Demons of Domesticity

Modern Economic and Social History Series

General Editor: Derek H. Aldcroft

Titles in this series include:

Insuring the Industrial Revolution
Fire Insurance in Great Britain, 1700–1850
Robin Pearson

Lives of the Philadelphia Engineers
Capital, Class and Revolution, 1830–1890
Andrew Dawson

Arms and the State
Sir William Armstrong and the Remaking of British Naval Power, 1854–1914
Marshall J. Bastable

The Seaside, Health and the Environment in England and Wales since 1800
John Hassan

Lancashire Cotton Operatives and Work, 1900–1950
A Social History of Lancashire Cotton Operatives in the Twentieth Century
Alan Fowler

Industrial Clusters and Regional Business Networks in England, 1750–1970
Edited by John F. Wilson and Andrew Popp

The Rise of Management Consulting in Britain
Michael Ferguson

The Myth of Mr Butskell
The Politics of British Economic Policy, 1950–55
Scott Kelly

Demons of Domesticity

Women and the English Gas Industry, 1889–1939

ANNE CLENDINNING

Nipissing University, Canada

ASHGATE

Published by
Ashgate Publishing Limited
Gower House
Croft Road
Aldershot
Hampshire GU11 3HR
England

Ashgate Publishing Company
Suite 420
101 Cherry Street
Burlington, VT 05401-4405
USA

Ashgate website: http://www.ashgate.com

British Library Cataloguing in Publication Data
Clendinning, Anne
Demons of domesticity : women and the English gas industry, 1889–1939. – (Modern economic and social history)
1. Women employees – England – History – 19th century
2. Women employees – England – History – 20th century 3. Gas industry – England – History – 19th century 4. Gas industry – England – History – 20th century
I. Title
331.4'8136363'0942

Library of Congress Cataloging-in-Publication Data
Clendinning, Anne, 1957-
Demons of domesticity : women and the English gas industry, 1889–1939 / Anne Clendinning.
p. cm. – (Modern economic and social history)
Includes bibliographical references.
ISBN 0-7546-0692-9 (alk. paper)
1. Gas industry—England—History. 2. Gas appliances—England—Marketing—History. 3. Gas industry—England—Employees—History. 4. Women—Employment—England—History. I. Title. II. Modern economic and social history series.

HD9581.G73E543 2002
338.4'768388—dc21

2002038375

ISBN 0 7546 0692 9

Typeset by DC Graphic Design Ltd, Swanley, Kent.
Printed and bound in Great Britain by MPG Books, Bodmin, Cornwall.

In memory of
Marjorie Clendinning
and
Beth Johnston

Contents

List of Figures

Modern Economic and Social History Series General Editor's Preface

Economic and social history has been a flourishing subject of scholarly study during recent decades. Not only has the volume of literature increased enormously but the range of interest in time, space and subject matter has broadened considerably so that today there are many sub-branches of the subject which have developed considerable status in their own right.

One of the aims of this series is to encourage the publication of scholarly monographs on any aspect of modern economic and social history. The geographical coverage is world-wide and contributions on non-British themes will be especially welcome. While emphasis will be placed on works embodying original research, it is also intended that the series should provide the opportunity to publish studies of a more general and thematic nature which offer a reappraisal or critical analysis of major issues of debate.

Derek H. Aldcroft
University of Leicester, UK

Acknowledgements

This study developed out of my interest in the history of women and work. In 1995, British Gas was undergoing the difficult process of privatisation and restructuring after 50 years of public ownership. One aspect of this corporate restructuring included the eventual elimination of the company's home service departments, staffed by women with home economics training. The idea for this book originated from conversations with women I knew who worked for British Gas. While they were concerned about their own futures within the company, the women also referred to the long tradition of public service to their customers that home service had represented. It was from Fiona Gould, Beth Johnston and Pat Davy, that I first learned of women's long history with the gas industry in England. From the moment I heard about them, I was fascinated with the Victorian lady demons, and their successors in home service, whose contributions to the history of women, work and business remained undocumented in a formal way. To the lady demons, past and present, who used their resourcefulness, common sense and good humour to create careers for women within the masculine world of business, I acknowledge your inspiration. This book is for lady demons everywhere.

Numerous individuals have assisted with the researching and writing of this book. This work began as a doctoral dissertation for McMaster University and I extend thanks to my supervisor, Richard Rempel, for his initial enthusiasm for this project. His encouragement, guidance and critical insight enabled me to complete this work. My thanks also go to the other members of my dissertation advisory committee, including Ruth Frager and Wayne Thorpe for their comments and astute criticisms, and Paul Fritz for his suggested readings on the history of consumption. I extend a special thanks to my external reader, Deborah Gorham, for her insightful and challenging evaluation of this work.

To the librarians, archivists and historians at the institutions where I worked, I offer my thanks for their invaluable assistance. At the outset, conversations with Brian Harrison, Sarah Milan and Penny Sparke helped me to develop a critical framework for the lady demon research, particularly with respect to the related histories of British feminism, women's work and domestic technology. David Doughan directed me towards some very useful archival sources located at the British Library and the Women's Library. This project would not have been possible without the assistance of Michael Hills, the director of the former London Gas Museum, then located in Bromley-by-Bow. I spent many hours in the museum's basement archive, and sincerely appreciated

Michael's generosity with the photocopying machine. His assistance by mail and fax continued until the museum closed and the collection relocated in 1999. Helen Ford, the head archivist at the National Gas Archive, and Anita Whitten, the librarian for the Institution of British Gas Engineers, have assisted with locating sources and documents, and I thank them for their time and patience in responding to my frequent questions. In this regard, I must also thank Julia Anne Lambert, curator of the John Johnson Collection, Bodleian Library, Alan Wakefield and Yvonne Oliver, from the Imperial War Museum and also S. Powlette at the British Library for assistance in gaining permission to reproduce photographs from their respective collections. At the University of Toronto, Jim Ingram from Photographic Services in Robarts Library assisted with the images from the *Architectural Review*. My appreciation extends to Guy Weaver, of the Radiation Company, for granting permission to reproduce advertisements under copyright. Sections of Chapter 3 appeared in the *Canadian Journal of History/Annales Canadiennes d'histoire*, 33, 1 (April 1998) under the title 'Gas and water feminism: Maud Adeline Brereton and Edwardian domestic technology'. Parts of Chapter 2 were published in *Women's History Review*, 9, 3, 2000 as '"Deft fingers" and "persuasive eloquence": the lady demons of the English gas industry, 1888–1918'. I thank the publishers of the above journals for granting their permission to reprint these portions.

Without the financial assistance of several funding agencies, I could not have undertaken the research required for this book. In that regard, I must acknowledge the Social Sciences and Humanities Research Council for supporting my doctoral studies. The history department at McMaster University provided travel bursaries that allowed for an extended research trip to London. Between 1999 and 2001, an SSHRC post-doctoral fellowship, held at the University of Toronto, facilitated further research in England and allowed for the necessary time to revise the manuscript.

Len White and Fiona Gould, both employees of the company formerly known as British Gas, have assisted with this project more than they will know. My gratitude extends to Fiona for introducing me to several former British Gas home service women. To Wendy Matthews, Grace Horrigan, Pat Davy and Pearl Thorpe (née Brown), I offer my thanks for their willingness to share their recollections, business papers and photographs. In this capacity, I also offer a special commemorative thanks to the late Kathleen Halpin and the late Elizabeth Warne, who had both agreed to be interviewed and generously offered their business papers. The conclusions that I have drawn from these sources are my own entirely, and I take full responsibility for the interpretation that I offer.

One relies so heavily on friends and colleagues for inspiration and assistance that it is hard to know where to begin in thanking them.

At McMaster University, Myra Baillie and Geoff Spurr offered support and commiseration, as and when required. Thanks especially to Myra for her comments on several chapters of the book, for suggestions regarding supporting references and for her unfailing encouragement during the final stages of the manuscript. Additional thanks extend to my post-doctoral supervisor, Lori Loeb, for her support and suggestions during the revision stage. David Santry, professor emeritus of chemistry at McMaster University and a former employee of British Gas, read and commented on sections of Chapter 4 that pertain to the history of the chemical industry. I thank him for his time and interest in this book. Sections of the manuscript have been presented at conferences in Canada, Great Britain and America. My gratitude to those who organised panels, commented on the various papers and offered suggestions, particularly Sascha Auerbach, Andrew August, Peter Bailey, Ken Cruickshank, Christopher Hosgood, Louise Jackson, Christopher Kent, David Manod, Daniel Ritschel and Margaret Walsh. I also thank the editors and readers at the *Women's History Review*, including Deirdre Beddoe, Clare Midgely, June Purvis and Jane Humphries, for their comments on the 'lady demons' article. And finally, a sincere thank you to Derek Aldcroft, the series editor, and to Tom Gray and the readers at Ashgate Publications for their comments on earlier drafts. My gratitude also extends to the members of the editorial staff for their guidance through the publishing process.

On a personal level, the members of the Gould-White household sustained this project from the very beginning. Their friendship, hospitality and, above all, endless sense of humour made for many interesting reprieves away from the archives. Friends in Hamilton, Windsor, Waterloo and, most recently, North Bay, offered encouragement throughout the duration of this enterprise; thanks especially to Suzanne Matheson, Alex MacKay, Margaret Shkimba, Jim Alsop, Matthew Hendley, Gordon Morrell and all of my colleagues in the history department at Nipissing University. To my sister and fellow McMaster graduate, Lynda, my thanks for her constant support, her kindness and her limitless generosity. Finally, I owe a great deal to Dermot Wilson. In a practical sense, his critical insight, editorial suggestions and technical assistance were crucial to the completion of this work. He has lived on a daily basis with the lady demons for as many years as I have, and, for his patience, his parenting and much more besides, I thank him. To our daughters, Imogen and Aurelia, you are the reasons that we need to know and respect the past. I dedicate this work to the memory of my mother, who encouraged my love of books, and to my sister-in-law, the late Beth Johnston, a lady demon in her own right.

List of Abbreviations

Throughout the book:

BCGA	British Commercial Gas Association
BISS	British Institute of Social Service
COS	Charity Organization Society
COSMITH	Committee Of Scientific Management In The Home
CPGC	Crystal Palace Gas Company
DORA	Defence of the Realm (Consolidation) Act 1914
EAW	Electrical Association for Women
EDA	Electrical Development Association
EMB	Empire Marketing Board
GLCC	Gas, Light and Coke Company
LCC	London County Council
LGB	Local Government Board
MARS	Modern Architectural Research Society
NCW	National Council of Women
NUGW	National Union of Gas Workers
NUSEC	National Union of Societies for Equal Citizenship
NUWS	National Union of Women's Suffrage
NUWSS	National Union of Women's Suffrage Societies
PEP	Political and Economic Planning group
SMGC	South Metropolitan Gas Company
WEF	Women's Employment Federation
WES	Women's Engineering Society
WEWNC	War Emergency Workers' National Committee
WGC	Women's Gas Council
WOGD	Women Operative Gas Demonstrators
WNHC	Workmen's National Housing Council
WSPU	Women's Social and Political Union
YWCA	Young Women's Christian Association

Throughout the annotation:

AAD	Archive of Art and Design, Victoria and Albert Museum
ATAOUFG	*A Thousand And One Uses For Gas*
BH	*Business History*
BL	British Library, Printed Books Collection
CAB	Cabinet Papers
EHR	*Economic History Review*
GJ	*Gas Journal* (known as the *Journal of Gas Lighting*

	until January 1917)
JGL	*Journal of Gas Lighting* (renamed the *Gas Journal* in January 1917)
LCC	London County Council
LMA	London Metropolitan Archives, Southampton Road, London
NAL	National Art Library, Victoria and Albert Museum
NGA	National Gas Archive, Warrington
RIBA	Royal Institute of British Architects
WHR	*Women's History Review*
WL	Women's Library, London Metropolitan University

Introduction

In December 1904, Miss Ethel Margaret Lovell-Wright of Northampton earned herself a handsome Christmas present. She took first prize in a championship cookery competition held at the International Gas Exhibition in Olympia, London, and received a gold watch from the organising committee.[1] To win the coveted timepiece, Lovell-Wright came first in several qualifying rounds, and was judged not simply on the quality of her cooking, but on her speed and efficient use of fuel. Because there were no restrictions on who might enter, many of the contestants were teachers of domestic subjects, experienced cooks and professional demonstrators. In fact, the culinary champion, Miss Lovell-Wright, was herself a professional lady demonstrator 'of many years standing' with the gas industry.[2] Mrs Mary Sutcliffe, a lady demonstrator with the Croydon gas company, judged the results of the competitions. As superintendent of the International Gas Exhibition's cookery programme, she was responsible for the two-week schedule of culinary demonstrations and competitions, which included the purchase of all the utensils and the food used by the competitors. For the discharge of these duties, Mary Sutcliffe also received an inscribed gold watch at the close of the exhibition.[3]

The significance of this commercial spectacle was twofold. This promotional event combined popular entertainment, in the form of the public exhibition, with advertising and consumer education. As the women bustled over their gas cookers, they demonstrated to their assembled audiences that coal gas, a volatile and toxic fuel, was potentially safe and easy to use. Promoted as a labour-saving domestic appliance, the gas-operated cooker had the power to revolutionise the Victorian home. The exhibition's competition, moreover, offered the gas industry's lady demonstrators a valuable opportunity for displaying their professional culinary expertise, competing with one another for the traditional emblem of corporate masculine respectability, the gold watch. The award served to recognise officially the promotional work that women had performed for the gas industry in the latter two decades of the Victorian era.

By the early twentieth century, the trade exhibition was a tried and tested means of attracting customers. From the mid-eighteenth century onwards, English natural philosophers, entrepreneurs and inventors had used the spectacle of the public demonstration to disseminate information about new scientific discoveries. By blending commercial entertainment and educational oration, experienced lecturers used novelty and showmanship not only to inspire audiences with curiosity, but also to pitch ideas at possible investors and to attract potential customers. Popular scientific performances continued into the nineteenth century, culminating in the greatest spectacle of Victorian science and industry, the Great Exhibition of 1851.[4] Hereafter, the trade exhibition became an effective and profitable means of promoting scientific information and its practical applications to mass audiences. As a promotional medium, it embodied the winning combination of entertainment, education, leisure and commerce.[5] The growing popularity of trade exhibitions was no doubt stimulated by contemporary discussions regarding working-class leisure and the need for genteel rational recreations.[6] Exhibition organisers congratulated themselves on the fact that their venues addressed this apparent social need, and simultaneously boosted trade and commercial enterprise.

Entrepreneurs and engineers in the gas industry readily engaged in this type of promotional activity using public lectures and, later, trade exhibition displays to promote the ever-expanding uses for coal gas in the English home. By the late 1870s, the majority of English cities and towns were served by at least one gas company, architecturally marked by the circular exoskeletons of gas holders that became a familiar feature of the Victorian urban landscape. Gas companies supplied the lighting for city streets and special civic events, as well as the indoor illumination of public and commercial buildings. Up until the mid-1880s and the advent of the Welsbach incandescent mantle, few private homes were fitted with gas lighting. That invention, along with the introduction of the penny-in-the-slot meter in the 1890s, made gas services more accessible for the lower-middle and working classes.[7] Gas could now be purchased by the penny-worth, and the gas companies adjusted their sales strategies accordingly.

Demons of Domesticity examines several related themes. First of all, it considers the process and the means that enabled British women to find an occupational place in the English gas industry and, as such, combines the history of women with that of business, labour and consumer culture. I am interested in several related elements of this occupational history. These include the reasons behind the expansion of sales careers for middle-class women in the gas industry from the late 1880s up to the late 1930s, and also the meanings that these women demonstrators attached to their new employment. I am here concerned with women's conscious efforts to define and assert a professional identity that served their needs as female white-collar workers, but one that also fulfilled the expectations of their employers, the highly stratified and predominantly masculine gas corporations. As early as 1905, gas industry spokesman Walter King, the editor of the *Journal of Gas Lighting*, speculated that the incorporation of women into the industry's promotional campaigns had significantly increased sales of appliances and services by appealing directly to female customers. The use of women experts to mediate between male producers and female consumers helped to domesticate, perhaps even feminise, gas technology, engendering sales and challenging gendered assumptions about the corporate sphere.

In general, the scholars working in the field of business history have concentrated on the history of corporations and enterprise, but tended to overlook the contributions of women. In an effort to correct the imbalance, feminist historians have introduced gender as a category of analysis in order to recover women's experiences and to understand how gender, as language and ideology, has shaped and defined the corporate world. For example, Wendy Gamber, Kathy Peiss and Angel Kwolek-Folland have noted that the traditional subject of business history, the firm, has been presented as a masculine enterprise, while the language of business history and entrepreneurship connotes male experience and essentialist masculine qualities.[8] In terms of our subject, with few exceptions, existing studies of the gas industry in Britain have been written by historians of business and labour. Business narratives document the expansion and development of the industry, describe major technological advancements and consider the effect of corporate restructuring at the managerial level, a process that transformed the industry from multiple small and local undertakings into larger regional or municipal companies that were then nationalised in 1949.[9] These histories generally adhere to the Chandlerian model of business history, which charts the adoption of specific business practices, such as scientific management, industrial restructuring and corporate mergers, as characteristics of successful enterprises.[10] Debates among historians of the gas industry have addressed the political and regional circumstances that facilitated the creation of municipally-owned gas undertakings during the nineteenth century.[11]

Labour studies by Joseph Melling and Derek Matthews focus on the experience of male industrial workers, the history of trade unionism and its effect on the gas industry, and the influence of governmental regulation as opposed to free-market capitalism.[12] For example, despite the overwhelming support by Victorian politicians and businessmen for free-market commercial development, some historians posit that economic state control and corporate welfare were evident in the nineteenth-century gas industry, with industrial copartnership upheld as an example of successful cooperation between capital and labour.[13] Although Wendy Gamber suggests that the history of women in business should also interest labour historians in their project of documenting work experience, traditionally, the history of labour has focused on workers engaged in manual labour and the effects of union solidarity and collective protest.[14] While a gendered analysis of labour history continues to enrich that discipline, labour studies of the gas industry have tended to overlook women's contributions. In general, the existing body of research compiles a masculine history, devoted to industrial issues, political policy, urban development, public utilities and services and the technical problems of large-scale production.

Among business historians, the adoption of a gendered analysis has been undertaken, for the most part, by historians of the United States. Margaret Walsh's study of female entrepreneurs in the long-distance transportation industry, Kathy Peiss's examination of American businesswomen and cosmetics, and Angel Kwolek-Folland's study of office workers in the insurance industry are all representative of this type of inquiry. The contributions of the above scholars not only add women to the history of business and enterprise, they also examine the gendered dimensions of the corporate sphere and women's efforts to create feminine and feminist spaces. Collectively, the above works add an integral component to business history and provide important clues for understanding the experiences and motivations of women in business. Joan Scott raises a word of caution: she notes that 'including women in business history requires more than documenting their exclusion, resistance and agency'. Although the recovery of female experience is important, Scott suggests that historians looking at women and gender need to consider 'how sexual difference is affirmed and produced through the creation and the service of consumer demand'.[15] Scott's point is useful for this study of women and the English gas industry. By introducing female sales representatives in the late 1880s, gas managers opened the door to a new profession for women domestic science experts, but, in so doing, they directed their campaigns to female consumers. This helped to feminise the market for domestic technology by underscoring the point that Victorian women were responsible for domestic labour, and, therefore, should be trusted with making important consumer choices for the home. It also appears, however, that the lady demonstrators (nicknamed 'lady demons') were mindful that their 'natural femininity' offered them distinct advantages in the business world. In other words, they were learning how to make gender work.

By recovering the experiences of women sales workers and their relationship to the development of advertising within their industry, I offer here a social history of the English gas industry from 1889 to 1939. Given the large number of gas companies, and partly because of the innovations of the largest urban corporations in the south east, this study places greater emphasis on the major undertakings that served London and the home counties. It documents the hitherto unexamined role that women played in the development of the gas industry by considering the expansion of sales and customer service occupations for women. Because the recruitment of women was an important aspect of the gas industry's evolving promotional methods, a history of women in the gas industry cannot afford to ignore the simultaneous development of its movement from itinerant trade fairs towards collective mass advertising. Thus, the history of the gas industry's advertising sector constitutes a secondary theme within this work. As we shall see, women were crucial to the production, dissemination and reception of corporate publicity. Consequently, women's relationship to the gas industry, as promoters, purveyors and consumers of gas services and material goods underpins the analysis presented here. Taking as fundamental this inextricable connection between domestic ideology, advertising and female consumerism, my research attempts to understand women's problematic relationship with the burgeoning 'world of goods' within the public and the private spaces of work and home. Examining the work undertaken by women in gas sales and marketing, this project re-evaluates several popular perceptions of the industry's overall public image, and the nature of its perceived rivalry with the electrical industry from the late Victorian years to the interwar period.

Demons of Domesticity contributes to the existing literature on the history of consumer culture by focusing on the second half of the production/consumption equation.[16] It considers the means and methods used by the gas industry to manufacture consumer consensus among its female customers from the late 1880s to the interwar years, assessing the overall social effect of gas technology on the English home. Victorian writers, including John Ruskin and Coventry Patmore, sentimentally characterised the home as a peaceful refuge from the outside world, overseen by idealised feminine 'angels' in the house. Despite this popular rhetoric, the home served as the focal point for a number of debates concerning female employment, infant health and hygiene, housing reform and urban sanitation.[17] The Victorian home, as a private sphere, was contested ground even in the late nineteenth century. A group of entrepreneurs and engineers within the gas industry actively addressed these issues, using advertising and public spectacles to position their industry in relation to these debates, while consciously attempting to increase the volumes of coal gas consumption via the sale and rental of domestic appliances.

From the 1880s, competition with the electricity industry motivated gas managers to invest seriously in various forms of promotional activities. It was not, as historians of the gas industry have assumed, because gas men seriously believed that electricity presented an imminent market takeover.[18] Instead, I posit that gas engineers worried about electricity because the public and the popular press appeared exceedingly vulnerable to the advertised claims of the electricians. Gas managers clearly understood that, if they were to maintain their competitive edge and improve their public image, they had to rely on more than gradual technological advancements and decreasing prices. Greater attention to the consumption of gas services, and the adoption of mass advertising and commodity spectacles, was crucial to their industry's economic survival. As Francis Goodall notes in his recent study of gas appliance manufacturers, the promotion of gas as a public service was complicated by the promotional activities of competing stove companies who rivalled one another for private sales and contracts with individual gas works. According to Goodall, the ironmongers and the stove makers who manufactured and sold gas fittings and appliances deserve much of the credit for the gas industry's continued success because they developed new applications for gas technology in the form of gas engines, heating stoves, cookers and domestic lighting.[19] There can be no denial that the technical and commercial innovations of gas appliance manufacturers increased fuel consumption; however, the commercial concerns of the appliance makers cannot be separated easily from those of the engineers who actually oversaw the gasworks. By the 1890s, in recognition of their shared enterprise, appliance makers and gas companies engaged in cooperative commercial activities, and, by the early twentieth century, had formed a collective advertising agency – the British Commercial Gas Association – to serve the entire industry. Its membership included gas engineers, ironmongers, appliance manufacturers and managers from private and municipally-owned gas companies.

By focusing on consumer activity and corporate selling strategies, this study, I hope, makes some contribution to that category of scholarship described as the history of consumption. In the 20 years that have followed the appearance of Neil McKendrick's, Colin Brewer's and J.H. Plumb's seminal essay collection, *The Birth of a Consumer Society*, an increasing number of publications have addressed the history of consumption and consumer activity as a necessary, but often unexamined, component of capitalism.[20] John Brewer and Roy Porter note that one of the distinguishing features of modern Western society, in addition to industrialisation and economic growth, has been the evolution of a 'consumer economy, with the consumers to go with it'. Observing that social stability depends, to a great extent, on the ability of political and economic systems to satisfy material expectations, Brewer and Porter suggest that to overlook the origins and evolution of consumer capitalism is to neglect a fundamental motivating force within modern Western history: the acquisition of material goods and services.[21] The success of consumer capitalism and the corresponding development of a mass consumer society has depended on the availability of goods to both satisfy and prompt people's increasing material expectations, stimulated by mass advertising and satisfied by available credit.[22]

The interdisciplinary publications devoted to the study of consumption demonstrate the thematic diversity of this subject. The history of consumption includes, for example, the history of retail sales, advertising, trade exhibitions and consumer credit, along with investigations into the social construction, manipulation and satisfaction of consumer desire.[23] The social history of leisure and popular culture also relies on the study of consumption as mass entertainment, which includes seaside excursions, commercial exhibitions and music hall performances. All were commodities aimed at diverse social and cultural groups.[24] Drawing on sociological and anthropological analyses of material culture, the history of consumption has moved far beyond the study of the retail trade by addressing questions related to the historical and cultural meanings of material goods and consumer behaviour. Consumption is both an economic and a social phenomenon: material possessions serve specific purposes, but also signify social relations.[25] According to Celia Lury, the utility of goods is 'framed by a cultural context' while even the most ordinary 'mundane objects of daily life', a cast-iron gas cooker for example, can assume cultural meaning.

As in the field of business history, feminist studies of consumer culture have considered the importance of gender.[26] Since the late nineteenth century and the expansion of the market for mass-produced consumer goods, retailers and advertising experts have agreed that the vast majority of shoppers were housewives shopping for themselves and their families. This, in part, was due to the household division of labour. As the household evolved from a unit of production to a unit of consumption, women assumed responsibility for acquiring things that might have been produced in the home in earlier years. By the end of the nineteenth century, as an increasing range of household goods and services became available in Britain – including canned foods, cleaning products and ready-made furniture – the domestic revolution was moving hand-in-hand with the modern consumer revolution, and women figured prominently in a burgeoning consumer culture.[27] This connection between women and shopping raises some problematic issues. Victorian female shoppers were often stereotyped as weak and irrational beings, seduced into overspending or shoplifting.[28] Moreover, the Victorian advertising industry made women into commodities, presenting them simultaneously as both the subjects and the objects of the capitalist commercial exchange system.[29]

Because of women's subordinate social, economic and cultural position and the patriarchal nature of the organisation and semiotics of mass consumption, some feminist scholars believe that women were more vulnerable than men to the exploitative forces of commercial culture. Herein lies the debate over whether consumer culture has liberated or victimised women. This debate is framed by the development of the Victorian advertising industry that alleged determined feminine spending patterns for personal and domestic items, the latter being for display in the middle-class home that was itself an indicator of conspicuous consumption.[30] Several historians have observed critically the ironies of a social liberation that depended on mass advertising and the spending of men's money. In the same manner, critics have argued that the freedom to wander around a shopping district or a department store was hardly a means of claiming public space for women when those areas had been specifically designed by retailers to imitate the comfort and protection of that quintessential feminine space: the middle-class Victorian home.[31]

In contrast to these analyses, however, a growing number of studies have demonstrated that women politicised their consumer activities, using buying power as a means of social activism via cooperatives, boycotts and rational shopping. Late nineteenth-century women participated in the making of a modern urban culture via their social interactions in public commercial spaces, such as department stores, restaurants and city streets.[32] For example, as Erika Diane Rappaport has demonstrated, the feminisation of London's West End was characterised by the expansion of public services for women, which included rest rooms and ladies' clubs. The appearance of the wandering female *flâneuse* and the urban lady guide signified women's new self-assurance in metropolitan retail districts.[33] Public anxiety regarding women's growing economic influence as consumers coincided with the apparent declining power of men's work as producers (as the advent of the mechanised assembly line and the application of Frederick Taylor's principles of scientific management de-skilled the production process). In contrast, the work of consuming, undertaken predominantly by women, remained 'un-Taylorized and self-regulated' and, therefore, still required decision-making skills.[34] Advocates for domestic education for girls and women underlined the importance of women's prudent consumer activity and the necessity of training women to be skilful shoppers and money managers. From approximately the late Victorian era, shopping became an essential aspect of homemaking in homes where women were typically the producers of goods and services which were then consumed by men, children and themselves. Although women generally purchased goods for the home, such as foods or furnishings, they worked on, or with, those goods, thereby blurring the clear delineation between consumption and production.[35]

Following the above arguments, consumption was therefore not necessarily a thoughtless, passive and trivial diversion, but a critical component of capitalism which granted women a certain degree of economic and social power. In some cases, such as the women's cooperative movement, the act of shopping itself reflected support for specific social causes, including women's suffrage, trade unionism and housing reform. Noting the theoretical debates regarding women and the history of consumption, Mica Nava posits that female consumer responses cannot be gauged by commercial propaganda, and that the female desire for commodities should not be read as evidence of absolute corporate indoctrination. Instead, consumerism should be understood as a constructed discourse, 'through which disciplinary power is both exercised and contested'.[36] For example, by the early 1900s, British women's increased enthusiasm for penny-in-the-slot gas cookers did not simply reflect the acceptance of a persuasive industry-wide advertising campaign: this acceptance of new technology was fed by contemporary debates surrounding the apparent physical deterioration of the urban poor and the efficacy of domestic training for girls and women. The need for increased access to domestic technology figured prominently in public discussions concerning working-class housing reform and women's employment both inside and outside the home. Female consumers and female sales representatives participated in these debates, while their words and their actions openly contested the patriarchal meanings of home, work and leisure. Victorian women who entered sales jobs in the gas industry – working at trade exhibitions, making the rounds of customers' homes or demonstrating appliances in company showrooms – quickly realised that existing and fluctuating definitions of femininity and masculinity had an effect on their evolving profession. As Rappaport has shown, gender and class identities were produced and contested, not only in the private sphere, but also in the public commercial sphere of street and shop.[37] Since their work encompassed both the domestic and the commercial realm, the gas industry's female sales staff present us with a unique example of the ways in which Victorian and Edwardian women negotiated gendered definitions of public and private behaviour.

Commercial practice blurs the assumed distinctions between male producers and female domestic consumers, between the separation of male public and female private space. Because women assumed responsibility for domestic spending and, as retail managers increasingly noted, women preferred to buy from other women, feminine contributions to the history of consumption warrant detailed study. Here I will argue that the gas industry's continued success increasingly depended on the loyalty of female consumers in the home and the promotional and public service work of its lady demons. Although the women functioned as sales personnel, their work was multifaceted; selling gas was not like selling silk or soap, but was a far more complex process, as consumer choice depended on the individual's access to that utility service. While the industry spokespeople lobbied architects, builders and politicians, the women's home service staff aimed to educate and empower female consumers, persuading them to make rational choices by demanding the most affordable and efficient utility system for their homes. Naturally, in the interests of the gas industry, that meant coal gas. While several economic historians have briefly mentioned the gas industry's efforts to attract female consumers and modernise its public image, no single study explores the social, political and gendered dimensions of that advertising process, or considers the roles played by women as the buyers and sellers of domestic technology.[38]

The lady demons – professional, educated and middle-class – were the first women employed in the gas trade. To date, their contributions have escaped the notice of historians of that industry. Similarly, scholars of women and work have neglected the lady demons and their attempts to create a role for professional female personnel within a predominantly male corporate world.[39] It remains surprising, for example, that the efforts of the gas industry's public relations specialist Mrs Maud Adeline Cloudesley Brereton have been absent from the historical record, particularly given her known connections to political progressivism.[40] In many respects, the lady demons established a conduit between male producers in the workplace and female consumers in the home, between the public and the private spheres of work. Their mandate to assist and educate female consumers represents an early example of customer service, a now familiar aspect of modern retailing, but one which existing histories of consumption have practically overlooked.[41] While the demonstrators were agents of the gas companies, many firmly believed that modern domestic technology would significantly reduce domestic labour and enable women to devote their time and energy to more important and rewarding activities outside the home. Labour-saving domestic technology gained importance during the interwar years as the servant shortage meant that more middle-class women kept their houses without the benefit of outside help. Feminist studies of domestic technology, in both the American and the British context, have shown that household mechanisation had the opposite effect: the availability of specialised 'labour-saving' domestic equipment and cleaning products actually increased the amount of time that women spent on housework because it raised expected standards of cleanliness.[42] By some accounts, the corporate home economists are chiefly to blame for promoting an image of domestic perfection that depended on continuous consumption.[43] Into the post-World War II era, women realised that the liberating potential of domestic technology had thus proved a false hope. Nevertheless, this should not negate the efforts of the gas industry's home service women who saw a direct correlation between hot water, decent meals and women's health.

As an occupational group, the lady demons of the gas industry offer an interesting case study that combines the history of women in business and the study of domestic consumption. Chapters 1 and 2 of this book consider how English women found a place in the predominantly male gas industry, noting the varied effects of the Victorian domestic science movement, technological innovations in the gas industry and the rise of mass advertising through the trade exhibition. During the Edwardian years, the lady demons formed special sales departments within the major gas companies, calling directly on customers in their homes to offer personal instruction in the proper use and care of gas stoves. Part social workers, part salesgirls, the lady demons moved between the traditionally gendered and separate spheres of work and home, consciously constructing a professional image that reconciled the social tensions of this new occupation.

Chapter 3 examines the lady demons' constrained attempts to reconcile definitions of Victorian femininity with their personal work experiences in the public sphere, as showroom demonstrators and door-to-door sales representatives. Their efforts to reach female consumers were assisted by the industry's innovative publicity collective, the BCGA, established in 1912, which was possibly the first industry-wide advertising agency in Britain. As editor-in-chief of the BCGA promotional material, Maud Adeline Cloudesley Brereton recognised the liberating potential of gas technology in terms of reducing women's domestic labour. She lobbied architects and builders on behalf of the gas industry, but she also urged her colleagues to look beyond the pounds and pence of profit and consider the social importance of their industry in terms of raising overall standards of living. Chapter 4, the years of World War I, considers how the gas women's carefully constructed professionalism was simultaneously challenged and strengthened by the influx of female replacement workers in the gas industry. Throughout World War I, when gas women cooperated with government agencies to assist on the 'domestic front', their efforts further legitimised their professional status within the gas industry, consolidating their public role as spokeswomen for their customers.

Chapters 5 and 6 examine the gas industry between the wars. During these most turbulent decades, due to its intensified commercial rivalry with the electricity industry, the gas industry undertook its most costly promotional schemes. These include most notably the redesign and modernisation of company showrooms, the production of documentary films and the construction of a model housing project in central London. At the same time, the industry relied heavily on its female sales personnel to help modernise its corporate image and maintain domestic sales by appealing to women consumers in the home. In addition to this corporate agenda, however, the work of the home service women continued to have a social welfare dimension. Through the Women's Gas Council, established in 1935, women in the gas industry interacted with a number of interwar feminist groups. They helped to exert pressure on the national government – under Stanley Baldwin and, later, Neville Chamberlain – to extend the school meals programme, raise the building standards for working-class housing, issue family allowance payments to women and create better facilities for infant nursery schools. Their efforts coincided with those of the 'new feminists', notably Eva Hubback and Eleanor Rathbone, whose social welfare reforms recognised and accepted women's 'equal but different' status as related to their domestic and maternal responsibilities.[44] Although they represented the sales agenda of their employers, home service women still challenged prevalent assumptions regarding the economic value of women's domestic labour in the home and feminine professionalism in the business world. Miss Lovell-Wright's gold watch, awarded for her expert sales performance and her culinary abilities, suggests that the gas industry and the home service demonstrators well realised the commercial potential of domestic proficiency, in both the workplace and the home. It was this awareness of women's fundamental importance to the corporate world that prompted a relatively small group of women to mediate between the producers and the consumers of domestic technology.

Notes

1 'Last day of the Exhibition', *Journal of Gas Lighting*, 20 December 1904, 898–899.
2 'Results of the quick cookery competitions', *JGL*, 6 December 1904, 761; 'The lady lecturer and teacher', *JGL*, 31 January 1905, 283.
3 Institution of Gas Engineers, 'Special attractions and entertainments', *Report of the Advisory Committee of the International Gas Exhibition, 1904* (London, 1904), 29–30.
4 Richard Altick, *The Shows of London* (Cambridge, Mass., 1978), 370–377; Simon Schaffer, 'The Consuming Flame: electrical showmen and Tory mystics in the world of goods', in John Brewer and Roy Porter (eds), *Consumption and the World of Goods* (London, 1993), 489–526; Thomas Richards, *The Commodity Culture of Victorian England: Advertising and Spectacle, 1851–1914* (Stanford, Calif., 1990).
5 Richards, *Commodity Culture*, chap. 1; Jeffrey A. Auerbach, *The Great Exhibition of 1851: A Nation on Display* (New Haven and London, 1999).
6 Hugh Cunningham, *Leisure in the Industrial Revolution, c. 1780–1880* (London, 1980); Peter Bailey, *Popular Culture and Performance in the Victorian City* (Cambridge, 1998), 21–26.

7 Stirling Everard, *The History of the Gas, Light and Coke Company, 1812–1949*, reprint of 1949 edition (London, 1992), chaps 9, 18; Trevor I. Williams, *A History of the British Gas Industry* (Oxford, 1981), chap 5; Hugh Barty-King, *New Flame: The Illustrated History of Piped Gas and the Way It Changed the Lives of Those Who Lived and Worked in Britain, 1783–1984* (Tavistock, 1984), chaps 9, 10.

8 Wendy Gamber, 'A gendered enterprise: placing nineteenth-century businesswomen in history', *Business History Review*, **72** (Summer 1998), 190; Kathy Peiss, '"Vital industry" and women's ventures: conceptualizing gender in twentieth-century business history', *Business History Review*, **72** (Summer 1998), 218–214. See also Angel Kwolek-Folland, *Engendering Business: Men and Women in the Corporate Office, 1870–1930* (Baltimore, 1994); Kathy Peiss, *Hope in a Jar: The Making of America's Beauty Culture* (New York, 1998); Margaret Walsh, *Making Connections: The Long Distance Bus Industry in the United States* (Aldershot and Burlington, VT, 2000).

9 Everard, *History of the Gas, Light and Coke Company*; Barty-King, *New Flame*; Williams, *History of the British Gas Industry.*

10 Alfred D. Chandler, Jr, *The Visible Hand: The Managerial Revolution in American Business* (Cambridge, Mass., 1977).

11 J.F. Wilson, 'Ownership, management and strategy in early north-west gas companies', *Business History*, **33**, 2 (1991), 203–221; John T. Wilson, *Lighting the Town: A Study of Management in the North West Gas Industry, 1805–1880* (London, 1991); Robert Millward and Robert Ward, 'From private to public ownership of gas undertakings in England and Wales 1851–1947: chronology, incidence and causes', *BH*, **35**, 3 (1993), 1–21.

12 D.A. Chatterton, 'State control of public utilities in the nineteenth century: the London gas industry', *BH*, **14**, 2 (1972), 166–178; Derek Matthews, 'Laissez-faire and the London gas industry in the nineteenth century: another look', *Economic History Review*, **39**, 2 (1986), 244–263.

13 Joseph Melling, 'Industrial strife and business welfare philosophy: the case of the South Metropolitan Gas Company', *BH*, **21**, 2 (1979), 163–179; Mary Mills, 'The gasworkers of South London', *South London Record*, 3 (1988), 6–16; Mary Mills, 'Comment: George Livesey and profit sharing: a comment on some recent literature', *BH*, **33**, 4 (1991), 87–90.

14 Gamber, 'Gendered enterprise', 194–195.

15 Joan W. Scott, 'Conceptualizing gender in American business history', *Business History Review*, **72** (Summer 1998), 249.

16 Francis Goodall, *Burning to Serve*: *Selling Gas in Competitive Markets* (Ashbourne, Derbyshire, 1999); Goodall considers the commercial aspects of the gas industry, but concentrates on the efforts of the appliance manufacturers.

17 Annmarie Adams, *Architecture in the Family Way: Doctors, Houses, and Women, 1870–1900* (Montreal and Kingston, 1996).

18 Everard, *History of the Gas, Light and Coke Company*, 262–263; Barty-King, *New Flame*, 134–136.

19 Goodall, *Burning to Serve*, 11, 68–71.

20 Neil McKendrick, Colin Brewer and J.H. Plumb, *The Birth of a Consumer Society: The Commercialization of Eighteenth-Century England* (Bloomington, Ind., 1982). See also Brewer and Porter (eds), *Consumption*

and the World of Goods; Gary Cross, *Time and Money: the Making of Consumer Culture* (London and New York, 1993); John Benson, *The Rise of Consumer Society in Britain 1880–1980* (London and New York, 1994).

21 Brewer and Porter (eds), *Consumption and the World of Goods*, 1–7.

22 Cross, *Time and Money*, 154–155.

23 Peter N. Stearns, 'Stages of consumerism: recent work on the issues of periodization', *Journal of Modern History*, **69** (March 1997), 103–117.

24 Peter Bailey, 'The politics and poetics of modern British leisure', *Rethinking History*, **3**, 2 (1999), 131–175; Bailey, *Popular Culture and Performance*, chap. 4.

25 Celia Lury, *Consumer Culture* (New Brunswick, NJ, 1996), chap. 2.

26 Joan W. Scott, 'On language, gender and working-class history', *International Labor and Working-Class History*, **31** (Spring 1987), 1–13.

27 Lury, *Consumer Culture*, 124; Margaret Horsfield, *Biting the Dust: the Joys of Housework* (London, 1997), chap. 9.

28 Elaine Abelson, *When Ladies Go A-Thieving: Middle-Class Shoplifters in the Victorian Department Store* (New York, 1989); Rachel Bowlby, *Just Looking: Consumer Culture in Dreiser, Gissing and Zola* (New York, 1985), chap. 2.

29 Victoria de Grazia with Ellen Furlough (eds), *The Sex of Things: Gender and Consumption in Historical Perspective* (Berkeley, 1996), 7. See also Lori Anne Loeb, *Consuming Angels: Advertising and Victorian Women* (New York, 1994), 26–32; Penny Sparke, *As Long As It's Pink: The Sexual Politics of Taste* (London, 1995), chap. 2.

30 de Grazia with Furlough (eds), *Sex of Things*, 7; Sparke, *As Long As It's Pink*, 22–24; Lury, *Consumer Culture*, 124–132.

31 Margaret Beetham, *A Magazine of Her Own? Domesticity and Desire in the Woman's Magazine, 1800–1914* (London, 1996), 148–154; Robin Dowling, 'Femininity, place and commodities: a retail case study', *Antipode*, **25**, 4 (1993), 295–319; Bowlby, *Just Looking*; Benson, *Rise of Consumer Society in Britain*, chap. 8.

32 Margaret Finnegan, *Selling Suffrage: Consumer Culture and Votes for Women* (New York, 1999); Mica Nava, 'Modernity's Disavowal: Women, the City and the Department Store', in Mica Nava and Alan O'Shea (eds), *Modern Times: Reflections on a Century of English Modernity* (London, 1996), 38–76; Belinda Davis, 'Food Scarcity and the Empowerment of the Female Consumer in World War I Berlin', in de Grazia with Furlough (eds), *Sex of Things*, 287–310; Kathy Peiss, 'Making Up, Making Over: Cosmetics, Consumer Culture, and Women's Identity', in de Grazia with Furlough (eds), *Sex of Things*, 311–336; Erika Diane Rappaport, *Shopping for Pleasure: Women in the Making of London's West End* (Princeton, NJ, 2000).

33 Rappaport, *Shopping for Pleasure*, chaps. 3–4.

34 Nava, 'Modernity's Disavowal', 56–57.

35 Lury, *Consumer Culture*, 123.

36 Mica Nava, *Changing Cultures: Feminism, Youth and Consumerism* (London, 1992), 165–166, 168.

37 Rappaport, *Shopping for Pleasure*, 12.

38 Leslie Hannah, *Electricity Before Nationalisation: A Study of the Development of the Electricity Supply Industry in Britain to 1948* (London, 1979), 186; T.A.B. Corley, 'Consumer marketing in Britain 1914–60', *BH*, **29**, 4 (1987), 73; Sue Bowden and Avner Offer,

'The Technological Revolution That Never Was: Class, Gender and the Diffusion of Household Appliances in Interwar England', in de Grazia with Furlough (eds), *Sex of Things*, 267.

39 Meta Zimmeck, 'Jobs for Girls: The Expansion of Clerical Work for Women, 1850–1914', in Angela John (ed.), *Unequal Opportunities: Women's Employment in England 1800–1918* (Oxford, 1986), 153–177; Lee Holcombe, *Victorian Ladies at Work* (Hamden, Conn., 1973). Possibly due to the pioneering work of electrical engineer, Caroline Haslett, the role of women in the electrical industry has received slightly more attention. See Suzette Worden, 'Powerful Women: Electricity in the Home, 1919–40', in Judy Attfield and Pat Kirkham (eds), *A View From the Interior: Feminism, Women and Design* (London, 1989), 128–147.

40 Michael Freeden (ed.), *Minutes of the Rainbow Circle, 1894–1924* (London, 1989), 204.

41 Alison Adburgham, *Liberty's: A Biography of a Shop* (London, 1975); Alison Adburgham, *Shopping in Style: London from the Restoration to Edwardian Elegance* (London, 1979); W. Hamish Fraser, *The Coming of the Mass Market, 1850-1914* (Hamden, Conn., 1981); Bill Lancaster, *The Department Store: A Social History* (London, 1995). Studies of American consumer culture have devoted more attention to the topic of customer service and the role of in-store demonstrators. See Susan Porter Benson, *Counter Cultures: Saleswomen, Managers and Customers in American Department Stores, 1890–1940* (Chicago, 1986).

42 Ruth Schwartz Cowan, *More Work for Mother: The Ironies of Household Technology from the Open Hearth to the Microwave* (New York, 1983); Caroline Davidson, *A Woman's Work Is Never Done: A History of Housework in the British Isles 1650–1950* (London, 1982); Christina Hardyment, *From Mangle to Microwave: The Mechanization of Household Work* (Cambridge, 1988); Suellen Hoy, *Chasing Dirt: The American Pursuit of Cleanliness* (Oxford, 1995); Pamela Walker Laird, 'Progress in separate spheres: selling nineteenth-century technologies', *Knowledge and Society: Research in Science and Technology Studies*, **10** (1996), 19–49; Horsfield, *Biting the Dust*.

43 Dolores Hayden, *The Grand Domestic Revolution: A History of Feminist Designs for American Homes, Neighbourhoods, and Cities* (Cambridge, Mass. and London, 1981), chap. 13; Horsfield, *Biting the Dust*, 147–150.

44 Brian Harrison, *Prudent Revolutionaries: Portraits of British Feminists between the Wars* (Oxford, 1987).

CHAPTER 1

The Victorian Kitchen Revolutionised

> What can a woman be worth
> Who knows simply nothing of cooking,
> Who handling the beauties of earth!
> All helpless is looking?
> For food to be pleasant and good,
> Depends on the way that you treat it;
> If carelessly roasted or stewed,
> 'Tisn't prudent to eat it.
> It may be as soft as new bread.
> It may be as tough as old leather;
> May lie just heavy as lead,
> Or light as a feather.
> So life may be burdened with pain,
> Or quite a succession of pleasure,
> If men a good cook can obtain.
> They call her a treasure.[1]

On 9 May, 1873, the reclusive Queen Victoria made a rare public appearance. She attended a cooking lecture. Accompanied by her daughters, the princesses Christian and Beatrice, Her Majesty was received ceremoniously by the commissioners of this International Exhibition, then being held at the exhibition grounds near the Albert Hall. The queen and princesses were ushered into the exhibition's School of Cookery to observe the preparation of a savoury omelette by a qualified male chef and four kitchen maids. The school's official lecturer, J.C. Buckmaster, provided the commentary for this culinary spectacle, speaking in the best tradition of the Victorian orator.

Buckmaster's four-minute presentation was an abbreviated version of the lengthy two-hour public lectures he was accustomed to delivering at the International Exhibition. In those few minutes, however, Buckmaster outlined the purpose of the cookery school: to show 'the best and most economical methods of domestic cooking' and the 'variations to which kitchen utensils may be fairly applied without injury'. He commented on the merits of egg dishes as nutritious, easy to prepare and affordable, even 'for the poorest of Her Majesty's subjects'. In particular, Buckmaster underlined the importance of cleanliness and the perceived relationship between domestic order, family harmony and good character, since 'dirt and virtue can never dwell comfortably together'.[2] Throughout this dialogue, the professional chef in attendance used a gas cooker to whip up a parsley omelette. The golden delicacy was offered to the queen and princesses; they each tasted it, pronounced it 'very good' and departed the School of Cookery to tour the remainder of the exhibition. Queen Victoria, with a cooking staff of 45, attending a cooking class? Absurd, perhaps; however, Buckmaster's lectures, which elevated cookery to a matter of national importance, were the sensation of the exhibition.[3] Crowds filled the auditorium twice-daily, with visitors even paying an additional sixpence over the general one shilling admission fee to secure a front-row seat and a taste of the finished products (see Fig. 1.1).

The 1873 International Exhibition, especially its School of Cookery, embodied two fundamental preoccupations of mid-Victorian society: the idealisation of home and the exaltation of industrial progress.[4] Though apparently contradictory, these two sentiments were compatible. The home was the natural refuge from industrial society and could potentially mediate the dehumanising effects of urban expansion and industrial change.[5] The home was never an isolated sanctuary, however, and the Victorians welcomed technological advancements that made new standards in domestic comfort and efficiency possible. The use of coal gas for domestic lighting and cooking was one such innovation of the late Victorian period.

1.1 J.C. Buckmaster and the Cookery Exhibition

Appliance manufacturers and gas companies would eventually adopt the format of Buckmaster's cookery classes to market their products. Although male chefs and gas managers initially did the demonstrating, the companies quickly realised that lady demons were more effective promoters, particularly when trying to reach a broad range of consumers. Several contributing factors by the late 1880s influenced the corporations' decision to hire women as sales specialists. These included the expansion of the domestic science movement, technological change within the gas industry, and the legitimisation of advertising as a respectable business practice. This chapter considers these three factors in an attempt to discover how Englishwomen gradually found an important place in the Victorian gas industry.

Good 'plain cooking': the foundation of national health

Queen Victoria's cooking lesson provides an evocative image of her public and private roles. As monarch and mother of the nation, Victoria appeared concerned with the well-being of all of her subjects and, by extension, with the general health and strength of the nation. Indeed, the continued economic prosperity and political stability of the British Empire depended largely on the health of the working population. Buckmaster made this point in his opening remarks of the lesson by thanking the queen for her interest in the diets of even the lowest of her subjects. Such concern was not a particularly new phenomenon and aristocratic philanthropists, including the Earl of Shaftesbury and Baroness Angela Burdett-Coutts, had long made the connection between working-class poverty, eating habits and national health, a theme also taken up by the *Lancet*.[6] In 1861, the queen's own chef, Charles Francatelli, published a cookery book specifically for the labouring classes.[7] In addition to the topic of working-class cooking, the subject of dangerous food additives remained a contentious issue throughout the 1860s. Despite the passing of a food act in 1860, several members of the medical community, including Dr Henry Letheby (the public analyst for the City of London) and editors of the *Lancet*, exerted continual pressure for more effective parliamentary legislation. The medical practitioners wanted regulations that would have effectively criminalised the use of harmful additives and colouring agents that were then still commonly found in flour, beer, milk, tea, coffee and patent medicines. Their sustained efforts finally resulted in the 1872 Adulteration of Food, Drink and Drugs Act, which included some mechanisms for enforcement via food testing and inspections by borough officials.[8] As a means of educating the public about the new food standards, the 1873 International Exhibition included displays of pure and adulterated food products so that consumers could recognise the differences in colour, weight and quality.[9] By demonstrating an interest in the food of her people, Queen Victoria overtly combined maternal sentiment with political design. In this respect, the royal outing to the International Exhibition, though brief, was not insignificant.

Without exaggeration, the queen was a recluse, making her appearance at the 1873 International Exhibition all the more reason for comment. Since the death of the prince consort in 1861 and Victoria's subsequent withdrawal from public life, the political press increasingly questioned the importance of a monarch that neglected public duties.[10] In some respects, the queen's participation in this trade exhibition was a tribute to her dear Albert's memory. The International Exhibition, designed to foster trade and good relations between Britain and her European neighbours, followed the tradition established by the prince consort with the Great Exhibition at Crystal Palace in 1851. Like the Great Exhibition, the International was a 'commodity spectacle' of new inventions and consumer goods, a tribute to industry and entrepreneurship and a means of advertising merchandise to the Victorian public.[11] Based on contemporary accounts, the focal point of the royal visit was the omelette lesson in keeping with the 1873 International Exhibition's thematic emphasis on food. Queen Victoria and the princesses were simultaneously royal patrons inspecting a national event, and an aging widow accompanied by her daughters obtaining some practical instruction in domestic economy. Both clearly signified the public and private images of Victoria's bourgeois monarchy.[12]

Certainly, in London throughout the spring of 1873, the subject of cookery fired the public imagination and Buckmaster's talks drew steady crowds throughout the International Exhibition's six-week engagement. No doubt, Buckmaster's oratory skill contributed to the popularity of his lectures. A familiar speaker on technical subjects, J.C. Buckmaster began his oratorical career in the 1840s with John Bright at the Anti-Corn Law League.[13] He followed Bright's advice and became a teacher, working at the Mechanics' Institute and giving occasional talks for the South Kensington Museum's department of science and art, under the direction of Sir Henry Cole. Although Buckmaster knew nothing of cooking, Cole persuaded the science teacher to be the commentator for a series of public lectures organised to complement the exhibition's food displays. Buckmaster's affable presentation, combined with the novelty of seeing a professional chef at work and the promise of a 'taste' for a sixpenny front-row seat, ensured the success of the lectures. Beyond their entertainment value, however, Buckmaster's lectures delivered practical instruction in economical cookery using simple but often unfamiliar ingredients, such as Italian macaroni, with the intention of introducing the British public to foods of other nations and of exploring foreign substitutes for 'those provisions whose price increased daily'.[14]

Based on his experiences at the International Exhibition, Buckmaster became an outspoken advocate for the good food cause which, he maintained, constituted the natural foundation of British social, domestic and moral well-being.[15] This position echoed existing views on the importance of home and family to national vitality. Yet, Buckmaster's cookery crusade went a step further. Since good plain cooking was the key to health and happiness, ignorance of the culinary arts signalled potential social disintegration. Changing patterns of work and consumption, sustained by 20 years of unparalleled economic prosperity, resulted in a degree of social mobility among the newly affluent middle classes. According to Buckmaster and Henry Cole, the women among these aspiring ranks no longer knew how to cook or manage their homes. By the same token, working-class women also needed domestic instruction to better utilise their scarce resources and correct the negative impact of women working outside the home.

The Victorian home and hearth

Buckmaster's lectures adhered to a particular format, not unlike that of the performances of earlier technical showmen.[16] He differed essentially, however, in that he was not specifically promoting a product as much as he was advocating good cooking as a means to individual health and national well-being. Yet, the International Exhibition contained a definite promotional element, given Buckmaster's ostentatious use of products displayed in the exhibition. Although the original intent of the cooking shows was public education, it was a select public.[17] Judging from the exhibition's one shilling entrance fee, and the additional sixpence for a front-row seat, the majority of the lecture patrons were affluent members of the middle classes. The reported crowds of 'fashionable ladies, sweetly surprised in discovering that a gridiron is quite different from a frying pan' next to 'stout middle-class housekeepers whose shrewd anxious glances betray their painful familiarity with the ways of plain cooks'[18] implied that any information gained was then passed on to cooks and servants. Buckmaster's culinary advice to the wives of English labourers and his subsequent criticism of waste and ignorance was appreciated by the attentive audience, but it was unlikely that many working-class women ever attended these educational lectures. Buckmaster's talks reflected and contributed to a Victorian preoccupation, on the part of the better classes, with food and cooking under the general subject of home management. In addition, his scathing anecdotes which pilloried the working man's diet added to the popular view that the social and economic ills of the labouring classes stemmed from a basic ignorance of thrift and domestic economy.

The 'cult of domesticity' and the development of distinct separate spheres for men and women are familiar themes in Victorian social and gender history.[19] The idealisation of home and family and the responsibility of women in the creation of that comfortable haven constituted a fundamental tenet of middle-class Victorian society. Novels, manuals and magazines suggested that this ideal had been firmly entrenched by the 1840s and corresponded with the expansion of the middle classes, the 'industrial bourgeoisie', between 1780 and 1830.[20] Defined in opposition to the labouring masses and the landed aristocracy and gentry, the middle classes adopted the ideals of laissez-faire capitalism, self-help and individualism, firmly rooted in an Evangelical code which demanded a reform of manners and morals, political and family life. Evangelical writers of the 1790s – notably William Wilberforce and Hannah More – established a connection between religious and domestic virtues, it being woman's special mission to provide a loving, Christian sanctuary from the evils of the outside world, thereby demarcating clearly the separate spheres of home and work, female and male.[21]

Corresponding changes in commerce and industry ensured the widespread adoption of this Evangelical discourse. The expansion of family shops to larger commercial and manufacturing enterprises often resulted in the physical separation of home and workplace, and consequently alienated many women from active involvement in family businesses.[22] While commercial families of the 1790s generally lived over their shops, their descendants, by the 1830s, resided in suburban homes with wives and children physically separated from the place of business. Absolved from commercial responsibilities, women were expected to focus on their home and children, and contribute indirectly to the business by providing a tranquil refuge for husbands, fathers and male children. According to gender studies of the middle classes, masculine self-respect depended on a rigid moral code of gentlemanly behaviour in public and on an orderly, financially secure household, the maintenance of which demanded of men 'a commitment to family life'.[23] This commitment was primarily both financial and emotional; the actual care of the home and family, and the management of domestic affairs rested with the middle-class housewife. Religious tracts underscored the perception that motherhood was women's 'natural' place in creation, professing that adult females needed families to give their lives purpose and contain their sexuality. Without home and family, a woman was redundant.[24]

Later generations of secular prescriptive authors, notably Mrs Sara Ellis and Harriet Martineau, echoed the sentiments of the Evangelical writers.

According to Catherine Hall and Leonore Davidoff, the domestic ideology of the 1830s and 1840s served as a buttress against the social and political unrest of these decades.[25] Moralist writers perceived the family as the basis of a stable society, while it was the duty of women to ensure social harmony through their moral influence within the home. Feminine virtue was manifest in all womanly activities, including the education of female children, the management of the housework and servants, and contributions to charitable work. Mid-century writers, including Lord Tennyson, John Ruskin and Coventry Patmore, popularised the sentimental image of woman as the 'angel in the house'.[26] The ostentatious decor of the prosperous Victorian home symbolised feminine spiritual influence and women's elevated moral status. Like the carefully contained parlour aspidistras, however, the activities and behaviour of these virtuous, gentle wives and daughters were circumscribed by rigid social conventions.[27]

Contemporary debates over the true meanings of femininity, masculinity and family were constructed around these definitions of sexual difference that firmly upheld the old Evangelical values. Social critics introduced new evidence in support of women's 'natural' social and intellectual inferiority, despite their exalted moral and maternal influence. Mid-century physicians and educators claimed that women were physiologically incapable of serious study given the size of the female cranium and the possibility that intellectual exertion would result in nervous hysteria.[28] Similarly, the 'limited energy theory' proscribed academic activities for adolescent girls since this interfered with the proper development of the female reproductive system, rendering young women unfit for their true profession: marriage and motherhood.[29]

The realisation of a domestic ideal that relied on servant help and leisure time exceeded the material and physical capacities of most middle-class families. Few women, even among the middle classes, sustained the image of the leisured Victorian matron, the attentive wife, virtuous mother and efficient manager of servants, as portrayed in the advice manuals and domestic magazines. According to Patricia Branca's study, the majority of middle-class women, those with family incomes between £150 and £300, had little extra money for lavish decorating and entertaining, let alone for a team of servants.[30] The employment of at least one servant, however – often a young maid-of-all-work – relieved the woman of the house from the most extreme drudgery and supplied a semblance of gentility. To keep at least one servant was a hallmark of Victorian middle-class respectability. At all costs, the 'ladylike wife and mother' did not engage in any paid employment, although her 'duties' might include charitable work among the less fortunate.[31] Women driven by financial necessity to contribute to the family income might do so covertly, by working at home. For example, fancy needlework and painting, traditional female accomplishments, were tacitly accepted forms of respectable female employment. Middle-class women performed these tasks in the home, even anonymously, and under the outward assumption that the work was for pleasure and not for profit.[32] Evidently, economies of scale determined individual domestic practice; nevertheless, overriding standards of female middle-class behaviour remained in place.

Female education effected the propagation of the Victorian domestic ideal. However, authors differed over the degree to which education should prepare girls for home life and motherhood. In her essay, 'Principles of Education', the prescriptive writer Elizabeth Sewell recommended against sending girls to school as it was incompatible with their natures and expectations in life. She briefly summed up the ideal Victorian girl as a model of feminine virtue and restraint: 'Girls are to dwell in quiet homes, amongst a few friends; to exercise a noiseless influence, to be submissive and retiring.'[33] While boys of the prosperous middle classes, businessmen, manufacturers and professionals with incomes between £800 and £1,000, left home for at least a portion of their education, girls remained behind and were trained by their mother or a governess.[34] The instruction offered to girls revolved around the acquisition of genteel accomplishments, including music, embroidery and French language, which were refinements that parents felt were necessary to attract prospective suitors.

Despite the alleged importance of the home and the female role within it, the formal acquisition of basic housekeeping skills received little attention until the late nineteenth century. Housekeeping manuals and advice books helped remedy this perceived shortfall in young women's education. Dena Attar attributes the growing number and increasing specialisation of mid-Victorian domestic manuals to the effects of industrialisation, urbanisation and social mobility (factors which redefined gender roles and altered material conditions).[35] The strict delineation between male and female spheres and the Victorian cult of the home inflated expected standards of comfort and cleanliness as visible signs of middle-class affluence and respectability. By the 1870s, the gentle 'angel in the house' had to be an efficient manager in her own right. As middle-class households became larger and employed more servants, the public values of methodical precision and order were increasingly applied to the private sphere to ensure that the home remained a peaceful refuge from the outside world. The mistress and her servants were responsible for maintaining this domestic perfection, but surreptitiously, since the home was theoretically a haven of peace and tranquillity, and not a workplace. Disruptive and noisy domestic chores had to be carried out according to precise work schedules when male householders were absent. For example, the housekeeping manuals aimed at middle-class women lamented about the annoyance which the annual spring cleaning caused for men, in terms of cold meals and rolled-up rugs. There was little reference to the exhausting work undertaken by the housewives and domestic servants during the annual cleaning and all that it entailed – everything from hauling down heavy drapes to beating carpets outdoors.

In addition, the mid-Victorian 'rationalization of housework' as explained in the housekeeping books emphasised the division of tasks into smaller units that could be repeated with greater frequency than in the past. As Leonore Davidoff notes, the span between laundering was reduced gradually from the quarterly wash of dirty linen, common in prosperous homes in the early nineteenth century, to the weekly wash of the mid-century middle-class home.[36] Alterations to housekeeping arrangements were reflected by changes in domestic architecture: rooms were designed for specific functions – such as eating, or caring for children – while areas of intensive labour, especially the kitchen, were distinctly separated from the front rooms. In fact, all signs of meal preparation, including the smell of cooking odours, were concealed in an effort to preserve the image of efficient domestic perfection.[37] The 'ladylike homemaker'[38] of the middle classes was expected to maintain an immaculate and comfortable home without apparent effort on the part of her servants, much less herself, but also without the benefit of practical training in household management. Apparently, despite women's 'natural' predilection to home and family, efficient housekeeping practices were not innate, but required cultivation and instruction, much the same as truly feminine behaviour was the product of correct education and a proper up-bringing.

Isabella Beeton wrote *Household Management* in 1861 because she felt that Victorian women desperately needed instruction in domestic affairs. The wife of journalist and publisher Sam Beeton, Isabella wrote for Beeton's *Englishwoman's Domestic Magazine*, a sixpenny paper consisting of household advice, serialised fiction, fashion news and needlework patterns. A newlywed woman of 25, Isabella was a novice housekeeper herself, but her practical suggestions and prudish notions on female behaviour made her a ready role model. On the basis of letters from her readers, Isabella attributed family discontent to the housewife's 'badly-cooked dinners and untidy ways'. In order to compete with the social attractions of the taverns, men's clubs and dining houses, 'a mistress must be thoroughly conversant with the theory and practice of cookery, as well as perfectly conversant with all other arts of making and keeping a comfortable home'.[39] Mrs Beeton wrote for the upwardly mobile middle classes, the wives of successful businessmen and professionals, faced with running larger homes than their mothers but lacking the requisite skills to manage several servants and larger budgets. Mrs Beeton also addressed the problems of modest households, and her information on plain cookery and invalid care could be applied to any income group.

Isabella's fundamental recommendations for successful household management depended not on the size of the family budget, but on the personal qualities of the mistress. She stressed the importance of early rising, personal cleanliness in the form of daily cold or tepid baths and hard work performed intelligently and thoroughly. The mistress set an example to those in her household, whether staff or family members, since it was her conduct that shaped the atmosphere of the home. Mrs Beeton could not overstate the importance of providing a good example for daughters so that they learned not only domestic skills, but also the true value of efficient housekeeping: 'Let the mistress of every house rise to the responsibility of its management; so that, in doing her duty to all around her, she may receive the genuine reward of respect, love and affection.'[40] Isabella's formula for household efficiency rested with the woman of the house. Her conduct and personality were paramount and, at least theoretically, unaffected by income levels.

Following Mrs Beeton's example, other household manuals upheld the pervasive ideal of the home as a comfortable sanctuary despite differences in living standards. More importantly, this image was potentially within every woman's reach through skilful domestic economy, no matter how modest the family budget. Mrs Eliza Warren's 1864 publication *How I Managed my House on £200 a Year* advocated cautious budgeting, living within one's means and a practical education for girls to instil in them a 'fondness for domestic association'.[41] She stressed the need for cooking, sewing, baking and washing skills. Like Mrs Beeton, Mrs Warren considered cleanliness to be the hallmark of respectability. She lamented that, with rising incomes and social pretensions, middle-class girls were increasingly unprepared to assume the domestic responsibilities that they inevitably faced, with ruinous implications for the Victorian home. Her solution was better, standardised training in the household arts for girls and young women, no matter what their economic and social background.

The apparent housekeeping crisis of mid-Victorian England, as perceived by writers Beeton and Warren, was equally fuelled by contemporary debates over the employment and education of working-class girls and women. In contrast to the prescriptive ideal of the middle-class ladylike homemaker, the model supported by the middle classes for the 'lower' orders was that of the 'good woman'.[42] She was a practical housekeeper responsible for her own cooking, childcare and housework.

Thrifty, prudent and methodical, the wife of a wage-earning husband, the good woman knew her social place, and accordingly trained her daughters to accept a similar lot. The image of the good woman was presented as the solution to many of the problems of working-class life, such as alcoholism, crime and high infant mortality. Self-help writer Samuel Smiles promoted this ideal for over 30 years, insisting that the home's moral atmosphere was dependent on 'good mothers' and that even the poorest dwelling could be a place of 'comfort, virtue, and happiness' under the influence of a 'thrifty, cheerful and cleanly woman'.[43] Despite the economic and class differences of the idealised ladylike homemaker and the good woman, both models emphasised cleanliness as an expression of feminine virtue and moral influence, reflecting the persistence of Evangelical rhetoric. Good housekeeping practices revealed a woman's character, no matter what her station.

Social commentators and clergymen articulated concern over the impact of the factory system and sweated piecework on the working-class home. The effects of women's work were most obvious in the northern mill towns where large numbers of married women, employed in the textile industries in Lancashire and Yorkshire from the 1830s onward, fostered a reversal in the economic position between men and women, thereby creating tensions over the gendered division of labour. Factory deputations lobbied for shorter hours and restrictions against women's work in an effort to save men's jobs and return women to their 'true spheres' of home and family. A northern factory inspector reported in 1861 that there was:

> ...no comfort in the dwelling of a working man whose wife is away from the home from 5.30 in the morning to 6.30 in the evening, except at mealtimes, for she is compelled to leave her children and her household to other hands; and having so little experience of her own, is quite unable to teach her daughter those attractive qualifications with which women keep their husbands from disreputable associations.[44]

As a means of getting by, female mill operatives employed other women, often neighbours or relatives, to wash their families' clothes and mind their small children. Whether women accomplished their household chores using hired help or masculine assistance, either domestic system was completely at odds with the idealised image of the working-class wife.[45] In addition, female industrial employment undermined the image of the respectable working-class man, who was perceived to be industrious, temperate, independent and, most importantly, capable of supporting his wife and children without female financial contributions.[46]

Although the homes of female mill operatives offered middle-class observers the most obvious example of domestic chaos, a far greater number of working women earned money at home via the outwork system of piece-rate manufacture. James Schmiechen's examination of the sweated industries reveals that the system was not a stage in factory production, but an entirely separate manufacturing process that relied heavily on unskilled and semi-skilled women and children.[47] Sweating was most common in the ready-made clothing industry: tailoring, shoe and bootmaking, dressmaking, shirtmaking and seamstress work was all undertaken by sweated female and juvenile workers. Sweated labour meant long, tedious hours, low pay and unhealthy surroundings since many women toiled in cramped workrooms or in the homes of others.[48] Under such circumstances, the working-class home was also a workplace, as in pre-industrial times, and hardly fitted the middle-class image of home as a sanctuary from industrial society. The working class rarely questioned this combination of family and work; it was a fact of life. The sweated nature of piecework meant that the entire dwelling was taken up with some miserable employment, which necessitated the use of small children in some menial capacity, like sewing on buttons, or occasioned their almost total neglect.[49] Like the women in the textile factories, women in the sweated industries often performed a double load of waged and domestic labour, undertaking domestic duties while the family slept.[50] Middle-class commentators expressed their horror at the standards of housewifery among working women and attributed the dirt and squalor not only to poverty, but also to simple ignorance and lack of domestic pride.

Catherine Hall points out the uneasy relationship between free-market doctrines in industry and the idealisation of home and family. The former depressed men's wages and encouraged the employers to hire women, both in the factories and as home-based pieceworkers, because women were cheaper. The resulting prosperity for the entrepreneurial classes made new standards in domestic order possible, thereby strengthening the ideal of home as private refuge, an ideal then internalised and aspired to by the working classes. Not surprisingly, working women lacked the financial means and the necessary time to fulfil such expectations.[51] Nevertheless, they were measured against the dominant standards of domestic practice because, as the household manuals suggested, efficient housekeeping was not dependent on money, but on the character and personal habits of the mistress. Within the sorry homes of the working poor, mothers had not the means, time or experience to instruct their daughters in proper household management and cookery.

The domestic science movement

Buckmaster's description of the working man's dinner reflected the growing sentiment that female waste and ignorance was the source of domestic hardship in the homes of the lower classes. In an 'amusing sketch', Buckmaster contrasted the French peasant's dinner with that of the British working man:

> In the former case, long before the dinner hour, a couple of ounces of meat, some crusts of bread, a few savory herbs and onions had been simmering in a little earthenware pipkin over a handful of charcoal, and the result is a nourishing and palatable mess. On the other hand, the wife of the English labourer thought herself in luck if she could rush out and buy perhaps double as much meat or bacon, a few minutes before 12 o'clock. She would then hastily place her purchase, supplemented by an onion, on a greasy frying-pan and serve up to her hungry husband a black, burnt chunk of wholly indigestible meat. This was considered a good dinner; what then must a bad one be?[52]

This was certainly not the first time that English working-class housewives had been accused of domestic mismanagement. Nor was it unusual for Victorian middle-class observers to accuse the lower classes of increasing their own economic difficulties with their wasteful and spendthrift habits. Prescriptive literature, such as that of Samuel Smiles, offered numerous recommendations for self-improvement.[53] According to an article that appeared in the *Lancet* in 1859, the French rag-picker dined better than the English mechanic because the English housewife 'throws one-third of her husband's dinner into the dust-bin'. Englishwomen, the commentary continued, did not know how to 'make much out of little' and instead of saving the kitchen stuff for soup and stews, working-class mothers were content to feed themselves and their children the usual ration of bread, butter and tea.[54] Likewise, the type of ineptitude that Buckmaster associated with working-class cookery was equally prevalent, he maintained, in the kitchens of the rich and even the middle classes. To remedy the situation, he recommended that girls and women of all social groups take compulsory training in cookery and household management. Housewifery courses might complement the academic and artistic education offered to the girls of the prosperous classes and compensate for the poor example set for girls in working-class households.

According to Buckmaster and Cole, better training in the domestic arts seemed the obvious remedy to the apparent housekeeping crisis. They advocated the immediate establishment of a national cookery school and, in March 1874, the National Training College of Domestic Subjects opened its doors. Sponsored by the South Kensington Committee of Science and Art, the school was a direct result of the International Exhibition's cookery classes and was outfitted with the utensils from the preceding lectures. Queen Victoria and the renowned philanthropist Lady Burdett-Coutts were among the school's most prominent aristocratic patrons. Although Cole had envisioned a school for young women of all social classes, the first lady superintendent, Lady Barker, offered classes for ladies only, although these affluent students could bring their cooks. The two-week beginner's course cost three guineas; an extra guinea exempted a pupil from scullery duty, although this option forfeited the course certificate (see Fig. 1.2). During the first year, 176 students successfully completed the course and only 27 of them were professional cooks.[55]

By the autumn of 1875 and under the directorship of Edith Nicholls, later Mrs Charles Clark, the college expanded its courses to accommodate women of different economic groups. Mrs Clark actively promoted a scheme to reach working-class girls, especially those from London's poorer districts. In addition to the ladies' classes, the National Training College offered courses for women and cooks of households with weekly food budgets of between 20 and 100 shillings. A second course, cookery for artisans, was designed for women with a weekly food allowance of between seven and 20 shillings.[56] Recorded working-class budgets for the 1870s are scarce; however, judging from sample accounts from workers' homes in the 1850s and 1860s, a weekly food expenditure of between seven and 20 shillings would have been about

1.2 The National Training College of Domestic Subjects at South Kensington was the first domestic science college in the country and offered instruction to women of all social classes

average for semi-skilled workmen and their families.[57] Adolescent girls were taught plain cookery, intended for use in the artisan home or as training for employment as kitchen helpers and scullery maids. Under Mrs Clark's supervision, the National Training College sustained Cole's intent of making instruction available for all social classes, however dubious his motivations.

Late Victorian philanthropists upheld the regenerative effects of good cookery as sufficient reason for promoting housekeeping lessons for working-class girls and women. Regarding the moral and social value of cooking, Lady Burdett-Coutts confidently stated: 'If wives are clean, thrifty and good cooks, men will find their own homes so inviting they won't want to turn out to find comfort in public-houses or gambling saloons.'[58] The Universal Cookery and Food Association, co-founded in 1885 by J.C. Buckmaster and food writer and chef Charles Herman Senn, maintained a similar position. It sponsored free lectures to working- and middle-class wives and published twopenny pamphlets on artisan cookery. The association was also an employment registry for professional chefs, cooks and teachers of domestic subjects, and offered annual achievement awards to its most outstanding members. In his treatise on the nation's eating habits, author Henry Thompson spelled out the connections between proper food, physical development and the health of the British 'race'. He was particularly concerned with infant and child nutrition, claiming that even among the affluent classes children suffered the ill-effects of poor nutrition. Such poor nutrition was not caused by insufficient quantity, but by eating the wrong foods, thus demonstrating an allusion to the new science of dietetics.[59] For late Victorians, good cooking was fundamental to the nation's physical and moral well-being.

The National Training College became the model cookery school as similar institutions opened in Liverpool, Leeds and Gloucester. It was not a coincidence that the advent of the private cookery college paralleled the national regulation of elementary schooling. Under the 1870 Forster Education Act, passed by William Gladstone's Liberal ministry, school management continued at the local level but visits from inspectors and curriculum guidance standardised primary education. The cookery colleges offered special courses to train teachers and lobbied heavily for the inclusion of domestic subjects for girls in national curriculums. Mrs Clark and the National Training College patrons, particularly Lady Burdett-Coutts, promoted domestic science as a way of improving standards of living and the general health of the entire population.

Formal instruction in domestic subjects was unheard of until the 1870 Education Act made elementary education compulsory for all children under the age of 12. Before that, the inclusion of needlework was justified, as it was a practical skill and because it taught girls to be quiet, to be occupied for hours at a time, and to be in mixed classrooms while teachers were free to concentrate on the more academic curriculum of male pupils.[60] Domestic subjects, such as plain cookery, laundering and housewifery, gradually entered the national curriculum between 1874 and 1894 with enrolment restricted to older girls. Ironically, just when working-class girls were ensured at least some form of national education, more class-time was devoted to non-academic subjects as school curriculums became increasingly gendered.

The primary lobbyists for domestic instruction in the new schools came from the domestic science colleges, especially those in the major urban centres of London, Liverpool and Leeds, which were the areas of greatest urban poverty. School board officials were hesitant to include domestic subjects because they envisioned problems with the organisation of timetables and school boards lacked the extra funds needed to equip practice kitchens. They also had difficulty finding qualified instructors. By late 1876, the London School Board had equipped only seven cookery centres in East and South London.[61] Girls from ten schools might attend in one week for a half day, with two separate groups per day. Outlying schools were equipped with a cooker. The costs of the classes were offset by selling the demonstration results to the children, meaning that the lessons had to result in individual baked goods, like scones or cakes, that sold cheaply. Thus fiscal restraints overrode the original intent of providing instruction in nutritious cookery for working-class families. Maintaining order in the domestic classes was practically impossible, given that there were 40 to 50 girls per class.[62] Often, it was the cookery mistresses who assumed the blame, and superintendents complained about the women's apparent 'inefficiency and inability to keep the classes in proper order and discipline'.[63]

The teaching of domestic subjects was contentious for many reasons. Victorian feminist educators, in particular, were critical of the subjects because they gendered the curriculum and cheated girls of valuable instruction time that could have been better spent. The debates about domestic subjects betrayed a social class bias since even feminist advocates of female education considered these subjects to be appropriate for working-class girls.[64] For example, in the 1860s, Bessie Rayner Parkes, editor of the *English Woman's Journal*, expressed her support for including the 'household arts' in the curriculums of industrial schools. A decade later, representatives of the domestic science colleges in England and Scotland lobbied the school boards to introduce domestic subjects for lower-class girls.[65] Board school mistresses were sceptical about the efficacy of domestic subjects. After 1880, when school attendance up to the age of 11 became compulsory, teachers faced crowded classrooms and an extended curriculum; cookery classes added to their already demanding workload. Those teachers who accepted the validity of domestic instruction still objected to the amount of time devoted to it at the expense of academic instruction. Parents of board schoolgirls complained that their daughters could learn housekeeping at home and that school should prepare them for something besides domestic service.[66] In particular, parents dreaded the laundry lessons since their daughters often returned home wearing damp pinafores that they had 'practice washed' in class.[67]

Yet, despite administrative difficulties and the questionable efficacy of the lessons, the instruction of domestic subjects broadened under the rubrics of technical education for girls. The extension of government grants to include domestic subjects, the 'payment by results' system, undermined the trustees' fiscal arguments to limit domestic economy classes. By 1893, the London School Board employed 105 home economy mistresses instructing 25,000 girls at 99 cookery centres. As more government-funded schools adopted domestic subjects, the demand for qualified cookery mistresses escalated. Graduates of the domestic colleges, like the one started by Buckmaster and Cole, filled these new teaching posts.

In an address before the Universal Cookery and Food Association, Buckmaster reflected on the impact of the 1873 International Exhibition on the domestic science movement. Its purpose had been 'to direct public attention to the importance and knowledge of food and cookery in the education of girls'.[68] Evidently, the cookery crusade that Buckmaster helped to launch some 20 years earlier had developed into a national cause involving educators, health officials and philanthropists. By the 1890s, the domestic science movement showed no signs of abating as middle-class social investigators and settlement workers noted the continued high levels of poverty among the urban working classes. Often, the strategies intended to reverse urban poverty, devised by housing experts and philanthropists, were calculated to reform the manners and behaviour of the working poor by promoting middle-class values of thrift and industry. For example, Octavia Hill, founder of the Charity Organization Society in 1869, hoped that her lady rent collectors encouraged working-class wives to adopt the habits and values of their social betters. According to Gareth Stedman Jones, household management classes, along with penny savings banks and teetotal working men's clubs, were some of the various means used to civilise and elevate the urban masses.[69]

Buckmaster, looking back over two decades, emphasised the International Exhibition's contribution to the development of domestic science. But the exhibition had another important function. As a trade show about food and the technology of food production, it had also introduced the public to technological advancements, and was therefore an important medium of advertising and sales. In this respect, the format of the 1873 International Exhibition and the participation of gas appliance manufacturers had a direct correlation to the later involvement of lady demonstrators as expert sales personnel.

Coal gas and the 'public good'

J.C. Buckmaster's 1873 cookery series proved to be an appropriate model for the gas company demonstrations some 20 years later. With this exposition of food and culinary techniques Buckmaster aimed to entertain, educate and advertise. For example, the School of Cookery's macaroni recipes were complemented by the working display of a pasta machine which the manufacturer had entered in the exhibition. The idea of both the machine and the product was to promote sales in England, especially since the manufacturer was based in Camden Town, not Italy.[70] By association, the cookery lectures served the added purpose of demonstrating recent innovations in domestic technology to Victorian audiences (the most obvious being the gas cooking stove). Although Buckmaster assured his audiences that all of the demonstrated foods could be successfully prepared on a kitchen fire, the advantages of the gas cooker were apparent in terms of cleanliness, temperature control and speed. It was no coincidence that the patent gas stoves of Sidney Leoni and Company of London were featured in the 1873 International Exhibition along with testimonials to their fuel efficiency.[71]

Despite the rapid expansion of the gas industry throughout the nineteenth century, the use of coal gas for domestic purposes was not widespread at the time of Buckmaster's 1873 cookery talks.[72] Early exponents of coal-gas fuel, including the Scottish engineer William Murdoch and his associate Samuel Clegg, recognised its potential for commercial lighting applications. In 1792, Murdoch lit his home at Redruth, Cornwall, with coal gas manufactured in a small iron retort in his back garden. A decade later, Murdoch illuminated Boulton and Watt's Soho iron foundry outside Birmingham; other commissions to light Midland cotton mills and factories followed.[73] Mill owners saw the advantages of gas lighting over oil lamps since gas burned brighter, thereby enabling the owners to extend their hours and productivity. Murdoch and Clegg continued to experiment and develop technology for on-site coal-gas production to fuel individual factories; however, Murdoch failed to grasp the wider implications of his inventions for public and domestic lighting.[74]

Of all the early experimenters with coal gas, and there were many in addition to Clegg and Murdoch,[75] Frederick Winsor appreciated the possibilities of this new fuel. He anticipated a system of central gas production and distribution to serve private homes, light public streets and meet the needs of industry. Winsor, a German who had emigrated to England from France after the Peace of Amiens in 1802, had conducted wood-gas experiments with the inventor Philippe Lebon. Winsor had greater faith in the potential of coal gas and devised a plan for centralised manufacture and massive distribution via underground pipes. In Europe, this scheme was met with general disapproval, given the lack of gas technology and the inconvenience of digging up the streets, so Winsor moved to London in the hope that English investors would be more receptive to his ideas.[76] England was also attractive as it contained the coal resources and mining and transportation systems to support such a venture on a wide scale. Winsor travelled to London to promote his idea and drum up investors with his live public demonstrations of gas lighting and cooking.

As Richard Altick has shown, the public exhibition was a popular form of entertainment and sales promotion with an extensive and curious history. Long before the 1873 International Exhibition or even the Great Exhibition at Crystal Palace in 1851, scientists and charlatans alike used public demonstrations to introduce new inventions and attract possible investors. Winsor adopted this promotional method and conducted a series of public demonstrations at the Lyceum Theatre, London, in 1804. According to Altick, Winsor hired an actor to lecture, because his English was poor, while he demonstrated, but the thespian was prone to drink and was not always available for the performances.[77] When possible, Winsor presented gas jets for lighting and showed off the Imperial Patent Light Stove, recommended for both cooking and heating. Like Buckmaster's lectures 70 years later, crowds flocked to Winsor's performances; however, the Lyceum gas show was more novelty than commerce and, in this respect, it closely followed a genre of scientific lecture designed to inspire awe and amazement initially, and attract financial investors later.[78]

Nonetheless, the Prince of Wales was sufficiently intrigued to invite Winsor to light the conservatory of his London residence, Carlton House, and then to mount a public display on Pall Mall. While visitors to the early indoor demonstrations complained of headaches and nausea due to the use of unpurified gas and improper ventilation, a situation hardly to Winsor's advantage, the outdoor lighting demonstrations had no such side effects and proved more persuasive. By 1808, Winsor had presented a paper to the Royal Society and convinced the Westminster parish authorities to install gas lamps on a section of Pall Mall. He soon acquired a committee of influential backers to promote his ideas for a national chartered gas company, resulting in the formation of the chartered Gas, Light and Coke Company.[79]

The technological history of the English gas industry has been documented elsewhere and, however interesting, is not the focus of this work.[80] What remains significant to this study is Winsor's original emphasis on coal gas as a public service that could substantially improve the lives of both the rich and the poor, in terms of comfort in their homes and safety on their streets. Indeed, gas did burn brighter than oil lamps and was favoured for that reason. Gas lights installed in St James's Park acted as a deterrent to thieves and prostitutes while the new lamps at St Mary's churchyard in Newington, South London, were meant to ward off the unscrupulous 'resurrectionists.'[81] Beyond its material benefits of gas, gas was touted as the smoke-free fuel even at this early date. Conversion to gas could potentially reduce the grimy fogs that were already choking Londoners in the early nineteenth century. The absence of coal soot also eliminated the 'necessity of the most wretched profession among men, that of our degraded and pitiable chimney sweepers', a filthy, hazardous job undertaken by young boys.[82] Although the hot, dirty working conditions at the gasworks and the sooty, smelly fumes of interior gas jets suggested otherwise, Winsor and his supporters presented coal gas as both hygienic and humanitarian. Despite the objections raised by their critics – for example, some members of the medical community – these dual images remained central to gas advertising well into the twentieth century.[83]

The gas industry underwent a period of rapid expansion following the 1812 establishment of the GLCC and within three decades London was served by a dozen gas companies.[84] From the 1860s onwards, the gas industry was subject to numerous parliamentary controls to regulate the quality, supply, price and the amount of dividends payable to stockholders. In essence, the controls were a trade-off, compensation for the virtual monopolies the companies exercised over their respective districts. One such regulation precluded the gas companies from manufacturing gas-making machinery. This clause, in conjunction with the gas engineers' preoccupation with the technical problems of coal-gas production, purification, storage and distribution, ensured the expansion of a manufacturing sector geared to the needs of this burgeoning industry.

For approximately the first 50 years, commercial lighting constituted the largest market for coal-gas production. By the mid-nineteenth century gas lighting had completely transformed all classes in Victorian society, as better-lit streets were safer for night outings and the illuminated theatres and concert halls served patrons of all classes. Although gas lighting was common for public use, it was still relatively unusual in the home. Hot, dirty and smelly, interior gas lamps required adequate ventilation, either by opening a door or window, to prevent the accumulation of sulphurous gas (the result of impurities in the gas and escapes from defective fittings). The manufacturers of gas appliances and fittings – for example, William Sugg and Company – worked to remedy these faults through the development of more efficient gas fittings. In 1868, Sugg patented the London Argand, an adaptation of the Argand oil burner.[85] Improvements to gas meters helped maintain uniform pressure at the works and also measured gas consumption with greater accuracy.[86] Nevertheless, the service to consumers' homes was hardly reliable since fluctuating gas pressure in the mains created an inconsistent supply, sometimes barely enough to keep the gas lamps illuminated. Consequently, the gas companies were often lampooned in the popular English press, criticised for their apparent complacency and indifference to their customers.[87]

During the 1870s, several commercial regulations affected the future development of the industry. Pending the approval of the Board of Trade, the City of London Gas Act (1868) sanctioned company amalgamations (a rationalisation process directed at reducing production costs and, theoretically, costs to consumers). Within 10 years, the GLCC absorbed six smaller London gas companies, while the South Metropolitan Gas Company extended its territory through similar takeovers.[88] The expansionary years following the Gas Act were also characterised by huge capital investments in new gas production technology by the GLCC, and, to a lesser extent, by the SMGC. One notable achievement was the construction of Beckton Works in Barking, named after GLCC director Simon Adam Beck, which initiated production in November 1870. Beckton Works enabled the GLCC to phase out its antiquated gas plants in Central London at Brick Lane, Curtain Road, Blackfriars and Westminster, thereby transferring the bulk of its gas-making facilities to the East End. Construction at Beckton continued throughout the 1870s since each corporate amalgamation increased the company's production requirements, necessitating the closure of older, inefficient works. By 1882, Beckton was three times the size originally planned with 12 retort houses, a labour force of around 2,000 men and a products works facility to manufacture tar and chemicals.[89]

Despite Beckton's state-of-the-art technology, the GLCC's huge capital expenditures and the rising cost of coal prices in the late 1870s kept London gas prices high. In contrast, smaller companies with less substantial capital investments, such as the SMGC, offered gas at a slightly cheaper rate, raising questions from consumers and the Board of Trade. In Birmingham, Mayor Joseph Chamberlain offered direct municipal control as the solution to discrepancies in gas services and prices. While this measure presented some concern for the private gas companies, it reflected, according to Francis Goodall, the 'civic consciousness' of some borough councils who upheld the view that consumers were better served by municipally-owned services.[90] The amalgamations and technological improvements of the 1870s failed to decrease gas prices at the rate anticipated. Dissatisfaction with the gas companies over prices and continued poor service contributed to a developing popular interest in electricity as an alternative source of light and fuel.

Gas versus electricity, and the advent of advertising

Historians of the gas industry maintain that competition with the electricity industry in the late 1870s instigated a burst of technological achievement in order to meet the new competitor head-on. According to Stirling Everard 'the coming of electricity as a competitor of gas lighting prompted the gas companies to explore the possibility of developing the cooking, space heating and water heating loads which hitherto had been very much neglected'.[91] Hugh Barty-King reiterates this point: 'the opinion of most in the gas industry at the end of the 1870s [was] that the days of gas-lighting were numbered, and that the best course for the industry was to shake itself out of its complacency and improve gas-lighting in order to make it acceptable.' To ensure its survival, the gas industry was forced 'to concentrate on perfecting alternative ways of using gas as a fuel'.[92]

This long-accepted view overstates the competitive threat which electricity actually posed at that time. True, while the inventions of Swan and Edison gave cause for concern, a number of gas governors felt secure in the knowledge that electrical technology did not present a serious, immediate challenge. Comparative experiments with electric light undertaken by William Sugg in the autumn of 1878 convinced the GLCC that the new technology was extremely expensive and produced a dimmer light, to the point that a member of board, Colonel William Makins, withdrew the motion that his company apply to parliament for permission to produce and supply electric light as well as gas.[93] Although several gas corporations made applications to local authorities to produce and sell electricity, the London corporations made no such move. Motions to this effect were presented to the GLCC board of governors in 1878 by both shareholders and company directors, but they were dropped, possibly because, as Walter King suggested in the *Journal of Gas Lighting*, electricity would not be an established institution for many years. Perhaps, the directors of the gas corporations felt confident that innovations in their industry could easily keep pace with anything electricity might offer.[94] No doubt Sugg's comparative experiments contributed to this perception. In 1882, Sugg read a paper before the Institute of Civil Engineers, stating the results of his tests on the illuminating power of gas and electricity. He noted that the members of the House of Commons had decreed gas obsolete and wired the

House of Commons with electricity, only to discover the inadequacies of electric lighting. To Sugg's satisfaction, the politicians reverted back to using the gas fixtures.[95]

This is not to say that advancements in electrical technology in the late 1870s had no impact on the gas industry. What worried British gas engineers more than any material achievements in electrical technology was the overwhelming enthusiasm with which the general public, and particularly the press, greeted this new American illuminant. For example, an article printed in *Cassell's Family Magazine* in 1880 entitled 'From Gas to Electricity' prophetically described the incandescent bulb as 'the light of the future' despite its then prohibitive cost.[96] The following October, *Chamber's Journal* ran a piece on electric light in New York, adding that in Britain 'the battle between electricity has fairly begun and it isn't hard to foresee to which victory will ultimately incline'.[97] A year later, *The Times* reported that at the opening of the rebuilt Savoy Theatre, the first public building to be lit entirely with electricity, the audience had cheered loudly in appreciation for the new technology. The original Savoy had been destroyed by a gas-related fire.[98] Reports of this type of popular enthusiasm for electrical lighting worried gas men far more than the actual achievements of the new industry.

Popular excitement over electricity reached a frenzy in 1882. A representative of Thomas Edison mounted an exhibition at the Crystal Palace, where thousands of spectators witnessed the highest efforts of the world's leading electric lighting engineers. The *Journal of Gas Lighting* was typically critical. The editor, Walter King, sourly condemned the Sydenham show as 'a huge advertisement' in the very worst sense: it was unscientific, all art and money, a spectacle of mere showmanship. In King's words, Edison's show did 'a vast amount of harm to impressionable people', as investors speculated in the new electrical companies while the value of gas shares plummeted.[99] Gas engineers responded with their own Crystal Palace show, which opened in December 1882. Sponsored by the Institute of Gas Engineers and jointly funded by gas corporations throughout the country, gas men hailed this International Electric and Gas Exhibition as 'the vindication for gas lighting'. It marked the gas industry's transition from product development to active product promotion at a time when Victorian businessmen in general were beginning to appreciate the benefits of mass advertising.[100]

A number of studies chronicle the development of Victorian advertising and the reasons for its expansion in the mid- to late nineteenth century.[101] For example, in their respective works, W. Hamish Fraser and Lori Anne Loeb attribute the growth of advertising in late Victorian Britain to rising middle-class incomes and the related phenomena of conspicuous consumption fostered by an expanded manufacturing sector that produced household goods, packaged foods, ready-made clothes and cosmetics. But selling gas was not like selling soap. Domestic gas installation was still considered an expensive luxury in 1882 and, despite the gradual increase in domestic gas lighting, coal was the primary fuel for heating and cooking. The 1882 International Electric and Gas Exhibition was followed by a succession of gas shows and lectures throughout Britain as corporations and appliance manufacturers alike consistently employed this method of sales promotion. While manufacturers extolled the benefits of gas lights, stoves and cookers through their public lectures and displays, it took several years for the gas authorities to discover that women made excellent and attractive showmen.

Gas cookery and its early demonstrators

Although Winsor's 1804 Lyceum lectures included illustrations of both gas lighting and cooking, the latter was still relatively uncommon even by the 1880s despite the gradual increase in domestic gas lighting. While the Aetna Iron Works, near Liverpool, manufactured primitive gas cookers as early as 1824, James Sharp, a Northampton gas engineer, was the first to publicise gas cookers and gas cookery.[102] In 1826 he was designing gas stoves, and by 1828 he had one installed in his own home. He staged public lectures and demonstrations in Northampton, and continued to do so after he moved to Southampton. During an 1836 lecture at the Leicester Mechanics' Institute, Sharp extolled the scientific benefits of gas over the competition (the traditional open hearth and the new coal range, then also at the experimental stage). Because gas burned cleaner than coal by any means, it eliminated the attendant problems of smoke damage to food and kitchen interiors. Gas was more economical than coal as the heat could be regulated easily. And since 'cooking by gas requires no attendance'[103] one cook might prepare a large meal for many people. Sharp's circular cooker was a monstrous two-tiered affair which featured a central boiler and several attached steamer compartments above and, below, an enclosed 4-foot by 2-foot oven of circular tin heated by a 15-inch gas ring. Designed primarily for commercial purposes, such as hotel kitchens, the Sharp cooker was commissioned by several large establishments including the Bath Hotel, Leamington. It appears from contemporary accounts that his smaller domestic models were not well received.

Alexis Soyer, a chef and culinary innovator, was another early supporter of gas technology. Having apprenticed in prestigious Parisian restaurants, Soyer later emigrated to London to join his brother who was chef to the Duke of Cambridge. After a series of engagements in aristocratic houses, Soyer became the head chef at the Reform Club. When he redesigned the club's kitchens in 1841, Soyer installed several small gas cookers in addition to the coal, charcoal and steam-heated cooking stoves.[104] Although the gas burners were a novelty, he praised their fuel economy and ease of use, and recommended similar installations for private homes.[105] In one celebrated gas demonstration, Soyer roasted a 535-pound baron of beef in an oven of brick and iron heated with 216 gas jets. Eight men carried the gigantic roast through the streets of Exeter, followed by a brass band and a crowd of 1,300 hungry onlookers. Despite this apparent enthusiasm for the merits of gas cookery, Francis Goodall suggests that Soyer's interest in gas technology was more a 'publicity-seeking search for novelty than a prolonged attempt to introduce gas to a wider market'.[106] While Soyer was no doubt promoting himself first and foremost, his stunts did draw attention to the use of gas for cooking, while the gas manufacturers were quite satisfied to accept this cheap and sensational publicity.

Although the public events described above suggest a growing interest in gas cookery, as do the numbers of gas stoves patented around 1850,[107] the general misgivings about the safety of gas cooking were difficult to dispel. For example, fearing the explosive and toxic qualities of gas appliances, despite the fact that the Crystal Palace itself was gas lit, the exhibition's royal commissioners ruled that no gas apparatus could be exhibited in operation. Disappointed by this decision, the Gasfitters Association mounted a supplementary display at the Royal Polytechnic Institute, Regent Street that prominently featured Sharp's enormous gas cooker.[108]

Nevertheless, the fears of the Crystal Palace commissioners reflected some genuine problems with early gas appliances. Imperfect gas purification and inadequate ventilation of early gas stoves meant that the gas often smelled strongly of hydrogen sulphide. Poor design and heat distribution resulted too often in food that was burnt from the gas jets, or soggy from being steamed in airtight compartments. The knowledge that gas was explosive and toxic if inhaled added to the perception that food cooked in a gas stove was equally poisonous.[109] In addition to gas's known toxicity, gas stoves were relatively expensive. Although its advocates stressed the cheapness of gas in comparison to coal, the initial outlay for a gas stove and the installation of fittings was far higher than the purchase price of a coal kitchener.[110] After 30 years, the gas companies offered customers the option of renting cookers and meters from the gas companies. By this time, however, the coal range had captured the cooker market, while open-fire cooking, whether in a traditional hearth or a small grate, was still common.[111]

By the mid-nineteenth century the 'combination' coal kitchener had grown in popularity especially in the Midlands and the south of England. But what made this cooker particularly popular was its movable panels which could be shifted to expose the fire or enclose it, thus producing the warmth and comfort of the open hearth and the convenience of flat cooking surfaces. It was this image of the open fire that contributed to coal's early advantage over gas. Given its connection with food, warmth and comfort, Doreen Yarwood describes the kitchen as the 'heart of the British home'.[112] The kitchen fire, especially the open hearth, was vital to the well-being of the household and was never allowed to go out. The cosy hearth with its attendant roast beef gently turning on a spit became a powerful image of English domesticity, whether in the grand country house or in the workman's cottage. While the combination kitchener with its glowing coals satisfied this sentimental ideal, the fully-enclosed gas stove did not.

In addition, the efficacy of the gas cooker was dubious regarding the preparation of that national favourite: roast beef. Henry Thompson's directions for roasting meat stressed the importance of a 'clear, bright fire' and basting 'frequently done with drippings from the pan below' using the open hearth and a mechanical device, the 'bottle-jack', to turn the spit.[113] This method of preparation had changed little since the sixteenth century and continued to be the gastronomic ideal as much for the Victorians as it was for the Tudors.[114] In contrast, cooking something inside the oven (or gas stove) was technically described as baking and, despite the provision of spits and hooks inside the gas oven to simulate roasting, people still claimed that the meat's taste did not compare to open-roasted beef. Cooks also complained that a gas stove tainted the drippings and that the critical basting stage was nearly impossible.[115]

Evidently, despite the efforts of gas inventors and chefs, like Sharp and Soyer, gas cookery made no significant gains on the domestic market until the 1880s. Although it had been 40 years since Sharp's first demonstrations, Buckmaster's utilisation of gas stoves in 1873 was still relatively novel and, despite the efforts of a few advocates, the prejudice against gas cooking remained unchanged. Attitudes began to shift with some technical innovations in the gas industry that affected the production and distribution of coal gas and the manufacture of appliances. These innovations were doubly significant given the gas companies' growing awareness of the advantages of advertising to attract customers and to cultivate a positive public image.

Conclusion

The cookery lectures at the 1873 International Exhibition set an important precedent in terms of both sales techniques and public education. The content, form and context of Buckmaster's talks foreshadowed the gas companies' later demonstrations, as they literally incorporated this promotional method. In addition, the 1873 lectures illustrated several important themes related to Victorian perceptions of class, gender and domesticity – perceptions which ultimately shaped the domestic economy movement and, by implication, the image and duties of the gas companies' lady demonstrators.

Buckmaster's talks suggest the growing impact of technology and consumption on the late nineteenth-century urban home. Given the developments in gas technology by the 1870s, this particular industry increasingly affected the Victorian home by setting new standards of comfort and efficiency. Gas lighting, as yet unchallenged by electricity, epitomised domestic modernity. By the 1880s, however, the highly publicised achievements in electric technology convinced gas engineers of the importance of consumer education. As we shall see in the following chapter, the lady demonstrators of the late 1880s served the somewhat separate agendas of those that supported the domestic economy movement and those that supported the manufacturers of gas appliances and fuel. At the same time, their developing profession contributed to new education and employment opportunities, particularly for middle-class women with aspirations to enter the masculine world of business and advertising.

Notes

1 C. Herman Senn (ed.), *Cookery Souvenir of the Universal Food and Cookery Association* (London, 1894), 23.
2 'The Queen at Kensington', *The Times*, 10 May 1873, 5.
3 'Cooking at the International', *The Times*, 23 April 1893, 4.
4 Geoffrey Best, *Mid-Victorian Britain, 1851–1875* (London, 1971), 74, 277–278.
5 Walter E. Houghton, *The Victorian Frame of Mind, 1830–1870*, reprint of 1957 edition (New Haven and London, 1985), 341–348.
6 'Cookery and diet', *Lancet,* I (1859), 338–339.
7 Charles Elme Francatelli, *A Plain Cookery Book for the Working Classes* (London, 1861).
8 John Burnett, *Plenty and Want: A Social History of Food in England from 1815 to the Present Day*, 3rd edition (London, 1994), 228–229.
9 'The International Exhibition', *The Times*, 2 May 1873, 7.
10 Adrienne Munich, *Queen Victoria's Secrets* (New York, 1996), 82–83.

11 Richards, *Commodity Culture*, 57–72.
12 Deborah Cherry, *Painting Women: Victorian Women Artists* (London, 1993), 129.
13 Dorothy Stone, *The National*: *The Story of a Pioneer College, National Training College of Domestic Subjects* (Lewes, Sussex, 1976), 5–6.
14 'Cooking at the International', *The Times*, 25 April 1873, 4.
15 Stone, *The National*, 25.
16 Richard Altick, *The Shows of London*; Schaffer, 'The Consuming Flame', in Brewer and Porter (eds), *Consumption and the World of Goods*, 489–526.
17 'The Queen at Kensington', *The Times*, 10 May 1873, 5.
18 'Cooking at the International', *The Times*, 25 April 1873, 4.
19 Catherine Hall, 'The Formation of Victorian Domestic Ideology' in Sandra Burman (ed.), *Fit Work for Women* (London, 1979); Leonore Davidoff, *Worlds Between: Historical Perspectives on Gender and Class* (New York, 1995); Catherine Hall and Leonore Davidoff, *Family Fortunes, Men and Women of the English Middle-Class, 1700–1850* (Chicago, 1987).
20 Hall, 'Victorian Domestic Ideology', in Burman (ed.), *Fit Work for Women*, 15. On the formation of the English middle class, see Harold Perkin, *The Origins of Modern English Society, 1780–1880*, reprint of 1969 edition (London and New York, 1991), chaps 6 and 7.
21 Hall, 'Victorian Domestic Ideology', in Burman (ed.), *Fit Work for Women*, 21–29; Davidoff, *Worlds Between*, 51–56.
22 Hall and Davidoff, *Family Fortunes*, 56–58.
23 John Tosh, 'Domesticity and Manliness in the Victorian Middle Class: The Family of Edward White Benson', in Michael Roper and John Tosh (eds), *Manful Assertions: Masculinities in Britain since 1800* (London, 1991), 48–51.
24 Hall and Davidoff, *Family Fortunes*, 114–115.
25 Ibid., 180. For a discussion of the impact of gender on political and social issues during the 1830s and 1840s, see Anna Clark, *The Struggle for the Breeches: Gender and the Making of the British Working Class* (Berkeley, 1995), chaps 11–13; Barbara Taylor, *Eve and the New Jerusalem: Socialism and Feminism in the Nineteenth Century* (London, 1983).
26 Catherine Hall, 'The History of the Housewife', in Ellen Malos (ed.), *The Politics of Housework*, (London, l980), 62.
27 Sparke, *As Long As It's Pink*, 31–49.
28 Deborah Gorham, *The Victorian Girl and the Feminine Ideal* (London, 1982), 149–151.
29 Joan N. Burstyn, *Victorian Education and the Ideal of Womanhood* (London, 1980), chap. 5.
30 Patricia Branca, *Silent Sisterhood* (London, l975), 55.
31 June Purvis, *A History of Women's Education in England* (Milton Keynes, 1991), 6–7.
32 Cherry, *Painting Women*, 82–83.
33 Elizabeth Sewell, 'Principles of Education', reprinted in *The Education Papers: Women's Quest for Equality in Britain, 1850–1912* ed. Dale Spender (New York, 1987), 144.
34 Purvis, *History of Women's Education*, 65.
35 Dena Attar, *A Bibliography of Household Books Published in Britain 1800–1914* (London, 1987), 12.
36 Davidoff, *Worlds Between*, 85.

37 Ibid., 87; Adams, *Architecture in the Family Way*, chap. 5.
38 Purvis, *History of Women's Education*, 7.
39 Sarah Freeman, *Isabella and Sam: The Story of Mrs. Beeton* (London, 1977), 186.
40 Ibid., 191.
41 Branca, *Silent Sisterhood*, 23.
42 Purvis, *History of Women's Education*, 7–10.
43 Ibid., 11.
44 Hall, 'History of the Housewife', in Malos (ed.), *The Politics of Housework*, 69; see also Sally Alexander, 'Women, class and sexual differences in the 1830s and 1840s: some reflections on the writing of a feminist history', *History Workshop Journal*, **17** (Spring, 1984), 139–140.
45 Hall, 'History of the Housewife', 69.
46 Keith McClelland, 'Masculinity and the "Representative Artisan" in Britain, 1850–80', in Michael Roper and John Tosh (eds), *Manful Assertions, Masculinities in Britain since 1800* (London and New York, 1991), 82–86.
47 James A. Schmiechen, *Sweated Industries and Sweated Labour: The London Clothing Trades 1860–1914* (Urbana, Ill., 1984), 40–42.
48 Ibid., 52.
49 Ellen Ross, Love and Toil: *Motherhood in Outcaste London 1870–1918* (London and New York, 1993), 152.
50 Clark, *Struggle for the Breeches*, 253–255.
51 Hall, 'History of the Housewife', 68.
52 'Cooking at the International', *The Times*, 25 April 1873, 4.
53 J.F.C. Harrison, 'The Victorian gospel of success', *Victorian Studies*, **1** (December, 1957), 155–164.
54 'Cookery and diet', *Lancet*, I (1859), 338.
55 Stone, *The National*, 11–14.
56 Ibid., 14–15.
57 Burnett, *Plenty and Want*, 169–173. According to Burnett, there was little change in the actual budgets of town-workers from the mid-century to the late 1870s and even up to World War I. Differences in living standards were determined by occupation, number of dependants and location. The semi-skilled worker with a marginally comfortable weekly wage of around 30 shillings might spend 14 shillings per week on food, mainly on bread, cheese, potatoes, meat, milk, sugar and fruit. Sweated workers, the worst-paid members of the labouring poor, might earn 12 shillings per week with a food budget of only three shillings per week; for a contemporary account see Maud Pember Reeves, *'Round About a Pound a Week*, reprint of 1913 edition (New York and London, 1980), 75–93. According to this study of Lambeth labouring families, weekly wages averaged around 25 shillings, with a weekly food budget of between 7 and 10 shillings.
58 Lady Burdett-Coutts quoted in C. Herman Senn (ed.), *Cookery Annual: Yearbook of the Universal Cookery and Food Association*, (London, 1902), 24.
59 Henry Thompson, *Food and Feeding*, 6th edition (London, 1891), chap. 1.
60 Dena Attar, *Wasting Girls' Time: The History and Politics of Home Economics* (London, 1987), 38–41.

61 LMA/LCC 1/ LCC /*Minutes of the London County Council Subcommittee on Cookery*, 20 July 1876, 78.
62 Helen Sillitoe, *A History of the Teaching of Domestic Subjects* (London, 1933), 37.
63 LMA/LCC 1/LCC/*Minutes of the London County Council Subcommittee on Cookery*, 11 July 1879, 194.
64 Jane Rendall, '"A Moral Engine"? Feminism, Liberalism and the English Woman's Journal', in Jane Rendall (ed.), *Equal or Different? Women's Politics 1800–1914* (Oxford, 1987), 126.
65 Helen Corr, 'The Schoolgirls' Curriculum and the Ideology of the Home, 1870–1914', in Glasgow Women's Studies Group (ed.), *Uncharted Lives: Extracts from Scottish Women's Experiences, 1850–1982* (Glasgow, 1983), 79, 83–84.
66 Annmarie Turnbull, 'Learning Her Womanly Work: Elementary School Curriculum 1870–1914', in Felicity Hunt (ed.), *Lessons for Life* (Oxford and New York, 1987), 96–97.
67 Historians of women's education have debated the reasons for the introduction of domestic subjects into national curriculums. For a critical account of domestic science teaching, see Attar, *Wasting Girls' Time*, 37–38; Anna Davin, 'Imperialism and motherhood', *History Workshop Journal*, 5 (Spring, 1975), 60–65. Historians of domestic subjects colleges adopt a more positive view of the discipline. See Neville Marsh, *The History of Queen Elizabeth College: 100 Years of University Education in Kensington* (London, 1986), 31–36; Margaret E. Scott, *The History of F.L. Calder College of Domestic Science, 1875–1965* (Liverpool, 1967), 14–21; Stone, *The National*, 20–25; Sillitoe, *A History of the Teaching of Domestic Subjects*, 37.
68 'Lecture delivered by J.C. Buckmaster, Esq. J.P. at St. James' Hall, 10 November 1892', reprinted in C. Herman Senn (ed.), *Cookery Souvenir, of the Universal Food and Cookery Association* (London, 1894), 131.
69 Gareth Stedman Jones, *Languages of Class, 1832–1982* (Cambridge, 1989), 192–193.
70 'The international exhibition', *The Times*, 2 May 1873, 5.
71 'International exhibition cookery', *Illustrated London News*, 17 May 1873, 458.
72 Davidson, *Woman's Work*, 63–66.
73 Everard, *History of the Gas, Light and Coke Company*, 14–16; Davidson, *Woman's Work*, 33.
74 Everard, *History of the Gas, Light and Coke Company*, 16–20.
75 Malcolm Peebles, *Evolution of the Gas Industry* (London, 1980), 7–9; Davidson, *Woman's Work*, 33; Barty-King, *New Flame*, 13–15.
76 Barty-King, *New Flame*, 20.
77 Altick, *Shows of London*, 371.
78 Schaffer, 'The Consuming Flame', in Brewer and Porter (eds), *Consumption and the World of Goods*, 490–494; Altick, *Shows of London*, 371.
79 Everard, *History of the Gas, Light and Coke Company*, 18–21; Barty-King, *New Flame*, 29–32.
80 Everard, *History of the Gas, Light and Coke Company*; Peebles, *Evolution of the Gas Industry*; Williams, *History of the British Gas Industry*; Wilson, *Lighting the Town*.

81 Barty-King, *New Flame*, 73, 76.
82 Everard, *History of the Gas, Light and Coke Company*, 18.
83 'Gas and water pipes', *Lancet* (1866), 139; 'Gas stokers', *Lancet* (1867), 707. The editor of this London medical journal claimed that, due to careless installation practices on the part of the large urban gas companies, leakage from gas pipes had contaminated municipal water supplies. The second article described the working conditions of stokers and called for an inquiry into the effects of the long hours and hot temperatures on the men's health.
84 Williams, *History of the British Gas Industry*, 26–27, 34; Chatterton, 'State control of public utilities', 166–167.
85 Williams, *History of the British Gas Industry*, 32.
86 Barty-King, *New Flame*, 50.
87 Goodall, *Burning to Serve*, 50–51.
88 Chatterton, 'State control of public utilities', 173–175.
89 Everard, *History of the Gas, Light and Coke Company*, 239–249, 264–266.
90 Goodall, *Burning to Serve*, 52–53.
91 Everard, *History of the Gas, Light and Coke Company*, 263.
92 Barty-King, *New Flame*, 135–136.
93 LMA/B/GLCC 34/GLCC, *Minutes of the Board of Directors*, 18 October 1878, 454; GLCC, *Minutes of the Board of Directors*, 23 November, 1878, 459.
94 'The lighting of the Fine Art and Industrial Exhibition at Worcester', *JGL*, 19 September 1882, 512.
95 William Sugg, 'Gas as an illuminating agent compared with electricity', paper read before the Institution of Civil Engineers (London, 1882).
96 'From gas to electricity', *Cassell's Family Magazine* (1880), 282–285.
97 'Electric light in New York', *Chamber's Journal*, 29 October 1881, 696.
98 *The Times*, 12 October 1882, quoted in Barty-King, *New Flame*, 136; see also 'The electric light in theatres', *Lancet*, I (1882), 188–189.
99 'The exhibition of gas apparatus and lighting at the Crystal Palace', *JGL*, 4 July 1882, 11.
100 'The encouragement of gas consumption', *JGL*, 9 May 1882, 830.
101 Fraser, *Coming of the Mass Market*; Loeb, *Consuming Angels*. See also Philip B. Meggs, *A History of Graphic Design*, 3rd edition (New York, 1998), chap. 11.
102 Williams, *History of the British Gas Industry*, 35; Davidson, *Woman's Work*, 66; Goodall, *Burning to Serve*, 29.
103 'A gas cookery lecture in 1836, reprinted from the *True Sun*, 1836', *Copartnership Journal of the South Metropolitan Gas Company* (August, 1925), 189.
104 Doreen Yarwood, *The British Kitchen: Housewifery Since Roman Times* (London, 1981), 95; Davidson, *Woman's Work*, 66; Sarah Freeman, *Mutton and Oysters: The Victorians and Their Food* (London, 1989), 169–173.
105 Goodall, *Burning to Serve*, 29.
106 Ibid., 31.
107 'Economic gas cooking stove', *JGL*, 13 January 1849, 3. Charles Ricketts patented a gas burning trivet suitable 'for humble dwellings of the urban working population'. Also see Hardyment, *From Mangle to Microwave*, 123.

Three gas stoves were patented in 1850 by Alfred King of Liverpool, James Sharp from Southampton and Ebenezer Goddard of Ipswich.

108 Hardyment, *From Mangle to Microwave*, 123; Yarwood, *British Kitchen*, 95; 'Apparatus for cooking by gas', *The Expositor*, 12 April 1851, 1.

109 Freeman, *Mutton and Oysters*, 117–118.

110 Ibid., 120. During the 1850s, the cost of a solid-fuel range varied from £1 and 18 shillings to £25 for a large model compared to £16 and £28 for a small and medium gas cooker respectively, plus a £2–6 installation fee for fittings and meters.

111 Davidson, *Woman's Work*, 60–63.

112 Yarwood, *British Kitchen*, 11.

113 Thompson, *Food and Feeding*, 76.

114 Yarwood, *British Kitchen*, 73.

115 Freeman, *Mutton and Oysters*, 120.

CHAPTER 2

Exhibitions and the Spectacle of Selling

> When the future historian comes to write an account of the world's progress in the latter half of the nineteenth century, he will, if he wishes to be faithful, give due prominence to one special feature which has contributed to that progress in a very marked degree. We refer to the important series of exhibitions which have followed the wonderful display made in the glass house in Hyde Park in 1851.[1]

As the author of the above quote suggested, the public exhibition became a familiar promotional technique for Victorian entrepreneurs and manufacturers, anxious to advertise the perceived connections between technological advancement, commercial goods and social progress. Further, while expositions presented new inventions and products to the public, they often provided the settings for intense commercial rivalries, as competing firms rivalled one another for the judges' awards of merit and, perhaps more importantly, favourable press coverage of their displays. Exhibitions were also the place where manufacturers could survey the state of their own industry vis-à-vis their rivals, and there was nothing as serious to the businessman as taking a good, hard look at the competition. This was exactly the mission of a contingent of gas engineers who on 28 February 1882 inspected the Electrical Exhibition at the Crystal Palace.[2] This exhibition was a promotional event designed to stimulate interest in England's fledgling electrical industry; Edward Johnson, a representative of Thomas Edison, had organised the public display. Johnson was an experienced electrical engineer as well as a promoter: he oversaw the operation of Edison's first telephone and lighting systems introduced to England in 1878. In addition, Johnson was responsible for the company's first central station in London, which opened in 1881, just months before the Crystal Palace electrical displays.[3] Evidently, the gas engineers had sufficient cause for concern.

The electrical exposition, held in the building's Central Nave, opened in late January; entrance to the lavish displays was included in the general admission price of one shilling. It must have been a wonderful sight: the gigantic greenhouse, Joseph Paxton's fantastic 'monument to the commodity',[4] radiant with the white light of electrical chandeliers in brilliant contrast to the grey, foggy winter of South London. According to contemporary reports, thousands of patrons visited the Crystal Palace that winter to view the dazzling rows of electric lights, the latest innovations from Europe and America, and to marvel at the endless assortment of electrical inventions, from motors and telegraphs to electropathic belts and hair brushes. The Electrical Exhibition offered its visitors a promising glimpse of a dazzling future, where new levels of unimaginable comfort and hygiene would be possible for all. The promise made at the Great Exhibition of 1851 seemed nearer fulfilment than ever before. Yet, despite the crowds and the excitement over electrical technology, the fascination with electrical gadgets and the rush to invest in the new electrical companies, the gas managers concluded, after their inspection of the Electrical Exhibition, that electricity posed no serious or immediate challenge to their own interests.

What could have induced the gas engineers to assume this nonchalance? Simply, it was the high cost of electrical technology and the conviction that electricity offered no greater competition than petroleum oil did. Reporting on the Electrical Exhibition before the Midland Association of Gas Managers, president R.O. Paterson admitted that electricity must be viewed as a 'rival illuminant' but, like oil, was not adaptable to the varied needs of an entire community. Paterson based his assumptions on the inordinate amount of equipment required to produce electrical power:

> One need only compare the long array of boilers, engines, dynamo machines, shafting etc. required to generate the electric current to do but a portion of the lighting of the Crystal Palace, with what would be required to supply, say, the Strand, to be convinced of the utter futility of the whole thing to supersede gas.[5]

In addition, Paterson cited the conflicting statements by electrical engineers regarding the cost of electric lighting. He also noted the questionable price estimates, based on 'theoretical evidence' taken from experiments performed under the most ideal conditions, as further cause for doubting the efficacy of electricity as a viable alternative to gas. Walter King, the outspoken editor of the *Journal of Gas Lighting* reiterated this position to the point where verbally attacking the electrical industry and its engineers became standard copy in the gas trade magazine. Having viewed the Edison show, King reassured his readership that gas engineers could return to their own work 'fully possessed of a renewed confidence in the future of gas lighting'.[6]

If the 1882 Electrical Exhibition did not persuade the gas companies to fear their new competitor, it did convince them that the electricity interests had one definite, if not dubious, advantage: they were better salesmen. Given the mounting public enthusiasm for electricity throughout the 1880s, the gas companies grasped the importance of self-promotion just as Victorian businessmen in general considered the commercial benefits of advertising. By the late 1880s, appliance manufacturers and gas companies were adding women to their sales teams, and discovering that they made convincing and persuasive sales representatives. While the cookery teachers were sympathetic to the commercial motives of the gas corporations, they undertook this promotional work for their own reasons. Advertising and sales offered new employment opportunities for women with home economics diplomas. Further, it extended the impact of the domestic science movement beyond the artificial realm of the classroom and directly into the Victorian home. This chapter traces the gas companies' active movement into the realm of advertising through the medium of public exhibitions and their use of lady demonstrators to sell their goods and services. To combat the negative image of coal gas and its association with soot, its smell and toxic vapours, the industry adopted the soft-sell approach. Using feminine persuasion to domesticate their product, gas managers sought to reassure consumers that coal gas was not simply a fuel for public and industrial applications. Lady demonstrators and lecturers reassured consumers, and particularly female consumers, that gas was safe and convenient for use in the home; this message had the important effect of encouraging gas consumption via a growing list of domestic applications, including using gas for lighting, cooking and heating.

From monopolists to traders

The 1851 Great Exhibition provided a blueprint for all the subsequent commercial exhibitions, from South Kensington's International Exhibitions of the 1870s to the Health and Smoke Abatement Exhibitions of the 1880s.[7] The electrical and gas companies were indebted to the 1851 Great Exhibition; they both attempted to recapture some of that commercial magic by holding their respective shows in the Crystal Palace. Even in the early twentieth century, the newspaper baron and proprietor of the *Daily Mail*, Alfred Harmsworth, titled Lord Northcliffe in 1905, imitated the Crystal Palace format for his Ideal Home Exhibitions, first held at Olympia, London, in 1908.[8] Public exhibitions also provided important venues for the articulation of Victorian national and imperial pride by presenting and reifying images of race, class and gender.[9]

Historians of Victorian Britain agree that the Great Exhibition of 1851 signified important social, economic and commercial changes. Following the unrest of the 1830s and 1840s, it inaugurated a period of unparalleled economic prosperity and relative political stability.[10] The exhibition paid tribute to the Victorian gospel of self-help and hard work made apparent by the superiority of English manufacturing; while also demonstrating the apparent success of peace via free trade and constitutional monarchy.[11] The Crystal Palace became the metaphoric ideal for mid-Victorian industrial capitalism. Jeffrey Auerbach offers a revisionist perspective of the Great Exhibition when he notes that although the Crystal Palace defined Britain as an industrial nation, design reformers, including Henry Cole, anticipated that the displays of foreign goods would highlight the perceived deficiencies in British design, which they believed were the result of Britain's inadequate provisions for industrial education. Further, Auerbach asserts that the Great Exhibition was not so much an expression of British industrial prowess, but a means of reviving the lagging economy of the late 1840s.[12] Despite earlier concerns regarding the moral and social consequences of Britain's industrial development, the 1851 Great Exhibition showed British spectators that the nation had been transformed from a rural pre-industrial society to an urban manufacturing economy.

In addition to its economic and political significance, the 1851 Great Exhibition marks the distinct break between traditional and modern forms of mass entertainment. Richard Altick contends that the Crystal Palace embodied the 'lofty ideal of rational entertainment' which included the opening of suburban parks and national monuments to the public.[13] The legacy of 1851 was seen in the new exhibition halls constructed for popular entertainment and commercial displays, including the rebuilt Crystal Palace at Sydenham, the Alexandra Palace at Muswell Hill and Earl's Court in Olympia.[14] With the gradual progress of popular education and rational recreation, the charlatans and freaks of the travelling shows commanded less credence, while the educational role of the public exhibition gained in prominence.

The 1851 Great Exhibition could also be described as the first international trade fair.[15] It illustrates unequivocally that the capitalist system had created a 'dominant form of exchange' and was engaged in the process of manufacturing a 'dominant form of representation' which later legitimised advertising as a progressive business practise.[16] In his semiotic analysis of the Great Exhibition and Victoria's Jubilee celebrations of 1887 and 1897, Thomas Richards suggests that the 1851 show established a language and format for representing the commodity. Its organisers incorporated the popular devices of Victorian spectacle – monumental architecture, fountains and elaborate scenery – to exalt manufactured goods, forever linking commodities to a technology of representation. In addition, the Great Exhibition invented 'a democratic ideology for consumerism', not of a classless society, but one in which 'everyone was equal in the sight of things', the assumption being that all people desired the same consumer goods even if they could not afford them.[17] The new 'rhetoric of representation' attributed to the Crystal

Palace transformed commercial practice by reifying the commodity exhibition. After 1851, the trade exhibition constituted the fundamental means of promoting consumer goods as it incorporated the winning combination of entertainment, education, leisure and commerce. Although the urban department store also fulfilled this role, particularly from the 1880s onwards, it never challenged the importance of the commercial trade exhibition, which still operated as an important venue where retailers could showcase their latest wares.[18]

For decades, gas enthusiasts recognised the promotional value of the public display. The German inventor Frederic Winsor first presented gas demonstrations in London's Lyceum Theatre. On later occasions, coal gas enthusiasts, including chef Alexis Soyer and gas engineer James Sharp, demonstrated gas cookery, and then paraded around the streets with gas-roasted sides of beef accompanied by brass bands and throngs of hungry followers.[19] The Institute of Gas Engineers mounted two displays in 1851: one at the Crystal Palace and the other at the Royal Polytechnic Regent Street. The latter clearly illustrates the commercial motivations of the gas engineers since the Great Exhibition commissioners forbade any machine or apparatus to be displayed in operation; however, one can only imagine the noise and physical danger to the public if such a rule had not been enforced. Two decades later, at the 1873 International Exhibition, the public orator J.C. Buckmaster drew attention to the gas stoves of the Leoni company that he used in a series of popular cookery lectures.[20]

By the late 1870s, following these intermittent displays, the gas companies and appliance manufacturers began to organise regular trade exhibitions to promote gas for purposes other than lighting. In 1877, W.J. Warner, the engineer of the South Shields Gas Company, sponsored an exhibition which apparently attracted around 30,000 visitors over its week-long engagement. Gas undertakings in Birmingham, Bradford, Halifax, Leeds, Wakefield, Rochdale and Newcastle-on-Tyne sponsored similar exhibitions within several months. The shows were mounted by both private and municipally-owned undertakings but with the common purpose of increasing gas consumption by countering public prejudice against gas appliances, particularly the gas stove. For example, after the Leicester town council purchased the town's gasworks in 1880, the new municipal company organised a trade exhibition so that consumers might actually see the advantages of using gas, instead of coal, for cooking.[21] The 1882 Worcestershire Industrial Exhibition, held in Bristol, featured a gas apparatus exhibition which included cooking stoves, lamps and engines. In an effort to familiarise the public with the production process of extracting gas from coal, the show also featured a model gasworks, supplied by Mr Hurt of Birmingham. The replica included a retort house and a gas holder.[22] Perhaps by presenting a miniature, sanitised version of a gasworks, the exhibition organisers hoped to impress upon the public the high level of technical expertise required to produce, purify, store and transport coal gas.

An article in the *Journal of Gas Lighting* praised the efforts of the gas undertakings with regard to the increasing frequency of promotional exhibitions, and congratulated the South Shields Gas Company for inaugurating the trend five years earlier. According to the author, the exhibitions had done a great deal to educate consumers regarding recent advancements in gas cooking stoves, engines and, thanks to the development of the inverted gas burner, improvements for domestic lighting.[23] However, in view of the comparable activities of the electrical inventors, there was little time for self-satisfied complacency. The following revealing quote suggests the perceived urgency of increasing sales through public exhibitions.

> Fortunately it appears to have become recognized (in theory, if not in practice) that it is now the duty of gas companies to discard the role of the monopolist, and take up that of the trader; and indeed that if gas managers at all desire the continued success of the concerns which are under their guidance, they must be prepared not only to exhibit their wares before the public eye, but to persist in extolling the advantages until its attention is attracted.[24]

The author, under the pseudonym of 'Owen Merriman', recommended the continued practice of 'gas apparatus exhibitions' but suggested the addition of free lectures on the uses of gas 'given in an interesting and popular style', and 'clearly illustrated' to encourage attendance and dispel misinformation. This article, which appeared shortly after the 1882 Edison display, indicates that the gas corporations were regular participants in trade exhibitions and were well aware of their benefits. Nonetheless, some felt the need to improve their sales pitch, perhaps by combining consumer education with entertainment through the medium of the popular lecture.

After the triumph of the 1882 Electrical Exhibition, something equally impressive was required, beyond a few eloquent spokesmen publicly fiddling with stoves. Something spectacular had to follow to counteract the 'advertising evil' of the electricians 'not because it is ... necessary in the view of scientific and experienced engineers, but because it will be politic with regard to the public, and will help restore confidence'.[25] According to Walter King, the 'artful displays of lighting in gorgeously decorated and furnished rooms' had a far greater impact on the consumer than any trade pamphlets by Swan and Edison. It was the spectacle that mattered. By proposing a gas exhibition of similar size and grandeur to the recent electrical display King urged the gas companies to act quickly. According to King, the absence of a special display of gas lighting to compare with the recent electric light show 'would be interpreted by the public and blazoned in the newspapers as an acknowledgement of weakness and defeat on the part of the representatives of an effete system'. The bombastic editor considered the 1882 Electrical Exhibition a challenge to which they must respond and if his colleagues failed to take action, they risked not only their market, but their reputations as businessmen.

As business writers like Walter King knew, the successful Victorian entrepreneur was undoubtedly a man of action. Contemporary prescriptive literature characterised the businessman as a gentleman soldier of commercial fortune. For example, among the essential personal qualities necessary for the successful business career, Samuel Smiles listed punctuality, honesty, attentiveness and accuracy in all matters. In addition, the businessman of the highest order required 'quick perception', the ability to execute plans efficiently and expeditiously, and like a military general, to organise and direct others with tact and firmness. According to Smiles, both Napoleon and Wellington were great generals precisely because they were also first-rate men of business.[26] By the 1880s, prescriptive literature and middle-class guidebooks characterised the businessman as the epitome of masculinity. The struggle for commercial success was often couched in the language of Herbert Spencer's 'social Darwinism' that emphasised the survival of the fittest, thus imbuing the rather sedentary and acquisitive pursuits of business with images of physical manhood.[27] In light of these popular associations, the Electrical Exhibition was potentially highly damaging to the nascent public image of the gas industry and its engineers. If they ignored this apparent commercial challenge, the gas industry and its promoters may have appeared weak, feminine and, therefore, unable to compete effectively in the corporate world. From King's point of view, then, the industry spokesmen had to respond to the electrical challenge with a strong commercial statement of their own.

From the outset, the 1882 Gas Exhibition was the largest publicity campaign the industry had ever undertaken. Indeed, it was the first collective exhibition ever staged by a group of independent and widely distributed organisations. The show was technically the responsibility of the CPGC which served the suburbs of south-west London; it assumed the financial management and risk of staging the show. However, the company directors appealed to the combined assistance of the gas interests to raise the necessary capital and ensure the success of the venture. The *Journal of Gas Lighting* editorials supported cooperative action to achieve the desired 'never-to-be-forgotten manifestation of the possibilities of gas lighting'; selfish apathy would have resulted in a mere 'gas-fittings bazaar, without unity, organization or purpose'.[28] The Gas Institute, a professional group of managers and engineers, nominated a representative committee chaired by George Livesey, the chief engineer of the SMGC, to assist the Crystal Palace gas managers with raising funds and staging the exhibit.

Despite Walter King's persuasive appeals, general support for the joint venture was not immediate. For example, the governors of the GLCC rebuffed Livesey's initial requests for subscriptions.[29] However, with the Electric Lighting Bill put to the House of Lords for a second reading and after Livesey's persistent appeals, the GLCC eventually reversed this decision. Corbett Woodall, secretary of the GLCC, was nominated to the organisational committee bringing with him a donation of £500 towards the forthcoming exhibition, thereby matching the contribution of the SMGC.[30] Donations from gas corporations throughout Britain brought the total to just over £2,000, a sum that practically guaranteed success.

Scheduled to open in late November 1882, the gas engineers rushed their exhibition. Possibly, by following soon after the Edison show, they meant to provoke a positive comparison with electricity while the previous exhibit was still fresh in the public's memory. In addition, since the electrical technicians were invited to participate, this gave the competition little time or resources to mount yet another lavish show. Nevertheless, the gas engineers had difficulty in meeting their own deadline. For a month, the CPGC ran apologies in *The Times* for the postponed opening. Indeed, when the Lord Mayor officially opened the International Electric and Gas Exhibition in mid-December 1882, many of the exhibits had not been installed. Further, the electrical lights and displays confined to the north half of the building were still unfinished by opening day to the obvious satisfaction of the gas companies.[31]

The aggrandisement of the gas industry and its fundamental contribution to the comforts of Victorian life was carefully orchestrated in the numerous displays. In addition to the old gas jets along the palace roof, the South Nave was fitted with gas lamps 30 feet above the floor, which virtually flooded the building with light. The palace's decorative theme rooms – the Pompeiian, French and Birmingham Courts – which were theatrical in themselves, were particularly spectacular when lit with the latest incandescent gas lamps and burners from manufacturers in England and the continent. Rows of stalls exhibiting gas apparatus, lamps, stoves and engines filled the South Nave and the Western Corridor. To match the decorative rooms of the previous electrical show, a suite of rooms furnished to middle-class taste illustrated the domestic applications of gas, with particular emphasis on the kitchen. Judging by the space devoted to the exhibition, and the number of participants, it was an ambitious event.

The organisers, investors and participants of the Sydenham show, which closed in May 1883, hailed it a marvellous success. As a commodity spectacle, this exhibition brought to the consumer a new awareness of the possible uses of gas in the home; it was one huge advertisement. Moreover, the show reassured the investing public, the stockholders of both the private companies and the members of the municipal corporations that the uses of gas were still growing. By the spring of 1883, gas stocks regained former values while the market speculation in overvalued electrical shares declined dramatically, given the exposure of some dubious patents.[32] The triumph of the organisers was marred slightly, however, by public accusations that the judge's awards of merit had been unfairly granted to Livesey's personal friends, thus creating a long-standing feud between the SMGC's chief engineer and George Bray, a prominent gas fittings manufacturer from Leeds.[33] In addition, there were suggestions that toxic gas fumes had killed the tropical plants inside the Crystal Palace. The rumour was substantiated by the revelation that exhibitors had wastefully overused gas as they literally attempted to outshine one another, despite the cooperative premise of the exhibition. However, the gas interests denied any responsibility for damage to the vegetation, since the building had more than adequate ventilation. They attributed the nasty rumours to embittered electricians.[34]

Despite the enthusiasm of journalists and electrical engineers for the new industry, British electrical technology lagged behind that of America, Germany and France. Foreign companies, specifically Siemens (1881) and Edison (1882), sponsored England's first two power stations, but neither turned a profit and both closed in 1884.[35] After these slow beginnings, the British electrical industry made some important gains in the 1890s. R.E.B. Crompton's experimental power station in the West End electrified a new development of terraced mansions. It proved to be more successful than the earlier ventures, but Crompton had the advantage of serving an elite Kensington clientele. Entrepreneurs complained that the regulations of the Electrical Lighting Acts of 1882 and 1888 discouraged private investors. However, as Leslie Hannah suggests, a greater deterrent was the financial losses after the 1882 electrical 'bubble' and the inability of the electrical industry to compete with the established gas industry.[36]

The gas industry's 1882–83 exhibition marked an important turning point in terms of commercial practice. Like many Victorian businessmen, the gas managers were recognising the need to advertise. In general, the leading members of the Gas Institute realised that they had to promote their industry actively to develop and maintain their market and public confidence. No longer monopolists, like any other trader, they had to sell their services and products. It mattered little whether electricity was a real or a hollow competitor if the public believed that gas was outmoded, or soon to be. Astute businessmen, like Walter King and George Livesey, contended that the 'commodity spectacle' was the best means of challenging the electrical interests, and the public exhibition became the gas industry's principal promotional strategy.

By admitting that advertising constituted an essential commercial practice, gas managers were among a growing number of Victorian entrepreneurs and retailers who were anxious to extend their markets and quash their competitors. Historians of material culture generally agree that the development and legitimisation of advertising as a business coincided with the years of Victoria's reign.[37] The queen herself became a popular motif in advertisements for everything from cocoa to velvet; even her Jubilee celebrations were important commercial events, heavily sponsored by merchants and manufacturers. But despite steady expansion from the mid-century onwards, the business of advertising still carried disreputable connotations. Originating with the sandwich-board men, the billposters and the barkers described by Henry Mayhew, advertising was perceived as both a questionable practice and a public nuisance.[38] According to some Victorian social commentators, notably Thomas Carlyle, self-aggrandisement or 'puffery' was a degrading and wasteful practice; quality goods would sell themselves without gimmicks, slogans and gaudy images.[39] However, a growing number of professional advertising agents and manufacturers thought otherwise; both groups were willing to risk some degree of criticism if it meant increasing product sales. Moreover, the Great Exhibition of 1851, which had focused unprecedented attention on the commodity, was practically an advertisement for advertising, despite the lofty political pretensions of Prince Albert.

By the mid-nineteenth century, the cost of advertising was considerably less than ever before. Improved printing practices after 1848 lowered the price of wall posters and contributed to the great boom in public billposting; hereafter, all available urban space, on hoardings and buildings, was plastered with posters. As Chancellor of the Exchequer, William Gladstone's reformist fiscal policies, notably the repeal of the paper duties in 1860 and 1861, provided another important boost to the advertising industry.[40] Consequently, the number of newspapers and consumer magazines grew, as did the amount of printed space per issue devoted to commercial advertising.[41]

As advertising prices decreased, the most ambitious companies invested vast amounts in self-promotion; however, the majority of manufacturers resented this growing trend. The purveyors of soaps and patent medicines were the most extravagant in this respect. For example, by 1875, Thomas Barrett's advertising budget at A & F Pears exceeded £100,000, spent on everything from paintings by John Millais to endorsements from Lillie Langtry.[42] James Beecham invested £95,000 in advertising in 1884, while Thomas Holloway, another celebrated chemist, spent £50,000 in 1883. Businessmen were pressed to advertise in some form since it appeared that, in any given product area, the first firm to advertise experienced so dramatic an increase in sales that its competitors were forced to follow suit.[43] As more companies were tempted to advertise, reservations about the respectability of the practice were slowly overcome. Increasingly, its opponents levelled their criticisms, not at the practice of advertising but at its most tasteless manifestations; this included the erection of poster billboards in Trafalgar Square or the projection of magic lantern images on the National Gallery.

In view of the growing dependence on advertising, Walter King's support for more self-serving promotional techniques was not surprising. As we have seen, the International Electric and Gas Exhibition of late 1882 was in direct response to Johnson and Edison's show earlier that year. Despite the rush of manufacturers to the newspapers and the popular journals, the gas industry's preferred advertising medium was the public exhibition. Individual appliance companies, like the Main Co. and John Wright & Co., sponsored newspaper advertisements for their stoves and heaters. Similarly, William Sugg and Company, a leading manufacturer of gas lamps, lighting apparatus and gas stoves, was also an active promoter. Descriptive pictorials of Sugg's 'Aladdin' Reading Lamp appeared in the *Illustrated London News* in 1887.[44] Nevertheless, the appearance of printed advertisements for gas services was relatively unusual before 1900.

In contrast to the infrequency of its press campaigns, the gas interests regularly participated in the growing number of commercial and educational exhibitions. For example, the gas appliance manufacturers entered wares in the South Kensington International Exhibitions in the mid-1870s; this was soon followed by the practice of mounting exhibitions devoted exclusively to gas apparatus.[45] Touting the cleanliness and comfort of gas as compared to solid fuel, the gas corporations and manufacturers were prominent supporters of both the Manchester Smoke Abatement Exhibition, held in the spring of 1882, and the International Health Exhibition staged in London in 1884. The industry used both of these events to construct a positive public image of gas as reliable, affordable and safe. Given consumers' generally low opinion of the Victorian gas companies and gas engineers, this attempt to reform the industry's image was not only ambitious, but essential. According to Sarah Milan, nineteenth-century consumers viewed the gas companies with contempt because of the high incidents of gas leaks that resulted in explosions or accidental deaths. Popular fiction lampooned the gas engineer as a fraudulent charlatan who overcharged the public despite the poor and unreliable service. Furthermore, the smoke pollution and waste associated with the extraction of combustible gas from coal created a highly negative impression of the industry – one that was associated with the worst effects of industrialisation. Consequently, the adoption of gas technology in the home was construed by some Victorians as the invasion of the industrial world into the domestic.[46] Even though improvements to gas meters and mains helped regulate the gas supply, and the industry attempted to train gas fitters and supervise installations, the image of the disreputable gas companies remained largely unchanged by the 1880s.

By linking conversion to gas lighting and cooking with public health, improved housing and national well-being – the organising themes of Smoke Abatement and International Health exhibitions – the gas industry attempted to elevate the acquisition of gas appliances from a convenient luxury to a critical necessity. In an attempt to eradicate the associations that coal gas had with noxious fumes and smoky interiors, published testimonials by physicians and scientists substantiated the gas engineers' claims that the use of gas appreciably reduced smoke pollution and was a safe, non-toxic and inexpensive fuel for cooking. The Smoke Abatement Committee tested gas cookers of various manufacture and size and offered its positive endorsement.[47] Victorian manufacturers, particularly of food, drink and patent medicines, frequently emphasised the restorative and health-giving qualities of their products and sought testimonials to this effect. The gas interests went further. By incorporating the crusade for national health into their promotional pitch they assumed the role of benevolent capitalists motivated by both economic and humanitarian concerns.

The business of selling

One of the most decisive marketing strategies that the gas companies adopted was the decision to rent out appliances to their customers. While the trade exhibitions introduced new technology to Victorian consumers, the rental payment scheme circumvented the obvious difficulty of paying for a large appliance in one lump sum. In the 1880s, gas cooking stoves were not cheap. The purchase price of a cooker averaged between £5 and £7 or about one-tenth of a year's wages for a working-class man in steady employment. Only the affluent middle and upper classes were able to afford them; moreover, it was a considerable outlay for householders who simply wished to try gas cooking on an experimental basis. The CPGC were the first to implement the rental system during the 1860s, in an attempt to remedy this economic difficulty.[48] Several years later, the governors of the SMGC also resolved to adopt the practice of hiring out stoves to its customers, with the intention of increasing the consumption of gas by encouraging the general use of the gas cooking stove. The rental rate charged was just 10 per cent of the cost of the stove per annum; however, the governors were confident that the inexpensive rental costs would be offset by an increase in gas sales.[49] Livesey put this suggestion to the directors in June, which was the low season for gas consumption and a time when housewives would have been most inconvenienced by the disadvantages of cooking over a coal fire.

In 1880, the Leicester Gas Company elected to rent cookers to its customers to promote gas consumption. Although local ironmongers accused the gas company of unfair competition, the Leicester governors responded that they were not intentionally undercutting the tradesmen, but simply trying to increase consumer demand for their product, something which the tradesmen had been unable to do since none of them dealt exclusively in gas appliances.[50] By the mid-1880s, a growing number of gas companies throughout the country had adopted the rental system with positive results.[51] In his 1882 annual report to the SMGC shareholders, company secretary and chief engineer George Livesey attributed the past year's increased gas consumption, despite a mild winter, to the 'considerable addition of gas cooking stoves let out on hire'.[52]

The hire-purchase system was not unique to the gas companies. Enterprising furniture salesmen used this method of credit as a way of stimulating sales in the 1870s and 1880s, particularly in rural areas.[53] Sewing machines had been financed by monthly instalment payments as early as the 1860s; by the 1880s, the Singer Sewing Machine Company employed over 30 collectors calling for weekly instalments in London's East End, a district infamous for its clothing sweatshops.[54] However, according to W. Hamish Fraser, the hire-purchase system applied to only a few items before 1914, including sewing machines and pianos. His study reveals that in the latter decades of the nineteenth century Victorian retailers were less inclined to offer their customers credit and increasingly demanded cash payments. The no-credit policy maintained a steady cash flow, and eliminated tedious book-keeping and the difficult task of repossessing goods in default of payments. The new department stores and multiple shops strictly denied credit sales. In compensation, these progressive retailers, like Thomas Lipton or Jesse Boots, guaranteed their customers lower competitive prices.[55]

As retailers, the gas companies were ideally placed to offer instalment payment schemes. They insisted on advance payment for stove rentals. If a customer defaulted on a payment, the gas service was discontinued, rendering the stove useless. Householders could choose between three payment options: they could purchase a stove outright in cash; hire-purchase their cooker by quarterly instalments paid in advance at the company offices; or rent a stove from the company, again by advance quarterly payments made with the company cashier. Usually, maintenance costs were included in the quarterly rental fee, which averaged between two and three shillings, depending on the company. Some undertakings, like the SMGC, hired out gas heaters for hot water at a quarterly rate of three shillings and four pence.[56] Service coverage varied as well. The SMGC payment scheme included all maintenance and repair costs for its cookers, while the GLCC specified that some services were extra and payable prior to repairs.[57] These options were irrelevant to the consumer who had decided to use gas since the gas companies held a virtual monopoly in their respective districts.

Repossession of the appliance followed any lapse in payment, and the large companies like the SMGC and the GLCC employed stove inspectors to undertake this work when necessary. In addition, because all gas meters, fittings and rented stoves were considered the property of the gas companies, these items were legally protected from distress under the terms of the Gas Act. The following case of Mr T.A. Hardy illustrates this point. A Mr Mulford rented a stove from the GLCC, but when his property rent fell into arrears Mr Hardy seized and auctioned the stove. The GLCC brought Mr Hardy, the auctioneer, before the bench claiming that the stove was a mere fitting for gas and, as the property of the company, was not liable to distress for rent or to be taken in execution thereof. The gas companies asked for this protection as essential in this aspect of their trade and, although the initial ruling was not in the GLCC's favour, the company won a later appeal.[58] In a sense, this ruling insured the gas companies' goods against unlawful resale, a guarantee that ordinary retailers with hire-purchase agreements did not possess. The gas companies' hire schemes were successful because they used the terms of the Gas Act to protect their rental appliances. In addition, by demanding quarterly, and not weekly, advance payments, they offered somewhat exclusive credit arrangements that required fewer, but larger, payments than most instalment schemes. Because this type of payment plan was still beyond the majority of the working population who operated on slim weekly budgets, the rental clients were more solvent financially and, therefore, were at a reduced risk of default. In cases of lapsed payments, the companies simply shut off the gas first, and retrieved their property later.

From its implementation, the hire system worked exceedingly well. Although company reports indicate that there were stove returns, either from default or dissatisfaction, the number of cooking stoves hired out increased exponentially throughout the 1880s and 1890s. In June 1889, George Livesey reported that the SMGC had 10,000 cookers on hire; a year later, the number reached 13,864. Livesey estimated that approximately 19 per cent of their customers took advantage of the hire system, while a 'considerable number' of patrons owned their cooking stoves outright.[59] The directors of the CPGC estimated that they had

1,171 cookers on hire, while only 258 consumers had purchased theirs.[60] Evidently, in conjunction with the public exhibition, the hire system increased domestic gas consumption and helped overcome the traditional prejudice against gas cookery. By the late 1880s, the gas companies were beginning to add a new dimension to their promotional strategy. They used lady demonstrators (lady demons), often graduates of the cookery schools, to promote their wares at the trade exhibitions.

Lady demons and the travelling show

From evidence in gas company records and industry publications, the lady demons made their public debut in mid-1888. While it remains unclear which company or individual first decided to place women on the exhibition platform, it appears that the London gas companies and the appliance manufacturers initiated the use of lady salesmen around the same time. Beginning in the summer of 1888, the SMGC staged regular exhibitions of gas apparatus, featuring cookery lectures, in all parts of the company's South London district. According to Livesey, the added publicity resulted in 'a great demand for stoves and in the dissemination of much practical information amongst consumers'.[61] Livesey's report does not specify whether the lecturers were male or female, but the omission of this detail may indeed suggest that women were doing the talking. Earlier accounts of illustrated lectures in the gas press advertised the name of the male chefs who addressed the crowds; however, it is probable that the SMGC engineer responsible hired female cookery teachers for these in-house lectures because female lecturers were cheaper and readily available from the domestic science colleges. It is also possible that Livesey did not wish to advertise to his directors that the company was hiring women for publicity purposes, a move that might have appeared controversial. In a later report to the board of directors, Livesey notes that the experiment with illustrated cookery lectures was repeated throughout the spring and summer of 1889. Although the report still does not specify the gender of the cookery demonstrators, Livesey opines that the events met 'with great success' and the SMGC received orders for over 200 stoves per week 'for many weeks in succession'.[62] Other companies were more forthcoming in allowing women to receive some of the credit for the accelerating demand for gas cookers. In the spring of 1889, the Richmond Gas Co. of London and Warrington – manufacturers of cooking stoves, meters and gas fittings – mounted an exhibition of gas appliances at the Workington town hall. Mrs Sophie Thwaites, a domestic economy teacher from Liverpool, offered a course in practical cookery in connection with the Richmond appliance display.[63]

Around this same time, the CPGC reported that it had enlisted three women – the cookery author Mrs Alting-Mees, and teachers Miss Stelfox and Miss Agnes Kelman – to give free cookery lessons on gas stoves at parish halls within their district.[64]

The purpose of the lessons was simple: to increase gas consumption through gas cookery. This is confirmed by the CPGC governors' explicit instructions that 'the stoves to be used in the lectures are to be those let out on hire by this company and that none other will be allowed in the lecture halls'.[65] While it is difficult to imagine the cookery teachers smuggling in unauthorised cookers, some lady demons were under contract with particular appliance manufacturers.[66] Presumably, the CPGC took care to engage teachers who were able to promote the make and models of appliances that the company was capable of supplying. The lessons were designed to instruct women in the methods of gas cooking as a means of encouraging them to rent a gas cooker, and company engineers distributed blank hire forms to the audience after the demonstrations. According to a report presented by the CPGC's chief engineer, the majority of those in attendance, an average of 150 women per lesson, already had gas cooking stoves; however, the customers needed instruction regarding the use and care of their new appliances. Nevertheless, the engineer reported an increase of more than 50 per cent in stove rentals compared to the same month in the previous year. In addition, 53 new gas stoves were hired out for May 1889 with another 85 for June.[67] This increase suggests that the lessons simultaneously assisted current customers who already had gas appliances and won over new customers to the company.

But, despite the evidence, the CPGC directors were unconvinced that the cookery lessons directly translated into sales. According to company records, stove rentals increased steadily throughout the entire year, and not simply when the company offered the spring and summer lessons. In January 1890, cookers on hire totalled 1,501; two years later, in January 1892, the numbers reached 2,235, indicating a steady rise despite the infrequency of the cookery lessons. Although it was reported that Miss Kelman's lessons were always well attended, the CPGC directors discontinued the annual spring cookery lessons.[68] They made this decision at the very moment when gas companies outside London were expanding the role of their female cookery advisors. Geography may account for this difference. Presumably, consumers in London could obtain information on gas appliances and rental schemes at the growing number of company showrooms, another marketing strategy initiated by the metropolitan companies around 1890.[69] Smaller companies lacked the necessary capital and staff to outfit and supervise permanent showrooms, and depended more extensively on local trade exhibitions and lady lecturers as their means of sales promotion.

If the London companies were ambivalent about employing women to sell gas stoves, the appliance manufacturers and rural gas companies were not. Literally dozens of small exhibitions organised by the gas appliance manufacturers toured the countryside. No exhibition was complete without a lady demonstrator to show off the best uses of the gas stove through her culinary expertise. The gas exhibitions were a form of travelling show. For example, during the winter of 1892, the Richmond Gas Co. toured Sheffield in January, Carlisle in February and Blackpool in March. Miss Golding, of London, delivered twice-daily cookery lectures at all three exhibitions, which usually ran for one or two weeks.[70] During that same winter, Richmond sales crews appeared in Darlington, Newcastle-on-Tyne, York, Middlesbrough, Southport, Putney, Llandudno, and the list goes on.[71] Every show presented free cookery lectures by lady demons, including Madame Gothard of London, Miss Owen of South Kensington and Mrs Bennett. On a few occasions, a professional chef was engaged, or Mr E.M.T. Richmond presented his essay 'Gas as a Public Servant'. Not to be outdone by Richmond's sales blitz, the Davis Gas Stove Co., John Wright & Co. and the Fletcher Russell Corporation all mounted similar travelling exhibitions that year, complete with cookery demonstrations by qualified female experts. A photograph of a Davis Gas Stove Co. crew 'on the road', taken at a 1905 gas exhibition in Bognor, presents two unnamed lady demons posing with their male colleagues (see Fig. 2.1). The placement of the appointed women, who stand on-stage with the appliances behind and

2.1 Two lady demons were among the representatives from the Davis Gas Stove Co. who attended the 1905 gas exhibition in Bognor

to either side of the seated men, suggests that the lady demons were part of the exhibition itself: temporary and portable. In contrast, the seated men in the photograph's foreground were permanent sales representatives and gas engineers with fixed and stable positions in the industry. However, women were soon lobbying for a recognised position within the industry by advocating permanent jobs for women in gas sales departments.

The appliance shows were often hosted by the local gas authorities and mounted in town halls throughout England. The exhibitions' organisers considered the class and gender of audiences when determining their programme. Afternoon demonstrations often presented 'high class cookery' designed for middle-class female consumers; 'plain cooking' demonstrations occurred in the evenings in an attempt to draw a mixed working-class audience. The exhibitions indicated that selling gas and appliances was a collective effort since the shows were co-sponsored by the local gas boards, tradesmen and the appliance manufacturers. For example, the local Miners' Institute sponsored a gas exhibition at Parkgate and featured the popular cooking talents of Mrs Sophie Thwaites, the same cookery teacher that had demonstrated for the Richmond stove company on previous occasions.[72] This method of advertising must have been both expensive and labour-intensive, considering the cost of transporting heavy gas appliances and publicising the events, in addition to the effort of setting up and presenting the displays on a fortnightly basis. However, it must have achieved results, since manufacturers continued this practice and sales increased. For example, in Sheffield, the number of stoves hired out rose from 400 to 1,200 in only one week, after the Richmond display and Miss Golding's persuasive talks.[73]

In some cases, the initiative to present cookery lectures originated with the women themselves. In 1894, Miss Ida Cameron approached the directors of the GLCC with an offer of free instruction for their customers, given at her London cookery school, in the use of gas stoves. Cameron was an experienced instructor. A graduate of the Liverpool Training Centre of Cookery and a former cookery mistress in a London board school,[74] Cameron was a familiar lecturer for the Davis Gas Stove Co. The governors agreed to Miss Cameron's proposal, purchasing tickets for the classes at one shilling each to be distributed among their female customers. This indicates that, in some cases, the initiative to hire women came from the domestic economy experts themselves. A year later, Cameron was engaged by the GLCC to run a cookery series for the company somewhere in London's East End, suggesting that she successfully promoted her professional services by securing another engagement with the London company.[75] According to Agnes Kelman, who had worked for the CPGC in 1889, lady demons were in high demand and could earn 'really large sums of money' working as freelance lecturers. Apparently, in addition to the work she did for the gas companies and appliance makers, Miss Kelman organised her own speaking engagements by renting 'town halls, theatres, music halls, skating rinks, drill halls and old churches' for her cookery demonstrations which attracted paying audiences of up to 1,000 people.[76]

Cameron and Kelman may have been inspired by the example of other businesswomen successfully earning a living through the promotion of the domestic arts. For example, Mrs Agnes B. Marshall, the proprietor of a successful cookery school on London's fashionable Regent Street, conducted public cookery performances for audiences of up to 600 people, apparently entertaining her audiences while offering practical instruction. While the demonstrations were a means of advertising her school, Mrs Marshall was also selling domestic equipment, as her premises featured a large retail showroom that stocked everything for the Victorian kitchen, from fancy moulds, to cutlery and gas cookers. Regarding the latter items, customers were assured that Mrs Marshall had tested and cooked with all of the leading brands.[77] Given this example, other women may have set themselves up in the demonstration field, albeit on a more modest scale, but, evidently, this was a new source of employment for educated women with entrepreneurial spirit.

By the 1890s, women had found a place in the English gas industry as appliance demonstrators, albeit on a contract basis. Several factors account for this development. One reason was the impact of competitive advertising. The CPGC and the Richmond Gas Co. experimented with lady demons and, given their evident success, the competition followed suit. Although the use of female demonstrators was a novel practice, it did not originate with the gas companies. Like the hire-purchase payment schemes, the use of lady demons was another technique that may have originated with the sewing machine industry. Apparently, from the early 1860s, Isaac Singer employed attractive young women to operate sewing machines in the window of his New York shop. As Singer's agencies spread throughout America, Britain and Europe, each store maintained a staff of at least one salesman, a competent mechanic and a lady demon.[78]

The importance of using lady demons should not be underestimated. While the women provided practical instruction in the use of the item, whether it was a sewing machine or a cooker, they simultaneously and publicly feminised, and domesticated, new technologies. In her study of the Paris garment trades, Judith Coffin convincingly illustrates how the sewing machine companies feminised their technology by directing their advertising and their sales techniques at women and girls in the home.[79] The gas industry adopted a similar approach, increasingly using lady demons to cultivate a demand among female consumers. Initially, however, this marketing tactic meant a change of direction. Prior to the

1880s, the majority of gas cookers in use were located in the kitchens of public institutions, such as clubs, hotels and hospitals, and were used almost exclusively by professionally trained male chefs. For example, recall that a professional male chef had demonstrated the use of a gas stove at the 1873 International Exhibition in Kensington while a male lecturer provided the commentary. In its report, *The Times* contrasted the 'venerable chef du cuisine, clad all in white like a bridegroom' with the four 'buxom kitchenmaids' in 'coquettish mob caps'.[80]

In the above instance, it was very clear that the men were in charge of using the technology and disseminating the information about it, while the women offered assistance and, more importantly, sex appeal. From the mid-1880s onwards, however, the gas companies and appliance manufacturers focused their attentions on domestic consumers by directing their sales campaigns at private homeowners and housewives. In order to transform the unfamiliar gas cooking stove from an industrial tool, operated by men, into a familiar domestic appliance, used daily by middle-class housewives and their plain cooks, the gas companies had to make gas cooking stoves appear safe and simple to operate. No doubt, gas promoters hoped that the sight of their lady demons using the stoves would convince reluctant female consumers that they too could master the art and science of gas cooking. By using women to sell the cookers to other women, the gas companies and manufacturers developed an effective means of gender product identification. It was part of the process of feminising the gas cooker by shifting its association away from the public institution of the restaurant or the hospital, and into the private scullery of the middle-class home. The use of lady demons might also be interpreted as another means of countering the public's negative associations of the gas industry with the worst aspects of industrialisation.

The appliance manufacturers and the gas companies recognised that, by the 1890s, shopping had become an increasingly feminine activity for both buyers and sellers. Each woman's habit of consumption, as she made the rounds of the fashionable West End shops, served as a status symbol of her husband's wealth. According to Judith Walkowitz, the expansion of the mass market and the growth of the department stores in the late nineteenth century increased this consumerist activity to a wider range of middle-class women, while it established new social patterns for the act of shopping itself.[81] Cafes, teashops and rest rooms catered to the needs of the feminine shopper. Eventually, in an effort to attract female shoppers into their emporiums, department store owners, such as Gordon Selfridge, included lounges, writing desks and luncheon rooms within their commercial premises. Selfridge's advertisements urged female shoppers to spend their entire day in his store, depicting his shop as a self-contained social centre.[82] Within the department store, these feminine areas provided unescorted ladies with places where they could relax and enjoy themselves in public without fear of harassment. The afternoon high-class cookery demonstrations sponsored by the gas companies and designed to attract middle-class ladies aptly characterises this connection between female leisure, consumerism and entertainment. Further, the cookery show was simultaneously a public and a private space, like the department stores, where respectable ladies might enjoy an afternoon's amusement unescorted and undisturbed.

The replacement of male shop assistants with females contributed to the feminisation of material consumption.[83] Young, single middle-class women were increasingly employed as sales assistants in the new department stores and multiple shops. Retailers realised quickly that neatly dressed pretty girls made a favourable impression on their customers and were particularly understanding of the fashion needs of female shoppers. Lady Mary Jeune praised the patient and perceptive sympathy of shop girls for their often exhausting customers in all matters of fashion, colour and taste, stating that 'no man can ever understand all these little refinements'.[84] Shop girls had the added bonus of being cheap and tractable and, despite marginal improvements in hours and wages, the daily routine meant standing for 12-hour shifts with few breaks and no fresh air or physical exercise.[85] As sales assistants, the lady demons fared much better than their contemporaries in the large shops in terms of hours and status. Nevertheless, they still represented the movement to female sales personnel and the realisation on the part of businessmen that selling was an increasingly seductive activity.

Businessmen were not the first to grasp the publicity value of female promoters. Victorian Evangelical groups – particularly the Salvation Army, which was founded in 1878 – frequently used female speakers because, as Catherine Booth maintained, they were more successful 'in gaining the ear of the people'.[86] The public spectacle of female salvationists attracted capacity crowds for the sheer novelty of seeing women speak in public. Doubtless, the same type of curiosity that filled town halls to see female Evangelicals also accounts for the capacity audiences of the lady appliance demonstrators at their evening performances. In contrast to the highbrow cuisine of the matinées, the evening shows of 'plain cookery' were designed to appeal to mixed audiences, both in terms of class and gender. The *Journal of Gas Lighting* frequently contained comments on the number of men present at these evening demonstrations, with the implication that male customers attended to observe the public spectacle that the women offered and not because they had need of culinary advice. Perhaps the men had hoped for on-stage 'buxom kitchenmaids'. In much the same manner that the female Evangelicals sold religion, the lady demons were feminine boosters for the gas industry.

But the promotion of gas appliances was a specialist business and required trained personnel. Improperly used, gas technology was potentially dangerous, as one unfortunate demonstrator proved to her audience while showing a stove at the Lostwithiel Cookery School.[87] The lady sustained facial injuries after an explosion when she went to light the stove; apparently, the gas had not been properly turned off from the previous lecture. Information concerning the proper handling and safety of gas stoves was critical to the sales pitch. Physicians regularly expressed their concerns about the safety of gas stoves, citing an incident where a customer had suffered serious injury as a result of an explosion that occurred as she opened the oven door. The injured party received a settlement of £50.[88] Not surprisingly, the gas companies wished to avoid accidents of this nature, either to its demonstrators or its customers, given the resulting litigation and negative publicity.

The domestic science colleges, which were established and respected institutions by the 1890s, supplied these qualified lady demons. In most cases, the cookery graduates were more than willing to promote gas appliances for the manufacturers and corporations since this provided an additional source of employment to the board cookery schools. By working through the public lecture circuit, experts in the domestic subjects might extend their professional credentials and broaden the general impact of the domestic science movement. Headmistresses of cooking schools offered published testimonials for particular stoves.[89] Cooking mistresses doubled as lecturers on the exhibition circuit.[90] Increasingly, the cookery instructor was a trained professional whose opinion carried weight, much like the doctors who endorsed Victorian patent medicines. But, unlike the bogus referrals found in the medicine advertisements, the cookery teachers personally and publicly endorsed these appliances. Although the editorial staff at the *Gas World* worried that cookery teachers were unqualified to judge the technical merits of particular stoves,[91] the demonstrators assessed appliances on the basis of the criteria that mattered most to female consumers: how easy they were to use and maintain.

Faith in the lady demons was bolstered by Marie Jenny Sugg's popular book, *The Art of Cooking By Gas*.[92] Published in 1890 by the wife of William Sugg, the book was both technical and practical. It presented detailed descriptions and illustrations of various gas appliances and their practical uses in the home along with pages of recipes tested by Mrs Sugg in gas cooking stoves. Apparently, Mrs Sugg was the daughter of a French chef and a culinary expert in her own right.[93] Not surprisingly, she spoke warmly of how William Sugg's award-winning Westminster Kitchener was suited to the cooking needs of the 'average family'. The GLCC adopted Sugg's Westminister as its standard rental stove. In 1897, Lillie Richmond published a similar cookery manual advocating gas in the home. Although

Miss Richmond aimed 'to suit the tastes and requirements of all classes', the pages devoted to servants and their misuse of gas appliances, as well as the recipes in French for garnishes and hors d'oeuvres does suggest that the book was directed at affluent middle-class readers. Still, it was a popular book and had reached its fourth edition by 1903, with sales over 30,000.[94] Both of these publications furthered the perception that women were able and effective spokespeople for the gas industry.

The domestic science movement had an important effect on gas sales in terms of creating a market for new technology in the home. The enormous increase after 1860 in books, magazines and compendiums devoted to household management directly reflects the increased complexity and specialisation of middle-class housekeeping and the perceived need to assist women in meeting the higher expectations of mid-Victorian England. The domestic science movement developed out of this apparent 'housekeeping crisis'.[95] It helped foster a more informed middle-class female consumer who was increasingly familiar with issues such as home management, infant health and hygiene. From the 1890s onwards, middle-class women were more aware of the benefits of modern appliances, including the gas stove. For example, household manuals, like Mrs Beeton's *Household Management*, included comparisons of cookers, furnishings and lighting arrangements. As Christina Hardyment notes, a comparison of the 1883 and 1893 editions reveals the increasing acceptability of gas cooking. In the former edition, the editors praised the cleanliness and efficiency of gas cookers, but cautioned that few cooks knew how to use them safely and with good results. In the latter edition, ten years later, the account of gas cooking was more enthusiastic, possibly because of the increased number of gas cookers in use and the declining prejudice against this method of cooking.[96] The official approval which gas stoves received in the 1893 edition coincides with the gas industry's exhibition blitz of 1892–93. The widely-advertised free cookery lectures by lady demons during the blitz addressed the 1883 editors' earlier misgivings concerning the safety and accessibility of gas stoves. The 1893 positive endorsement then constituted the best form of mass advertising, given the manual's extensive readership.

As noted in the previous chapter, philanthropists and educators anticipated that home management instruction would indirectly improve the health of the labouring classes. Lady Burdett-Coutts and J.C. Buckmaster, ardent supporters of domestic economy training, firmly upheld this view, as did Edric Bayley, a prominent member of the London School Board.[97] Better cooking on more scientific cookers provided another possible means to this end. Such an alliance between the gas industry and the domestic economy advocates gave the gas corporations an unprecedented opportunity. As school officials struggled to equip their housekeeping centres, appliance manufacturers and gas corporations offered to install and maintain classroom cookers in exchange for their quarterly rental fee of between two and three shillings, depending on the company.[98] From the mid-1870s, the Fletcher Russell Corporation supplied the gas cooking stoves to the London cookery centres. They were usually their standard No. 140 model, a squat, solid caste-iron box that resembled a safe with a grill and four burners. Several lady inspectors reported complaints against the Fletcher stoves regarding their poor ventilation and emission of noxious fumes. Children and teachers complained of headaches, drowsiness and nausea, and, in some cases, girls were too ill to continue with their afternoon schoolwork following the morning cookery class. As a result of these complaints, the London County Council Sub-committee on Cookery instructed the appliance manufacturer to immediately provide proper pipes in all cases of these stoves to ventilate the fumes, otherwise the school board would seek replacements.[99] Obviously, the corporations met the school board's demands, since they hoped that if girls learned how to cook on gas stoves, they might influence their families to hire cookers, or at least, they would choose gas cookers for their own kitchens in the future. Conversely, future sales were threatened by the early bad impressions of gas technology.

Gas technology for the working classes: the 'penny-in-the-slot'

Historians of domestic technology identify the pre-payment meter, patented in 1888 by W.E. Price, as one of the most significant innovations in Victorian gas technology in terms of the creation of new markets.[100] Its inception coincided with another important innovation. In 1884, while conducting experiments with thorium and cerium, Austrian metallurgical chemist Carl Auer (later Baron Auer von Welsbach) discovered a process whereby a chemically treated fabric held over a gas flame produced a brilliant white incandescent light. Auer's experiments were greeted with scepticism by the English gas trade press, and it took several years for him to perfect the chemical formula of thorium and cerium that was used to soak the mantles. By 1891, the chemist had successfully developed a robust upright mantle that gave 50–60 candle power for up to 3,000 hours; the improved Welsbach mantles were mass produced by Auer's Incandescent Gas Light Company, based in Westminster.[101] The Welsbach incandescent mantle, and its later imitators, revolutionised gas lighting by producing a brighter and cleaner gas light for both public and domestic gas installations. The Welsbach system was expensive, however, particularly because of Auer's aggressive defence of his exclusive patents. Limited distribution also affected the price, given Auer's decision to market his mantles and burners through local ironmongers and not the English gas suppliers, as the tradesmen sold their wares with a higher mark-up than the gas companies.[102] Eventually, however, Welsbach's company resolved its differences with the gas undertakings by agreeing to market its lighting systems through the new gas company showrooms.

It is of significant importance that the introduction of the incandescent mantle coincided with the development of the pre-payment gas meter, first patented in 1888. Although the first proposals to adapt this technology to gas lighting and appliances were greeted with some scepticism, experiments with the pre-payment meters by the Ramsgate Gas Company soon convinced the larger undertakings of the efficacy of the new technology. Birmingham and Liverpool soon offered its customers the option of pre-payment meters, followed by London's two largest companies (the SMGC and the GLCC, in 1892 and 1894 respectively).[103] As its use increased, gas men realised that the penny-in-the-slot meter opened new markets by enabling them to sell gas to the lower-middle and working classes by providing a more affordable payment system than the usual advance quarterly instalments. In 1892, the year the company launched the penny-in-the-slot meter, the SMGC offered to install the penny mechanism on any meter or stove in their jurisdiction and gave a small bonus to any employees who obtained new customers for this service. The company supplied and maintained the fittings. In return, however, slot meter customers paid slightly more per cubic foot of gas than 'regular' advance payment customers.[104]

The penny-in-the-slot meter had an immediate effect on gas sales. George Livesey estimated that the SMGC would gain over 6,800 new customers for the year ending June 1893 as a result of the new payment scheme.[105] The majority of the slot meter customers were members of the lower-middle and working classes. According to Livesey's report the increased consumption of gas resulted from 'the demand of weekly tenants who, with scarcely any exceptions, did not use gas until the introduction of the coin meters'.[106] Landlords and tenants favoured the hire system, since neither party invested permanently in the gas fittings or appliances as the gas company assumed the installation and repair costs.[107]

The philanthropic housing corporations and their movement to re-house the 'respectable' working classes also affected the demand for pre-payment meters. In 1895, the GLCC signed a 21-year agreement with the Metropolitan Association for Improving the Dwellings of the Industrious Classes to supply pre-payment meters to all of their buildings.[108] Between June and December 1901, the SMGC reported the installation of 7,130 penny-in-the-slot meters in London model tenements.[109] The increasing popularity of the automatic meter can also be gauged by the growing number of workers assigned to the task of emptying them – a job that was done on rotation every four–six weeks. To give an impression of the numbers of penny-in-the-slot meter customers, the SMGC employed 90 collectors in 1901 at an annual salary of £60.[110] Although we know relatively little about their personal circumstances, it is possible that the collectors occupied the same situation as other white-collar workers in the lower-middle class, including clerks and shop assistants. Caught between the labouring classes, below them, and the more affluent members of the business and professional middle class, white-collar workers often internalised middle-class social values and aspirations of upward mobility, but lacked the income to fulfil these ideals.[111] As outdoor representatives of their employers, the companies stipulated that the collectors had to dress respectably. Despite their low salaries and ambiguous class status, the collectors often wore black coats and bowler hats (see Fig. 2.2).[112] The men travelled their districts on foot, checking customers' meters to ensure that the mechanisms were operating properly. They then emptied the meters, tallied the coins and carried away the heavy coppers in canvas bags for later deposit in the company accounts at designated bank branches. Although their salaries were low, the collectors occupied positions of trust because they were responsible for the company's money and public image, given their direct contact with customers on a regular basis.

2.2 An Edwardian gas company representative, in business attire and bowler hat, empties an automatic pre-payment meter in a customer's home

The pre-payment meter did have its disadvantages. Emptied every four–six weeks, penny-in-the-slot meters were easy targets for thieves, presenting temptations too great for some customers, and even for a few company collectors. The gas companies did not hesitate to prosecute offenders. For example, in October 1895, two 'boys', Edward Guerney and William Cox, were convicted of stealing money from one of the GLCC meters. Several months later, the GLCC charged Mr W. McCarthy with a similar offence, and the court sentenced that errant customer to 14 days' imprisonment and hard labour.[113] That same year, Enrico Poggi received the stiff penalty of nine months' imprisonment for stealing money from a meter. The following year, J.W. Reid and W. Merridew received respective sentences of 18 and 12 months' hard labour for similar offences.[114] It would appear that increasingly harsh punishment was meted out for stealing money from meters, perhaps in an effort to discourage this type of crime, bearing in mind that a prison term, even of several months, could have meant utter destitution for the families of the accused.[115] Meter inspectors were instructed to keep a strict watch on the meters and report any discrepancies. For example, a West Ham court ruled in favour of the GLCC in a case arising from a tampered meter valve which gave the customer considerably better value than the normal 25 cubic feet of gas per penny.[116] Before a Lambeth court judge, one SMGC gas inspector declared that meter thefts were increasing steadily, with his company experiencing at least 'four cases within the previous few weeks'.[117] In addition to the lost revenue, the companies worried that damaged meters, if undetected, could result in lethal gas escapes and explosions. In February 1896, the governors of the GLCC decided to engage two plain-clothed constables to detect people stealing money from the automatic meters.[118] However, this type of petty theft continued unabated, with at least one clever thief posing as a gas company collector to gain access to the customers' meters. In this particular case, the GLCC tried, albeit unsuccessfully, to recover the full amount of the missing money from the duped customers. The judge ruled that the company and customers should split the losses, a compromise favourable to the GLCC.[119] Pinching meter money was hardly grand larceny, but it was an annoyance for the gas companies and one that the customers usually paid for in the end.

Entrusted with so many pennies, company collectors occasionally helped themselves to the meter money as well. The case of Thomas Lovell is one of many. Lovell, a 28-year-old collector for the GLCC, failed to deposit his collection to a Brondesbury branch of the London South-Western Bank. He admitted to a company clerk that he was £16 short in his cash and could not pay it. Upon inspection, the account books revealed that he owed as much as £78 in total. With a warrant issued for his arrest, Lovell disappeared from his Kilburn flat, but was later located in Lincolnshire working as a general labourer; his wife and three children had already entered the poorhouse. Eventually, Lovell received six months' hard labour for embezzlement of his employers' money, a rather lenient sentence compared to the penalties given to those delinquent citizens who had simply forced open the meters. Lovell's light treatment may have owed to the fact that his prosecutors decided not to push the case.[120]

Despite the temptation that the gas companies' automatic coin meters presented to some customers and gas employees, the advantages to the lower-middle and working classes were still undeniable. Gas managers, like Livesey, congratulated themselves on the 'double satisfaction' of raising their revenues while 'adding to the comfort and safety of the wage-earning class'.[121] Livesey's 'double satisfaction' may also have stemmed from improved industrial relations between managers and gas workers following the labour unrest and the struggles with New Unionism in the late 1880s. In May 1889, under the leadership of Beckton Work's stoker Will Thorne, the National Union of Gas Workers and General Labourers successfully pressured the GLCC directors into accepting their demands for increased wages, an eight-hour shift for gas stokers and time-and-a-half pay for Sunday work.[122] The gas workers' settlement was one of several noteworthy victories for New Unionism between 1888 and 1889, along with the successful resolution of the London Dock Strike in August 1889 and the walkout of the London match girls at the Bryant and May factory.[123] The gas workers' victory was only partial, however, since Thorne's negotiations with the SMGC proved to be more difficult. While the GLCC maintained a grudging tolerance of the gas workers' union, George Livesey and the SMGC governors upheld a strict anti-union policy, addressing their company's labour disputes through old-fashioned, paternal capitalism. In July 1889, to avoid a strike at his works, Livesey agreed to the same terms and conditions as reached by Thorne at the GLCC. Later, when the union attempted to enforce Rule 16, that all people employed at the gasworks should be union members, the company responded by stating that it did not recognise the union, and preferred to employ non-union workers. To present an alternative to union membership, Livesey devised a copartnership scheme which offered bonus payments in the form of company savings accounts to some classes of workers.[124] Not surprisingly, Thorne advised the employees to reject Livesey's scheme and, in December 1889, 2,000 men at the Vauxhall works laid down their tools. Livesey immediately hired replacement workers, brought in special constables to protect 'free labour' and effectively locked out the strikers. In February 1890, after a two-month lock-out, the economic hardship experienced by the workers and their families prompted Thorne's union to suspend the strike. Although Livesey had agreed to reinstate all striking workers, he retracted just four days after a settlement was signed.[125]

The profit-sharing plan devised by Livesey in direct challenge to the gas workers' union offered an important alternative in labour relations. This plan was later adopted by other gas corporations and companies outside the gas industry.[126] But were Livesey's intentions as self-serving as they appeared? Contemporaries and historians have failed to agree. Labour activist, John Burns, publicly denounced Livesey's copartnership scheme as 'the golden link to bind men to their employers'. This comes as no surprise, given Burns's involvement with the Social Democratic Federation, his founding role in the National Union of Gas Workers and General Labourers, and his evidence to the 1892 Royal Commission on Labour. Nevertheless, the SMGC governor had his supporters, including George Holyoake of the Mechanics' Institute.[127] By the mid-1890s, with the earlier labour crisis behind them, managers like Livesey boasted confidently about the beneficial impact of the gas industry on the working classes, both in terms of home comforts and stable employment.[128]

The cookery crusaders, including J.C. Buckmaster, praised the gas industry's efforts to increase the popularity of the gas stove, considering its potential to raise culinary standards and general health among the working classes. Helen Bosanquet, an influential professional social worker with the Charity Organization Society, echoed this sentiment. Although she unsympathetically attributed working-class poverty to poor housekeeping, improvident spending via credit and an inability to plan beyond the immediate future, Bosanquet admitted that the poor diet of the working classes was related to inadequate housing conditions, not simply ignorance of domestic management:[129]

> Here is a family living in two rooms, with no trace of store cupboard, pantry, or larder. True, there is perhaps a cupboard in each room, but these only just suffice to hold, one the coal, the other the crockery. It is clear that there can be no stock of edibles; straight from the shop to the table is the only plan, and probably the healthiest in houses which are so closely packed with humanity.[130]

From her observations of working-class life, Bosanquet knew that the great multitude of Londoners lived in cheap lodgings 'poorly adapted to culinary purposes'. While a woman could balance a saucepan, a frying pan or a kettle over a small open grate, it was difficult to prepare a decent meal with so little means. Moreover, she added that even where a generous landlord had put in a cooking stove, or, in the model dwellings where cookers were the norm, 'the time and skill necessary to the art are often wanting'. Prepared food, hastily purchased from the local fried-fish shop or the eel-pie shop, was undeniably simpler for the working-class wife.

Better housing and proper instruction in the safe use of domestic appliances could change this pattern. As the number of pre-payment cookers in working-class homes increased, so did the need for instruction in their proper use and care. Consequently, the task of the gas industry's lady demons assumed both commercial and social dimensions.

Despite Bosanquet's admission that most working-class homes were ill-equipped for meal preparation, she still promoted the benefits of teaching women simple cooking and the elementary laws of health, particularly regarding the care of infants and children.[131] Even after two decades of sporadic government-funded domestic subjects training for girls and women, the continued poor health of the labouring classes gave cause for public alarm. Charles Booth's extensive research indicated that the living standards of the London poor had changed little in 25 years, despite the material prosperity of the late Victorian era. The calm resolution of the London Dock Strike and the decline of New Unionism in the 1890s convinced the ruling classes that while the labouring poor constituted a serious social problem, they did not represent an imminent political threat.[132]

Earlier fears of social and physical degeneration re-emerged during the Boer War. General Maurice attributed the lagging campaigns in South Africa to the substandard physical condition of the British recruits.[133] Politicians, including the New Liberal Charles Masterman, equated the apparent 'racial degeneration' with imminent imperial decline, and the resulting Interdepartmental Committee on Physical Deterioration (1904) assumed the task of verifying these allegations.[134] After extensive testimony from witnesses, the commissioners reported that, despite external signs of Edwardian prosperity, a large percentage of the urban working population lived in constant want, if not abject poverty. In general, working-class housing was overcrowded, unsafe and unsanitary. Poor nutrition was common, particularly among children, women and the aged.[135] The study concluded that although the 'British race' was not in decline genetically, the negative impact of the urban environment on the health of the working population demanded an immediate response. Among its many recommendations, the report of the Interdepartmental Committee on Physical Deterioration listed improved training for girls and women in domestic subjects, infant healthcare and hygiene.[136] The crusade of the Victorian cookery teachers – that being to make girls and women into better wives and mothers – was rejuvenated and politicised by Edwardian fears of racial decline and imperial disintegration.[137]

The mission of the gas companies

By the turn of the century, no gas exhibition or demonstration was complete without the lady cookery expert. While the alleged job of the demonstrator was to educate and inform the consumer, the performances contained an undeniably sensual element. Reports on these cookery demonstrations praised the pleasing appearance of the young women on stage in their crisp aprons and caps. Descriptions of the women's 'deft fingers' and 'persuasive eloquence' lent an almost sensual dimension to the performances, as did the references to the enticing smell of the food cooking and its olfactory impact on the crowd.[138] The final distribution of the cooked products, usually biscuits and cakes, completed the image that the demonstrations presented a general assault on the viewers' senses of sight, smell and taste.

Some managers feared that the cookery demonstrations were becoming a bit too entertaining. An article in the *Journal of Gas Lighting* noted that consumers left the demonstrations with more vivid recollections of the food prepared than the stove it was cooked on. The article's author suggested more technical information on the gas appliance itself and, especially, comparative facts on the costs of gas cooking vis-à-vis solid fuel should be added to the demonstrations. After all, the appliance, not the food, was of primary importance. However, once the attraction of the cookery demonstration was established, it was difficult to refashion the format from entertainment to technical education.[139] For example, to add another dimension to the usual cookery demonstrations, the organisers of the 1904 International Gas Exhibition, held at Olympia, London, organised amateur cookery competitions for women and girls.[140] Although they were intended to 'display the time and labour-saving merits of the gas-cooking stoves', the public competitions were, essentially, entertainment. Competitions for schoolgirls had the added bonus of proving the simplicity and safety of gas stoves by demonstrating that even children could successfully operate them. It was the spectacle, however, that remained paramount. Walter King went so far as to compare the cookery shows to other forms of popular entertainment:

> A boxing competition, a football match, or favourite music hall artiste, could not have brought into the lecture hall a bigger and more mixed audience than on Saturday evening; and had the hall been twice or thrice the capacity, we have not the slightest doubt that is would have been filled. People went away from the doors in numbers, with almost rueful faces, because they could not get in.[141]

The exhibition organisers offered substantial prizes to the competitors: a gold and a silver watch for the two adult finalists and cash awards for the children.[142] Increasingly, entertainment and advertising were overshadowing the technical and educational function of the cookery demonstration. Ironically, professional cookery teachers and lady demons were not restricted from entering the competition, and, consequently, they often took home the top prizes.[143] Unfortunately, available sources do not indicate whether the general public realised that many of the adult contestants were also professional demonstrators, or indeed whether this was perceived as a conflict of interest. Apparently, the gas suppliers and appliance manufacturers had staged a perfect advertisement for their services and merchandise, while simultaneously rewarding their female speakers for their promotional abilities.

Could the expertise of the cookery teachers be engaged more profitably? The *Journal of Gas Lighting* addressed this question in an article entitled 'The services of the lady lecturer and teacher'. Reflecting on the immensely popular cookery demonstrations and competitions at the 1904 International Gas Exhibition, Walter King questioned whether the gas companies made full use of the lady cookery experts who were prepared to assist them. Beyond the usual lectures and cookery classes, the writer advocated that the gas companies hire qualified women full-time to undertake systematic visiting, inspecting and advising to consumers in their homes. In addition, the lady demons might establish contacts with organisations connected to churches, chapels and factories that catered to the needs of mothers and young women. The experts could also offer free cookery classes to girls and women where instruction of this nature was not provided by the local school. The article noted the growing numbers of domestic subject graduates able to undertake this work.[144]

The home visit format which Walter King had suggested was standard procedure among settlement workers, district nurses and COS officials. Renowned charity workers, including Helen Bosanquet and district nurse Margaret Loan, have written about the efficacy of this practice and the proper etiquette that it required.[145] Despite the intrusive qualities often attributed to the COS, Bosanquet was sensitive to the public view of the visits as an invasion of working-class privacy. She recommended that visits should not be made 'unless on some definite errand; or unless acquaintance has been previously made with them; or lastly unless there is some special reason for believing that the visit will be acceptable'. These criteria were general guidelines, however, sufficiently broad and therefore unlikely to restrict the visitors in their charity work. Borrowing from the philanthropic model, the *Journal of Gas Lighting* stressed the importance of grace, tact and diplomacy, even comparing the lady demons to missionaries when King asked: 'why the educated and refined lady charged with the duty and mission of the gas suppliers should not be as welcome as the clergyman or the district visitor cannot be divined'.[146] Evidently, King failed to comprehend that, in many cases, Edwardian working-class women resented the stream of official visitors that routinely marched through their homes.[147] Add yet another 'lady visitor' to the representatives from the parish, the school board or the housing authority, who regularly ventured into the homes of the labouring people.

Should one more visitor matter, particularly when the gas companies were simply safeguarding their own property? While providing an educational service and fostering good public relations, the lady demons could save the gas companies money. Although following the hire system led to increased company revenues, service costs absorbed much of the profits. King lamented the rising costs of repairing hire cookers, given the 'terrible state of many samples of returned stoves seen at different works'.[148] Improper stove use and customer negligence contributed to this situation. However, the lady demons, in their multiple roles of inspector, house-to-house instructor and canvasser, could offer personal assistance and instruction to customers and their servants in the home.

Editorial response to King's article was immediate; the majority of the letters were from cookery teachers, who were familiar names on the exhibition lecture circuit.[149] While all of the women agreed that lady demons should be retained full-time with the gas companies, they differed over how the ladies might carry out their new roles. The topic of home visits was controversial, and for good reason. Miss M.E. Betts of the Davis Gas Stove Co. favoured surprise stove inspections with annual prizes for the housewife with the cleanest cooker throughout the year. In contrast, Mrs Godfrey Sutcliffe, the chief organiser of the cookery lectures and competition at the 1904 International Gas Exhibition, wrote that based on her own experience it was not wise to call unexpectedly, however much 'tact' one might possess. She cautioned: 'Ladies, as a rule, do not like taking others into their kitchens; and cooks strongly object to it.' Through advertising, customers could be invited to make use of the lady demons' expertise, either in the company showrooms or in the customers' homes. Home visits were useful, but a lady never intruded uninvited. Miss M.S. Birkett of Southport objected to home visits entirely, opining that canvassing 'house-to-house' was 'more suitable' for men. Instead, Miss Birkett advocated closer cooperation between domestic economy schools and gas corporations.[150]

In addition to the home visit, the content of the lectures presented another topic for debate. Several lady demons, including Miss Birkett, advocated the revision of the lectures' content to include more technical information on the care and operation of gas stoves.[151] Mrs A.M. Collins, of Tollington Park, also recommended that the lectures be made more practical. She suggested a cleaning demonstration in addition to the cookery show at the end of every lecture. Mrs Collins insisted that the lecturer 'should take the stove to pieces and clean it in full view of the audience'. She stated, confidently: 'I think there is nothing undignified in this.'[152] Some demonstrators might disagree, however, as cleaning the stove was a dirty, disagreeable task, usually relegated to a domestic servant, if the household had one.

The debate over 'to clean, or not to clean' may have stemmed from the demonstrators' possible concerns regarding their desire to disassociate themselves from the popular image of domestic servants that was reinforced everywhere by the illustrated press and advertisements. As Victorian manufacturers used lady demons to show off their stoves, visual representations of women, by the early 1900s, were increasingly incorporated into the trade literature and printed advertisements. But the portrayals of femininity found in the advertisements clearly contradicted the corporate image that the lady demons actively sought to create, in terms of class background, education and occupational status. Not all appliance makers used images of women in their advertisements, and many did not advertise at all. However, those that did – specifically the Wilson Stove Company and the J.J. Coll Company – adopted a gendered language of visual representation which was developed by advertising agents and illustrators. Advertisements for other household products, like furniture polish and soap, suggest obvious parallels. Therein, women were often depicted as naive, sexual and childish; frequently, they appeared as domestic servants, with inappropriately short skirts and exposed ruffled petticoats.[153] In the same way that the illustrations of 'lovely seaside girls' represented 'material comfort, sexual excitement and hard selling',[154] the image of the flirtatious female servant was another popular advertising symbol. Her sexual ambiguity promoted a variety of products, including food, cigarettes, furniture polish and, eventually, cooking stoves.

Recent scholarship on women in Victorian advertising offers some insight into the commercially constructed ideal of femininity and domesticity. Advertisements reflected the ideology of separate male and female spheres, with women at the centre of family life entrusted with the health, happiness and moral well-being of men and children. This sentimental portrayal of Victorian femininity contrasted with the hedonistic consumption of the late nineteenth century, as middle-class homes became the showcases for the 'paraphernalia of gentility'.[155] With Victorian women increasingly responsible for the domestic management of the home, the conspicuous display of tasteful affluence also became a feminised activity.[156] As middle-class women constructed the perfect domestic interior, they were likewise pressured to aspire to physical perfection themselves.[157] Evidently, given women's increased participation in the mass market, late Victorian advertisements frequently portrayed women as both the consumers and the objects of consumption.

The image of the female domestic servant in late Victorian advertising simultaneously asserts and contravenes this ideal of virtuous consumption. As a commodity, the maid herself was a symbol of respectability, as the presence of a household staff, even a single maid-of-all-work, indicated middle-class status. Nevertheless, the maid also represented social deviance. Associated with sexual promiscuity, servant girls frequently appeared in Victorian pornography as being characteristically available and easily corruptible by male employers. The decline into prostitution was considered to be such a serious occupational hazard that middle-class philanthropists established homes and clubs for female servants in the hope of preventing this apparent downward spiral.[158] The cute maids of the late Victorian and Edwardian advertisements embodied both of these themes. Set in respectable middle-class domestic interiors, they exuded a playful sexuality with teasing eyes or exposed petticoats and legs. For example, a popular advertisement illustration for 'Heartease Cigarettes' showed a female domestic, in a short skirt and fashionable heeled mules, tapping discreetly at a parlour door, a tray of cigarettes in one hand. Her doe-eyed expression and slight smile are unmistakably sexual and hint at the sensual connection between the physical pleasures of sex, smoking and food.[159]

The appliance makers adopted this symbolic language of female representation. The graphic artist Dudley Hardy adapted his famous 'Gaiety Girl' figure to an 1896 advertisement for oil-fuelled cooking stoves manufactured by the Albion Lamp Co. of Birmingham. The shapely female cookery assistant, so identified by her white apron and flat cap, coyly upstages a portly chef as she bends over the simmering stove.[160] The woman's calf-length skirt and visibly shapely legs suggest an easy virtue, a Victorian stereotype often attached to domestic servants. In another example, a full-colour fold-out paper novelty, dated 1895 and distributed by the Wilson Stove Company, presents a young maid or cook with her mistress's gas stove.[161] Although a domestic servant, the young woman does not wear the traditional black and white uniform; instead she dresses in pink, with her neck exposed. A group of curious male onlookers peer into the kitchen: a soldier, a policeman and an unidentifiable rake in a top hat. Their noses look unmistakably red, either from the cold or drinking, and the men practically leer at the girl as if they were performing in some music hall farce. However, she stands at a distance, separated from them by the outside kitchen wall, safely assuming her air of teasing defiance. The squat iron cooker occupies almost half of the advertisement;

however, the tension between the woman and the three men draws the viewers' interest to this little drama, and not necessarily to the product.

A 1903 advertisement for the gas stoves of the J.J. Coll Company also presents a temptress in the kitchen (see Fig. 2.3). Showing a shapely leg, a pretty cook in a fancy apron and green dress, against a bright-red background, smiles demurely at the viewers while sampling a dish. She wears a chef's cap, similar to those required by cookery students, but this one sits casually askew on her curly head. She must be an accomplished cook, for the stove contains a multi-course meal, illustrating the complete versatility of the appliance. However, the feminine cook's demure smirk effectively undermines her culinary expertise. With no background setting, this image may have represented either a household cook or a youthful lady demonstrator. In either case, the manufacturer clearly adopted the conventional imagery of women, sex and food to promote its gas stoves to Edwardian consumers.

The image of the female cook found in advertising contrasted with the official self-image of the gas companies' lady demons. Possibly, the debates regarding the propriety of performing scullery work on stage indicated a genuine concern that, by doing so, they might be identified more closely with the flirtatious kitchen maids found on the trade cards. The general consensus among the demonstrators was that scrubbing stoves in public was acceptable because they were trained domestic experts: cleaning the stove properly was simply one aspect of the

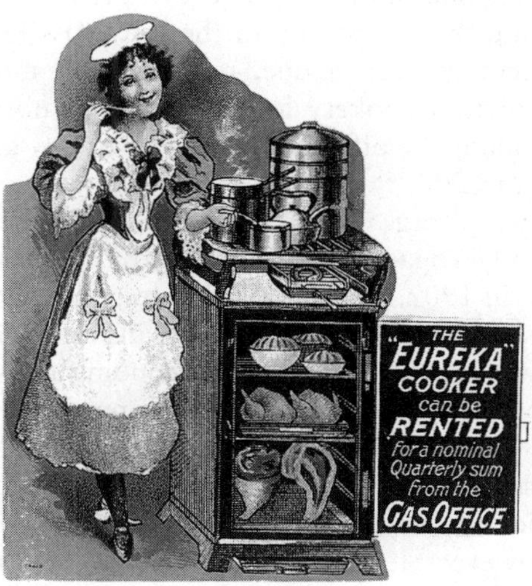

2.3 An attractive cookery demonstrator poses with a fully laden stove. Her flirtatious demeanour was at odds with the home service advisors' professional image

demonstrator's job. However, the women overlooked the fact that the male lecturers and chefs were not expected to scrub their stoves on stage. The notion that domestic tasks could be performed scientifically and efficiently points to another objective of the domestic science movement: the elevation of the occupational status of housewifery and domestic service via standardised training. Unfortunately, by accepting scullery work as part of the lady demon's job, women reinforced the gendered nature of domestic labour and foundered into the same dilemma as their contemporaries in the school system. Domestic labour would never be professionalised or widely respected as long as it remained exclusively women's work, particularly given the accepted perception that women's time was cheap.[162]

But even if the job demanded black leading a stove on stage, the lady demon must always be, in the words of Helen Edden, 'essentially a lady', because the class of consumers ranged from the highest to the lowest. In the tenements of London's East End, or the salons of South Kensington, the lady demon must never feel at a social disadvantage. In essence, she had to fulfil a long list of prerequisites. She had to be a domestic college graduate, familiar with all types of cookery and housewifery. In addition, she must be a technical expert on the operation and maintenance of gas appliances, although, actually, she was not required to fix cookers, as this was a fitter's job. She had to be a lady of excellent character, in possession of tact and good judgement, but of sufficient professionalism not to object to performing scullery duties in public. Realising the complexity of the job, Miss Edden stated that 'to gain the services of such a one, remuneration must be adequate'.[163] Miss Rose Brown, a cookery lecturer from Cambridge and author of housekeeping manuals, offered her services to any gas company for the annual salary of £200.[164]

Gas manager George Stevenson of Long Eaton explained that his company already engaged a full-time lady demonstrator: Miss Clegg joined his staff in 1902. She was hired in response to the huge increase in penny-in-the-slot meter customers in his district, many of whom were 'totally ignorant' of the 'proper and economical use' of gas stoves.[165] Stevenson's company opted for the regular appliance inspection; however, his customers were duly warned. Formal calling cards, tactfully worded, were issued several days in advance of Miss Clegg's visits, and the company reported few instances of hostility on the part of their customers. However, Stevenson did not recommend unannounced visits. The lady demon imparted primarily technical information: she always took the stove apart, explained how it worked and emphasised the connection between proper maintenance and the efficient operation of the appliance. If the customer wished, Miss Clegg would even cook a dinner and thus provide a complete personal home demonstration. Stevenson claimed that 'the experiment [that is, hiring a lady demon] proved a splendid investment', as maintenance costs decreased noticeably, while gas consumption increased, because customers used their stoves more frequently.

Following this editorial debate, the GLCC acted upon the women's recommendations. The board moved 'to engage the services of a lady to visit consumers at their houses and instruct them in the uses of gas-cooking stoves'.[166] They hired Helen Edden. A member of Buckmaster's Universal Cookery and Food Association and an experienced lecturer for the Davis Stove Gas Co., Edden had recently impressed audiences at the 1904 International Gas Exhibition. She accepted the job at £200 per year, which was double the earnings of cookery mistresses with the London School Board. By the end of the year, Edden had a staff of five.[167] Paid £1 per week, her assistants still earned above-average wages compared to female clerks and shop assistants.[168] In 1907, the SMGC employed lady demon Miss Wright at £120 per annum. Within a few years, Manchester and Birmingham also added lady demons to their staffs. Even some of the smaller gas companies, such as Enfield, Croydon and Hornsey began to employ them.[169]

Company records suggest that the lady demons enjoyed a certain degree of autonomy in the workplace. They formed completely new departments within the corporate structure and, although management arrangements varied, they were usually a sub-group of the inspectors or sales departments. For example, Miss Edden reported to GLCC chief inspector Francis Goodenough, but she managed her own staff of assistants.[170] Edden gave semi-annual reports before her employer's board of governors and was probably the first woman to attend their meetings. In addition to the occasional public lecture, the primary function of the lady demons was to visit and instruct customers in their homes in the proper use of gas stoves.

As Helen Edden's comments in the *Journal of Gas Lighting* implied, the GLCC ladies adhered to a strict code of etiquette and behaviour. The ladies issued calling cards in advance since surprise visits were considered an invasion of privacy that might annoy or alienate the customer. When making a call, the lady demons always wore hats and gloves, giving them the appearance of respectable middle-class district visitors. The GLCC advertised this customer service in a cookery book that they offered for sale to its customers (see Fig. 2.4). The accompanying photograph depicts a lady demonstrator neatly attired in Edwardian daywear. However, unlike most afternoon callers, she kneels beside a gas stove, gesturing with her hands as if responding to a question, while the lady of the house, hatless and in a blouse and a skirt, takes notes.

2.4 The Gas, Light and Coke Company, London, invited customers to consult its expert lady demonstrators with their household queries

Evidently, the illustration was meant to represent a home service visit, with the lady demon responding to the consumer. Given the constant walking involved in this profession, sturdy shoes were important. Also, no matter how many calls she might make in a day, the lady demon was never allowed to appear fatigued. Retired demonstrator Grace Horrigan recalled conversations with Maywyn Godby, an early GLCC cookery expert:

> In the early days they [lady experts] were told not to visit more than one house on a street especially in the better areas, like South Kensington. The company women were not to give the idea that they were calling door-to-door, or arriving at a customer's home in a breathless state.[171]

Women's efforts to define a corporate identity for themselves within the gas industry were influenced by Victorian conventions that determined class and gender behaviour. It is probable that many of the demonstrators were the daughters of middle-class families, with fathers in business and the lower professions, considering the cost of the required domestic college training. A two- or three-year course from a recognised domestic subject college cost approximately £25 per annum; it was, therefore, an education which required a certain amount of family assistance, even with the scholarships that some colleges offered.[172] Candidates for the domestic training colleges were carefully selected: especially during the late nineteenth century, the domestic college superintendents preferred women to be aged between their mid-20s and early 30s, with some experience of philanthropic work.[173] Just as the lady health inspectors – who were also drawn from the middle classes – felt that their class background added to their official credentials when dealing with working-class families, the gas industry's lady demons attempted to use their class status to their professional advantage.[174] The lady demons, however, found themselves in the rather difficult position of having to be entirely comfortable in the homes of the rich, the poor and everywhere in between. They could be neither condescending with the working-class housewives, nor presumptuous with their social betters, because, as salespeople, they could have been dismissed just as easily from the homes of either group. Symbolically, the lady demon is shown kneeling before the housewife, offering her services to both the customer and the corporation.

Clearly, regardless of their education and their social standing, the lady demons were engaged in the unladylike business of selling and cleaning gas stoves at a time when few women were considered to be technical experts. Just how much mechanical ability a lady needed to perform her job properly was open to interpretation. Wary that the ladies might usurp their jobs or work status, some male inspectors and stove fitters insisted that the women could not carry tools or be allowed to perform routine repairs of any sort. Yet, given their familiarity with the stove operations, the women were capable of undertaking minor repairs, such as tightening a stove valve or unclogging a blocked gas jet. Instead, however, the women were instructed to report any necessary repairs; a company fitter would then call on the customer to fix the stove. Inefficient though this may have been, it preserved a gendered hierarchy of job classification and status. The women did still occasionally undertake repair work. For example, one amused SMGC customer reported that a lady demon was mistakenly sent on a service call intended for a stove repairman. Much to the customer's satisfaction, the unnamed lady demon dismantled the stove, discovered the nature of the problem (a blocked gas jet) and cleared the mechanism with the aid of her nine-inch hat-pin.[175]

Offering instruction in customer safety was another integral aspect of a lady demon's job. Arguably, showing customers how to use the stoves safely and without injury was the fundamental lesson. Judging from the content of the company literature, however, economical cookery and cleaning procedures received priority. For example, an early brochure entitled *How to Use a Gas Cooking Range*, written by Helen Edden, stressed the need for cleanliness while information on the lighting of the gas jets received considerably less attention.[176] Edden's and Moser's *The Household Gas Cookery Book* devoted less than half of its 260 pages to gas safety.[177] Similarly, cookery experts and lecturers Amy Atkinson and Grace Holroyd listed cleaning procedures before safety precautions in their 1909 publication on gas cookery.[178] Unfortunately, having a spotless cooker offered no guarantee against an explosion. No doubt the gas companies wished to play down the potential danger of their appliances, and safety was conflated into a set of proscribed steps for proper use. While it appears that the lady demons were responsible for imparting some information on gas stove safety, the more conventionally feminine issues of cleanliness and domestic economy received company priority.

Conclusion

By the 1880s, the trade exhibition constituted the most important means of gas sales promotion, and one that appliance makers and gas suppliers used to their advantage. Although the gas entrepreneurs were not overly concerned about the immediate threat of electricity to their industry, they recognised that a more aggressive approach to marketing and sales was required lest the gas industry appear weak and unable to compete in the masculine corporate world. To help revitalise its trade exhibitions, the gas companies and appliance manufacturers looked to women as both the sales experts and the targeted consumers. Fulfilling the multiple roles of sales assistants, domestic science teachers and social workers, the lady demons were the first women employed by the gas companies. Appliance manufacturers engaged women on a contract basis to demonstrate their products, in an attempt to convince female customers of all classes that the gas cooker was a household appliance for everyday use, and not simply an industrial tool operated by professional cooks in public kitchens. Capitalising on the advertising appeal of cookery performances, the appliance makers and gas corporations used the soft sell of 'deft fingers' and 'persuasive eloquence' to break down prejudices against gas cooking. As the number of hire cookers increased, as a result of the introduction of the pre-payment meter, the individual gas companies enlisted lady demons on a permanent basis to educate and assist female consumers in their homes. They joined the ranks of the other outside gas workers, such as the fitters and the collectors.

The lady demons were not simply responding to the dictates of the corporations. Much of the initiative for this new occupation came from the cookery teachers themselves, as they sought to widen their employment opportunities through professional self-promotion. Representing the ideals of both the domestic science movement and

Victorian capitalism, and working in a completely new occupation, the lady demons moved between the public and private domains. Although they were businesswomen working outside the home, the lady demons promoted an ideal of Victorian domesticity that they personally did not represent. This tension was revealed in the debates concerning the proper conduct of female demonstrators; the implied assertion was that one can be both a working woman and a lady. In contrast to the male meter collectors, in their black coats and bowler hats, the lady demons' clothing and public conduct reflected their unique occupational status and their need to distinguish themselves from other people in public places and in their customers' homes. Not surprisingly, the guidelines for the lady demons' behaviour during work reinforced several idealised traits of middle-class womanhood, including tact, respect for the feelings of others, good judgement and modest dress. In this regard, the demonstrators' occupational image upheld Victorian feminine values rather than negating their claim to respectable womanhood. According to Angel Kwolek-Folland's study of the insurance industry in America, it was a position the gas demonstrators shared with other middle-class women entering the masculine business world during the early twentieth century.[179] By acting as the official representatives of the companies they worked for, the lady demons served as important conduits between the gas corporations and their consumers of all classes, fostering the development of customer relations, a now familiar aspect of product promotion.

As domestic experts and corporate agents, the lady demons blurred the conventional and gendered boundaries that separated the private domestic sphere from the public arena of business and work. We shall see in the following chapter that, during the Edwardian years and up to World War I, women assumed an even greater role in the industry's business of advertising and sales, thanks, in part, to the efforts of its female publicist: Maud Adeline Brereton. That the gas industry selected a woman in 1912 as chief editor of its promotional campaigns suggests the increased attention that some managers placed on the cultivation of the feminine consumer.

Notes

1 'Hobart (Tasmania) Exhibition', *JGL*, 19 March 1895, 607.
2 'The Gas Institute at the Crystal Palace', *JGL*, 7 March 1882, 412.
3 David E. Nye, 'Electrifying Expositions, 1880–1939', in Robert W. Rydell, Nancy E. Gwinn and James B. Gilbert (eds), *Fair Representations: World's Fairs and the Modern World* (Amsterdam, 1994), 144.
4 Richards, *Commodity Culture*, 17.
5 'The encouragement of gas consumption', *JGL*, 9 May 1882, 831.

6 'The Gas Institute at the Crystal Palace', *JGL*, 7 March 1882, 412.
7 Adams, *Architecture in the Family Way*, chap. 1.
8 NAL, AAD 9/1, *Ideal Homes Exhibitions*, introduction.
9 John M. MacKenzie, *Propaganda and Empire: The Manipulation of British Public Opinion, 1880–1960* (Manchester, 1984); Annie Coombes, *Reinventing Africa: Museums, Material Culture and Popular Imagination in Late Victorian and Edwardian England* (New Haven and London, 1994).
10 Best, *Mid-Victorian Britain*, 2, 3, 231.
11 Asa Briggs, *Victorian People: A Reassessment of Persons and Themes, 1851–67*, 2nd edition (New York, 1963), 40–43.
12 Auerbach, *The Great Exhibition of 1851*, 1,127.
13 Altick, *Shows of London*, 456, 472.
14 Ibid., 470.
15 Patrick Beaver, *The Crystal Palace, 1851–1936* (London, 1970), 12.
16 Richards, *Commodity Culture*, 3.
17 Ibid., 58.
18 The success of the *Daily Mail* Ideal Homes Exhibitions provides a perfect example of retailers using the public exhibition to further sales in addition to their department store outlets. See NAL, 1910 A.23 (45), London, *Ideal Homes Catalogue*, 8–23 April, 1910.
19 Williams, *History of the British Gas Industry*, 35; Davidson, *Woman's Work*, 66; 'A gas cookery lecture in 1836, reprinted from the *True Sun* (1836)', *Copartnership Journal* (August, 1925), 189; Barty-King, *New Flame*, 115; Freeman, *Mutton and Oysters*, 169–173.
20 'The Queen at Kensington', *The Times*, 10 May 1873, 5.
21 Goodall, *Burning to Serve*, 82–85.
22 'Gas exhibition in the west of England', *JGL*, 3 October 1882, 596.
23 'The encouragement of gas consumption', *JGL*, 9 May 1882, 830. The light from a flat flame burner was emitted upwards and horizontally; however, the development of the inverted flame burner produced a light that shone in a downward and more useful direction. This was improved by the addition of a sealed glass globe which produced a better and cleaner light than the neat flat flame burners; however, it was the incandescent mantle, developed during the mid-1880s by chemist Carl Auer, that eventually popularised domestic gas lighting in England. For an explanation of gas lighting fixtures, see P. Crawford Sugg, 'The rise of the gas appliance', *Gas World*, November 1984, 1–2.
24 'The encouragement of gas consumption', *JGL*, 9 May 1882, 830.
25 'Exhibition of gas apparatus and lighting at the Crystal Palace', *JGL*, 4 July 1882, 12.
26 Samuel Smiles, *Self-Help: With Illustrations of Conduct and Perseverance*, centenary edition with an introduction by Asa Briggs (London, 1958), 267–273.
27 Kwolek-Folland, *Engendering Business*, 45–47.
28 'Exhibition of gas lighting at the Crystal Palace', *JGL*, 4 July 1882, 11–12.
29 LMA/B/GLCC 36/*Minutes of the Board of Directors*, 28 July 1882, 269.
30 LMA/B/GLCC 36/*Minutes of the Board of Directors*, 4 August 1882, 271–272.
31 'The exhibition of electric and gas apparatus at the Crystal Palace', *JGL*, 19 December 1882, 1075, 1078.

32 'The electric light at home', *The Graphic*, 27 January, 1883, 58.
33 Goodall, *Burning to Serve*, 84–85.
34 'The close of the Crystal Palace Exhibition', *JGL*, 8 May 1883, 815–816.
35 Hannah, *Electricity Before Nationalisation*, 3–7. Although Michael Faraday discovered electromagnetic induction in 1831, the practical application of electricity took decades. By 1857, electric generators were being used at English lighthouses. The Siemens station was located in Godalming, Surrey, and the Edison plant at Holborn Viaduct, London. Early electrical enthusiasts included Lord Salisbury who experimentally lit his home with electricity.
36 Ibid., chap. 1. The early Electricity Acts (1882 and 1888) granted permission to undertakings applying to dig up urban streets to lay electrical wiring. To protect the public from the power of private monopoly, the acts regulated prices and provided for the purchase of electrical companies after 21 years, in the first act, and extended to 42 in the second act. Hannah asserts that the 'speculative and experimental' nature of the early industry discouraged investors more than the legislation.
37 Diana and Geoffrey Hindley, *Advertising in Victorian England, 1837–1901* (London, 1972); Leonard de Vries and James Laver, *Victorian Advertisements* (London, 1968); T.R. Nevett, *Advertising in Britain* (London, 1982); Fraser, *Coming of the Mass Market*; Loeb, *Consuming Angels*; E.S. Turner, *The Shocking History of Advertising* (Harmondsworth, Middlesex, 1965).
38 Nevett, *Advertising in Britain*, chaps 2–4.
39 Fraser, *Coming of the Mass Market*, 134–135.
40 H.C.G. Matthew, *Gladstone, 1809–1874*, 2nd edition (Oxford, 1988), 113.
41 Nevett, *Advertising in Britain*, 67–68.
42 Fraser, *Coming of the Mass Market*, 135.
43 Nevett, *Advertising in Britain*, 67, 71.
44 Barty-King, *New Flame*, 144.
45 'The encouragement of gas consumption', *JGL*, 9 May 1882, 830. The journal reported on gas exhibitions in both Bristol and Stockport in 1882. The Bristol show had 800 entrants and included a miniature working model of a gasworks supplied by Mr Hunt of Birmingham. See also 'Gas exhibition in the west of England', *JGL*, 3 October 1882, 596–597; 'Stockport gas exhibition', *JGL*, 7 November 1882, 828.
46 Sarah Milan, 'Refracting the Gaselier: Understanding Victorian Responses to Domestic Gas Lighting', in Inga Bryden and Janet Floyd (eds), *Domestic Space: Reading the Nineteenth-Century Interior* (Manchester and New York, 1999), 95–97.
47 Davidson, *Woman's Work*, 6, quoted from the *Report of the Smoke Abatement Committee 1882* (London, 1883), 17.
48 Davidson, *Woman's Work*, 67; Goodall, *Burning to Serve*, 92.
49 LMA/B/SMGC/Minute Book/*Meeting of the Board of Directors*, 11 June 1878, 447.
50 Goodall, *Burning to Serve*, 82–83.
51 Three out of the four London gas companies had adopted the hire system by 1882: the Crystal Palace Company in 1869, the SMGC in 1878 and the GLCC in 1882. LMA/B/SMGC, *Minutes of the Board of Directors*,

11 June, 1878, 447; 'Gas companies and the utilization of gas for cooking', *JGL*, 23 May 1882, 929; Davidson, *Woman's Work*, 67. Outside of London, the Gateshead Gas Company at Newcastle-upon-Tyne also introduced hire cookers.

52 LMA/B/SMGC/*Proprietors' Minute Book*, Directors' Report, 30 June 1882, 3.

53 Fraser, *Coming of the Mass Market*, 87.

54 Ibid., 176.

55 Ibid., 92–93, 111–121. For a discussion of the growth of department stores and multiple shops, refer also to Lancaster, *The Department Store*, 85–86, 104–105.

56 Barty-King, *New Flame*, 147.

57 LMA/B/SMGC/ *Proprietors' Minute Book*, Directors' Report for Year End 30 June 1891, presented 9 August 1891, 82; B/GLCC/36, *Minutes of the Board of Directors*, 19 November 1886, 38.

58 LMA/B/GLCC, *Report of Court of Appeal*, 26 July 1886. A similar court judgement ruled in the company's favour on 26 October 1886 when Herbert Smith & Co. seized and auctioned a GLCC rental stove.

59 LMA/B/SMGC/ *Proprietors' Minute Book*, Director's Report for year end 30 June 1889, presented 14 August 1889, 66; Director's Report for year end 30 June 1890, presented 20 August 1890, 74.

60 LMA/B/CPGC/Crystal Palace Gas Company, *Minutes of the Board of Directors*, 29 March 1889, 81.

61 LMA/B/SMGC/*Proprietors' Minute Book*, Director's Report for year end 30 June 1888, presented 15 August 1888, 56.

62 LMA/B/SMGC/*Proprietors' Minute Book*, Director's Report for year end 30 June 1889, presented 14 August 1889, 66.

63 'Gas exhibition', *JGL*, 23 April 1889, 778.

64 'Use of gas engines and stoves in Carlisle', *JGL*, 18 June 1889, 1139.

65 LMA/B/CPGC/ *Minutes for the Board of Directors*, 26 April 1889, 84.

66 Goodall, *Burning to Serve*, 87.

67 LMA/B/CPGC, *Minutes of Board of Directors*, 7 June 1889, 94.

68 LMA/B/CPGC, *Minutes of Board of Directors*, 26 February 1892, 115.

69 LMA/B/GLCC 38, *Minutes of the Board of Directors*, 5 April 1889, 310. The secretary and general manager advised the board to establish showrooms for the permanent exhibition of gas apparatus to promote increased consumption of gas. GLCC, 18 October 1889, 376; the GLCC opened its first showroom at Goswell Road at a cost of £630. Five years later, the GLCC invested £2,500 in its Kensington showrooms, not including the £1,500 annual rent for the property. LMA/B/GLCC 40, *Minutes of the Board of Directors*, 28 September 1894, 80.

70 'Gas exhibition and cookery lectures', *JGL*, 26 January 1892, 161; 'Gas exhibition', 2 February 1892, 215; 'Exhibitions of gas appliances', 23 February 1892, 355; 'Gas exhibition at Blackpool', 1 March 1892, 405.

71 For detailed descriptions of the exhibitions and cookery lectures see 'Exhibition of gas appliances at Darlington', *JGL*, 16 February 1892, 304; 'Gas exhibition', *JGL*, 8 March 1892, 453; 'Gas exhibition, Newcastle-upon-Tyne', *JGL*, 15 March 1892, 500–501; 'Exhibition of gas appliances', *JGL*, 22 March 1892, 549; 'Gas exhibitions', *JGL*, 29 March 1892, 593. As the above list suggests, numerous gas exhibitions and cookery performances were staged during the winter and spring months of 1892.

For lists of the spring lectures see 'Exhibitions of gas appliances', *JGL*, 3 May 1892, 811; 'Exhibitions of gas appliances', *JGL*, 10 May 1892, 862; 'Exhibitions of gas appliances', *JGL*, 24 May 1892, 978.

72 'Gas exhibition', *JGL*, 8 March 1892, 456.

73 'Gas exhibition', *JGL*, 2 February 1892, 215.

74 LMA/LCC/*Minutes of the London County Council Subcommittee on Cookery*, 11 May 1883, 461; *Minutes of the London County Council Subcommittee on Cookery*, 5 October 1883, 475.

75 LMA/B/GLCC 40, *Minutes of the Board of Directors*, 2 March 1894, 3; GLCC 41, *Minutes of the Board of Directors*, 21 June 1895, 201.

76 Goodall, *Burning to Serve*, 87–88.

77 Mrs A.B. Marshall, *Mrs. A.B. Marshall's Cookery Book* (London, n.d.), 20–25; Marshall includes several reviews of her cookery lectures that were staged in 1887.

78 Hardyment, *From Mangle to Microwave*, 45; Judith G. Coffin, *The Politics of Women's Work: The Paris Garment Trades, 1750–1915* (Princeton, NJ 1996), 79.

79 Coffin, *Politics of Women's Work*, 88–107.

80 'Cooking at the International', *The Times*, 25 April 1873, 5.

81 Judith Walkowitz, *City of Dreadful Delight: Narratives of Sexual Danger in Late Victorian London* (Chicago, 1992), 46–47.

82 Adburgham, *Shopping in Style*, 168; Lancaster, *The Department Store*, 53–55; Rappaport, *Shopping for Pleasure*, 154–177. Gordon Selfridge, an American formerly with the Marshall Fields Company in Chicago, opened his Oxford Street store in 1909. It catered for middle-class women by offering a writing room, a reading and rest room, post office and information bureau.

83 Walkowitz, *City of Dreadful Delight*, 49.

84 Lady Mary Jeune, 'The ethics of shopping', *Fortnightly Review*, 57 (January 1895), 126.

85 Ibid., 129. Lady Jeune sympathised with the shop assistants' requests for improved wages, regular breaks and better food.

86 Walkowitz, *City of Dreadful Delight*, 74–75.

87 'Gas exhibition at Lostwithiel', *JGL*, 5 March 1895, 503.

88 'The dangers of gas cooking-stoves', *Lancet* (November 1897), 1403.

89 'Parkinson gold medal gas cooker', *JGL*, 10 September 1895, 564. Matilda Lees Dods, principal of the Birmingham and Midland School of Cookery, endorsed the 'Parkinson gold medal gas cooker' claiming that it was used in all their cooking centres. Sophie Thwaites, of the International School of Cookery, Liverpool described the 'Parkinson' as 'easy to manage, very good in its results and most economical in the consumption of gas'.

90 For several examples of these sorts of endorsement see: 'Gas appliance exhibitions', *JGL*, 7 June 1892, 1082. Miss R.M. Watts of the Yorkshire School of Cookery demonstrated the stoves of C. Wilson & Co. of Leeds. She baked 50 loaves of bread which she then donated to local miners involved in a coal dispute. Miss Thompson of the Nottingham & District School of Cookery demonstrated the Main cookers in a Nottingham exhibition. 'Exhibitions of gas appliances', *JGL*, 12 April 1892, 678. Mrs Wilkinson, formerly with the National Training College in South Kensington, demonstrated for the Fletcher Russell Co. at Albert Hall in London.

91 Goodall, *Burning to Serve*, 84.

92 Marie Jenny Sugg, *The Art of Cooking By Gas* (London, 1890), 2–19.
93 Barty-King, *New Flame*, 146.
94 Lillie Richmond, *Cookery Recipes, with Special Hints on Gas Cooking*, 4th edition (London, 1903).
95 Attar, *Bibliography of Household Books*, introduction.
96 Hardyment, *From Mangle to Microwave*, 128.
97 Edric Bayley, *Education in London Board Schools* (London, 1888), 14. Bayley was an active supporter of technical education for boys and girls; however, with the latter, this usually meant domestic subjects like cooking, knitting and laundry. Bayley made frequent comparisons between the German and English systems of education, and feared that the inadequacy of English schooling would cause Britain to lose its industrial prominence as the 'workshop of the world'.
98 LMA/LCC/*Minutes of the London County Council Subcommittee on Cookery*, 15 July 1892, 238. The Commercial Gas Co. let stoves to five schools, at three shillings and sixpence per quarter; the agreement included normal stove maintenance. The GLCC rented stoves to 19 schools, at three shillings and nine pence per quarter, with a 10 shilling maintenance and repair charge; the SMGC let stoves to 17 schools at two shillings and sixpence per unit, with no specific charges for maintenance and repairs.
99 Goodall, *Burning to Serve*, 249.
100 Barty-King, *New Flame*, 157; Hardyment, *Mangle to Microwave*, 128; Davidson, *Woman's Work*, 36.
101 Goodall, *Burning to Serve*, 125–127; Sugg, 'Rise of the gas appliance', 1–2.
102 Goodall, *Burning to Serve*, 129–139. Goodall provides an excellent account of the Welsbach company's patent disputes with traders who wished to import continental gas mantles. He notes that the gas suppliers and the Welsbach company did not share similar interests, since the mantle producers wished to sell mantles and fittings at the highest cost possible, and using the least amount of gas to provide a strong light. In contrast, the gas companies wished to sell the fittings and mantles as cheaply as possible, to encourage overall increases to gas consumption.
103 Ibid., 101–103.
104 LMA/B/SMGC/*Proprietors' Minute Book*, Director's report for year end 30 June 1893, presented 16 August 1893, 100.
105 LMA/B/SMGC/*Proprietors' Minute Book*, Director's report for year end 30 June 1893, presented 16 August 1893, 100.
106 LMA/B/SMGC/*Proprietors' Minute Book*, Director's report for year end 31 December 1895, presented 12 February 1896, 122.
107 LMA/B/SMGC/*Proprietors' Minute Book*, Director's report for year end, 30 June 1894, presented 4 August 1894, 108. In June 1894, the SMGC averaged 300 new working-class customers per week, totalling 29, 705 cookers leased; *Proprietors' Minute Book*, Director's report for year end 30 June 1901, presented 14 August 1901, 178; by 1901, the SMGC served 116,000 working-class homes, with over 6,000 new customers in the last six months; 80 per cent had gas cookers. LMA/B/SMGC/*Proprietors' Minute Book*, Director's report for year end 31 December 1904, presented 30 January 1905, 30; in 1904, the company estimated 172,884 penny-in-the-slot meter customers and, of these, 150,113 hired penny-in-the-slot meter stoves.

108 LMA/B/GLCC 40, *Minutes of the Board of Directors*, 18 November 1895, 253.
109 LMA/B/SMGC/*Proprietors' Minute Book*, Director's report for year end 31 December 1901, presented 29 January 1902, 184.
110 LMA/B/SMGC/*Proprietors' Minute Book*, Director's report for year end 31 December 1901, presented 29 January 1902, 184. The GLCC paid its automatic meter collectors £75 per year, about equal to that of a senior clerk. See LMA/B/GLCC 40, *Minutes of the Board of Directors*, 23 November 1894, 184.
111 For discussions of white-collar workers and the lower middle class see Harold Perkin, *The Rise of Professional Society: England Since 1880* (London and New York, 1989), 96–99; Geoffrey Crossick, 'The Emergence of the Lower Middle Class in Britain: A Discussion', in Geoffrey Crossick (ed.), *The Lower-Middle Class in Britain, 1870–1914* (London, 1977); Christopher P. Hosgood, '"Mercantile monasteries": shops, shop assistants and shop-life in late-Victorian and Edwardian Britain', *Journal of British Studies*, 38 (July 1999), 322–352; Peter Bailey, '"White collars, gray lives"? The lower middle class revisited', *Journal of British Studies*, 38 (July 1999), 273–290; Geoffrey Spurr, '"Those who are obliged to pretend that they are gentlefolk": the construction of a clerking identity in Victorian and Edwardian London', unpublished PhD dissertation (McMaster University, 2001).
112 Apparently, the outdoor sales representatives of the South Suburban Gas Company in South London wore bowler hats, black coats and pinstripe trousers until the late 1930s. Interview with Pearl Thorpe (née Brown), retired home service advisor with the South Suburban Gas Company, London, 23 March 2001.
113 LMA/B/GLCC 40, *Minutes of the Board of Directors*, 21 June 1895, 201; GLCC 40, *Minutes of the Board of Directors*, 25 October 1895, 245.
114 LMA/B/GLCC 40, *Minutes of the Board of Directors*, 31 July 1896, 386.
115 George R. Sims, *How the Poor Live and Horrible London*, reprint of 1889 edition (New York and London 1984), 87.
116 LMA/B/LCC, *Notice to Consumers Re: Automatic Meters*, a special circular (GLCC, Horseferry Road, Westminster), April 1896.
117 *JGL*, 10 September 1895, 541.
118 LMA/B/GLCC 40, *Minutes of the Board of Directors*, 28 February 1896, 307.
119 LMA/B/GLCC, 41, *Minutes of the Board of Directors*, 5 February 1904, 178.
120 *JGL*, 1 October 1895, 677; *JGL*, 8 October 1895, 734.
121 LMA/B/SMGC/*Proprietors' Minute Book*, Director's report for year end 30 June 1901, presented 14 August 1901, 178.
122 Yvonne Kapp, *The Air of Freedom: The Birth of New Unionism* (London, 1989), 106.
123 Perkin, *Rise of Professional Society*, 111–112.
124 Melling, 'Industrial strife and business welfare philosophy', 167–170; Mills, 'The gasworkers of south London', 12–15.
125 Kapp, *Air of Freedom*, 117–118.
126 Christopher Moss, 'Industrial copartnership, labour unrest and the last Liberal government', unpublished MA thesis (McMaster University, 1983), 13–22.
127 *Labour Co-Partnership*, **1**, 4 (October 1894), 27. John Burns condemned the plan on the basis that it only applied to 'a few foremen' and was

designed to crush the unions. George Holyoake defended profit-sharing because, in conjunction with union activity, it offered the workers a means to acquire savings while fostering cooperation between workers and employers.

128 South Metropolitan Gas Company, *Co-Partnership Journal* (November 1908), 250; *Labour Co-Partnership*, 1, 4 (October, 1894), 27; LMA/Charity Organization Society/ *COS Minute Book*, 13 (December 1902–April 1908), address by George Livesey, 'Industrial Partnership and the Prevention of Distress', Special Meeting of the COS, 30 March 1903. George Livesey (1834–1908) epitomised the Samuel Smiles ideals of self-help, thrift and moral behaviour. The son of a gas engineer, Livesey spent his entire 60-year career with the SMGC. A strict Church of England teetotaller who forbade the sale of beer in the gasworks' canteens, he was an ardent supporter of the Band of Hope. Despite the many improvements to gas production which resulted from his technical expertise as an engineer, Livesey believed that his profit-sharing scheme, copartnership, was his most important contribution to the industry because it forced the workers to save by holding back a wage bonus in a company savings fund. Livesey believed that his plan offered a solution to employer confrontations with labour unions and presented a viable alternative to socialism among the working population.

129 Ross McKibbin, 'Class and Poverty in Edwardian England', in Ross McKibbin (ed.), *The Ideologies of Class: Social Relations in Britain, 1880–1950* (Oxford, 1990), 167–196.

130 Helen Bosanquet, *Rich and Poor* (London, 1899), 89–90.

131 Ibid., 92.

132 Stedman Jones, *Outcaste London*, chap. 16.

133 Ibid., 330–331.

134 *Report of the Interdepartmental Committee of Physical Deterioration*, Minutes of Evidence, 1 (1904), 13–14.

135 Burnett, *Plenty and Want*, 180.

136 Diana E. St John, 'Educate or domesticate? Early twentieth-century pressures on older girls in elementary school', *Women's History Review*, 3, 2 (November, 1994), 191–218.

137 Jose Harris, *Private Lives, Public Spirit: Britain 1870–1914*, 2nd edition (Harmondsworth, Middlesex, 1994), 241–245; G.R. Searle, *The Quest for National Efficiency: A Study in British Politics and Political Thought, 1899–1914* (Berkeley, 1971), 60–67; Richard Soloway, *Demography and Degeneration: Eugenics and the Declining Birthrate in the Twentieth Century* (Chapel Hill, 1990).

138 'The services of the lady lecturer and teacher', *JGL*, 24 January 1905, 218.

139 'Conducting exhibitions of cooking by gas', *JGL*, 17 January 1899, 126–127.

140 'Earl's Court gas exhibition', *JGL*, 11 October 1904, 1119.

141 'The last day of the exhibition', *JGL*, 20 December 1904, 898.

142 Ibid., 899. The schoolgirls received prize money of £2, 15 shillings and 10 shillings for first, second and third.

143 Ibid., 898–99; 'Results of the quick cookery competitions', *JGL*, 6 December 1904, 761; 'The lady lecturer and teacher', *JGL*, 31 January 1905, 283.

144 'The services of the lady lecturer and teacher', *JGL*, 24 January 1905, 218.

145 Bosanquet, *Rich and Poor*, 142–152.; Helen Bosanquet, *Social Work in London 1869–1914* (London, 1914), 56. See also Margaret Loan,

Outlines of Routine in District Nursing (London, 1905), 141, quoted in McKibbon, *Ideologies of Class*, 179–180.

146 'The services of the lady lecturer and teacher', *JGL*, 24 January 1905, 218.

147 Jane Lewis, 'The Working-Class Wife and Mother and State Intervention 1870–1918', in Jane Lewis (ed.), *Labour and Love: Women's Experience of Home and Family, 1850–1940* (Oxford, 1986), 111–113; Ross, *Love and Toil*, 204–209.

148 'The services of the lady lecturer and teacher', *JGL*, 24 January 1905, 218.

149 Between January and March 1905, the *JGL* printed 13 editorial letters and articles on the subject of the lady experts. Of these, 11 letters were from women undertaking this work. The following were among the list of contributors: Helen Edden, with the Davis Gas Stove Co. and a former lecturer at the 1904 Gas Exhibit, Ida Cameron, of the GLCC demonstrations, Mrs Sutcliffe, organiser of the Olympia cookery lectures, Mrs Charles Marshall and Miss Rose Brown, demonstrators with the Richmond Co. and Mrs Emilie Deane, who toured for the Willey & Co. stove works of Exeter.

150 'The lady lecturer and teacher: correspondence on the article published in last week's issue', *JGL*, 31 January 1905, 282–3; see the letters from Miss M.E. Betts; Mrs Godfrey Sutcliffe (formerly Miss Mary Ball); 'Miss M.S. Birkett writes', 14 February 1905, 419.

151 'Miss M.S. Birkett writes', *JGL*, 14 February 1905, 419.

152 Mrs A.M. Collins, 'The lady lecturer and teacher', *JGL*, 21 February 1905, 501.

153 Robert Opie, *Rule Britannia: Trading on the British Image*, (Harmondsworth, 1985), 84–85, 94–95; Hindley and Hindley, *Advertising in Victorian England*, illus. 4.1.

154 Richards, *Commodity Culture*, 228.

155 Loeb, *Consuming Angels*, 3.

156 Sparke, *As Long As It's Pink*, 44–45.

157 Beetham, *A Magazine of Her Own?*, chap. 10.

158 Pamela Horn, *The Rise and Fall of the Victorian Servant* (Dublin, 1975), 103–106, 133–136; Theresa McBride, *The Domestic Revolution* (London, 1976), 104–106.

159 Hindley and Hindley, *Advertising in Victorian England*, illus. 4.1.

160 Nevett, *Advertising in Britain*, colour plate, between pages 114 and 115.

161 Opie, *Rule Britannia*, 95.

162 Annmarie Turnbull, 'An isolated missionary: the domestic subjects teacher in England', 1870–1914, *WHR*, 3, 1 (1994), 82.

163 'The lady lecturer and teacher', *JGL*, 31 January 1905, 283, letter from Helen Edden.

164 'The lady lecturer and teacher', *JGL*, 28 February 1905, 584, letter from Rose Brown.

165 George Stevenson, 'Lady inspectors for cooking-stoves', *JGL*, 14 February 1905, 418.

166 LMA/B/GLCC 44, *Minutes of the Board of Directors*, 19 January 1906, 14.

167 LMA/B/GLCC 44, *Minutes of the Board of Directors*, 23 November 1906, 150.

168 Nora Phillips, Marian Edwardes, Janet Tuckey, E. Dixon (eds), *A Dictionary of Employments Open to Women* (London, 1898), 110–111, 122, 135–136. Annual starting salaries for female clerks averaged at £35–40; sales assistants

earned £10–60 depending on the type of shop and board arrangements, while board school cookery teachers earned around £60.

169 LMA/B/SMGC, *Minutes of the Board of Directors*, 8 September 1909, 91; 'Cookery hall', *JGL*, 14 October 1913, 129.

170 LMA/B/GLCC 44, *Minutes of the Board of Directors*, 19 January 1906, 14; also see GLCC, *Minutes of the Board of Directors*, 8 June 1906, 80; GLCC, *Minutes of the Board of Directors*, 26 October 1906, 135.

171 Interview with Pauline Horrigan, former head of the women's advisory staff, British Gas Council (St George Hotel, Colchester, Essex) 8 December 1995.

172 M. Mostyn Bird, *Women at Work: A Study of Different Ways of Earning a Living Open to Women* (London, 1911), 121–122.

173 Sillitoe, *A History of the Teaching of Domestic Subjects*, 111.

174 Miss E.M. Evans, 'The essential qualifications of a lady health visitor', *Journal of the Sanitary Institute*, transactions, **24**, 1903 (London, 1904), 310, 312, 317–318.

175 'The unfailing hatpin', *Co-Partnership Journal* (September 1915), 208. Although this particular incident was reported in 1915, the author suggested that lady demons had previously resorted to the hat-pin as a repair tool, and that, in 1915, fixing stoves was still not considered a demonstrator's job despite the growing numbers of men that had left the gas industry to join the armed forces. By 1918 the situation had altered completely with women undertaking a wide range of jobs in the gas industry, including stoking retort houses and loading coke.

176 John Johnson Collection/Gas and Electrical, Box 1/Bodleian Library/ Helen Edden, *How to Use a Gas Cooking Range* (Croydon Gas Company, n.d.). Instructions on how to clean the stove were italicised and underlined, while information on lighting the stove received no special attention. Users were instructed to turn off the 'taps' when the stove was not in use, but this critical information was only partially italicised.

177 Helen Edden and N. Moser (eds), *The Household Gas Cookery Book* (London, 1908) 6, 11.

178 Amy Atkinson and Grace Holroyd, *Practical Cookery with an Introduction on Cookery by Gas* (Leeds, 1909), 5.

179 Kwolek-Folland, *Engendering Business*, 179.

CHAPTER 3

Lady Demons and 'Well-Dressed Men': Gender and Sales in the Edwardian Gas Industry

Twelve little maidens all in a row
Whence do they come, whither do they go?
With radiant smiles and voices sweet
Tripping along with such dainty feet.

Twelve pretty maidens all in row,
What are they doing, I'd like to know?
Tripping along through snow and sleet,
In winter's cold, or summer heat.

So late to bed, so early to rise,
Working alike beneath gloomy skies,
Or fainting 'neath the sun's hot rays,
Without a look or word of praise.

Twelve little 'Demons' all in a row
Working hard for The Gas Light Co.!
Teaching the wives to bake and boil;
And to clean a stove without any toil.

They teach them all they want to know
From skinning a whale to stuffing a crow!
To blaze a kipper, or grill a mouse
This is why they trip from house to house.

So on they toil from day to day
With never a thought of rest or play –
Till a shaft is loosed from Cupid's bow
Which lures them from The Gas Light Co.

For Hymen's luminous torch burns bright
Changing the 'Demons' to Angels of light
In bridal array, far away they go –
Alas! and alack for The Gas Light Co.

[signed] L.P.

A 'Demon'[1]

The above verse appeared in the September 1911 issue of *Co-Partners' Magazine*, the official staff journal of the GLCC. The company magazine informed employees about the progress of its new profit-sharing scheme, introduced in 1909, and also reported the company's many social events, from the gasworks' cricket leagues to the children's Christmas parties. To foster a sense of community within the gas industry, and conciliation between workers and management, the journal covered major publicity events, like the national gas exhibitions, and reprinted the official minutes of the general stockholders' meetings. Although the GLCC was not entirely anti-union, the editors avoided any reference to trade union activity or employee discontent. *Co-Partners* was a carefully monitored company publication that promoted the ideal of cooperation between workers and management. It seldom gave voice to workers' grievances of any kind, except perhaps in the case of the 'Twelve Little Demons'.

The comic poem, signed 'A Demon', satirised the daily routine of the lady demons as they undertook their customer calls. Despite the demonstrators' constant efforts to construct a professional image of themselves, the poem's author presented the cookery experts as innocent 'little maidens' with honeyed voices and sweet smiles 'tripping along' on 'dainty feet'. In contrast to the social, corporate and even political imperatives of their work, the writer poked fun at herself and her colleagues as 'they trip from house to house' instructing housewives to perform unimaginable culinary feats, from skinning whales and stuffing crows to grilling mice. Obviously, the poem was meant to be a joke.

Nevertheless, behind the satire lay a smouldering element of discontent that made its appearance in the company journal unusual. The lady demon referred to the long hours and extreme weather conditions endured by the canvassing women with no apparent company recognition for either their efforts or their unwavering dedication. Perhaps the poet felt that the demonstrators might as well offer lessons in roasting rodents for all the governors knew of their work. She warned the company that the lady demons would readily abandon their professional positions and join ranks with the housewives they advised. Hymen, god of marriage, becomes a deliverer capable of transforming the professional public-serving 'Demons' to 'Angels' sheltered in the private sphere. Only then would the company have realised its irreparable loss.

The poem suggests that the lady demons, the first female employees in the industry, felt dissatisfied with a patriarchal corporate system that failed to acknowledge their professional contributions. Clearly, the demon poet was conscious of the social and occupational ambiguities of this new type of sales employment. However, she neglected to mention that, often, newly-wed lady demons merely exchanged one set of governors for another. Further, the domestic expertise of the young housewife may not have been any more appreciated by her new husband than it was by her old employer. As the poem suggests, gender ideology defined both the workplace and the home and the lady demons straddled the two, but not without apparent difficulty. The poet's contrasting images of the corporate 'Demons', faint with heat, their toil and the grime of the streets, and the idealised 'Angels', in gauzy bridal gowns suggested a longing for the latter, more conventional career choice, particularly during that sweltering summer of 1911.[2] She also reminded her readers that, apart from the occasional excitement of performing at a trade exhibition, the job of the demonstrator was often tedious, tiring and rather lonely.

Although lady demons became permanent staff members, around 1905, their occupational responsibilities remained open to question by gas managers and demonstrators alike. As the number of gas industry lady demons increased throughout the Edwardian period, so did corporate concerns about the integration of women into the gas industry. Because they were educated women from various levels of the middle classes, the demonstrators anticipated the expansion of their career opportunities, from simply demonstrating stoves to that of managing showrooms. Given their direct experience with female customers and their technical familiarity with gas appliances, the women saw themselves as specialised sales professionals. They used their higher education and occupational mandate to distinguish themselves from the female sales assistants of the high street department stores, and even from their fellow retail workers, the male gas showroom attendants. The construction of the lady demons' business personae reveals the manner in which class and gender intersected in the workplace. This chapter examines the development of a feminine corporate identity within the gas industry, in light of contemporary social and political events related to the Edwardian 'woman question'. At the same time, proposals for government intervention to improve public health, housing and child welfare provided a powerful impetus to the work of the lady demons. Discussions concerning their occupational responsibilities and their role in the sales process coincided with the founding of a national advertising organisation, the British Commercial Gas Association, a collective responsible for all of the gas industry's promotional material. The lady demons found a powerful ally in this organisation's chief editor.

During the twentieth century, the gas industry embarked on a programme of corporate modernisation in an effort to outpace the growing electrical industry. Increased attention to advertising, and particularly advertising directed at women of different classes, constituted an important aspect of this modernisation process. Despite its alleged conservatism, an image upheld by business historians and scholars of industrial design, the Edwardian gas industry must be given some credit for founding a marketing and public relations organisation that anticipated the importance of appealing both directly and indirectly to feminine purchasing power for consumer durables.[3] The BCGA, an industry-wide publicity group formed in 1911, targeted builders, architects and doctors, in the hope of convincing these professionals of the advantages of gas. Under the combined leadership of sales specialist Francis Goodenough and public health activist Mrs Maud Adeline Cloudesley Brereton – both members of the Edwardian 'Progressive movement' – the BCGA adopted the mandate already assigned to the lady demons: improved service and customer education. To this end, the BCGA propaganda addressed social issues of relevance to different classes, ranging from the servant shortage and domestic economy to child welfare and the need for improved working-class housing. In a recent article on the failure of electrical appliances to have an impact on the interwar English home, Sue Bowden and Avner Offer note that, in comparison to electricity, the gas industry made a 'strong commitment to advertising' to women of all classes.[4] This 'strong commitment' originated with the lady demons, but was strengthened by the organisation of the BCGA and the activities of Maud Adeline Cloudesley Brereton. Brereton remained an outspoken advocate for women's rights to better working and living conditions, and, by her example, she proved that women might be highly successful in the masculine world of business. Further, the adoption of public causes, such as women's rights, to promote gas effectively increased the public profile of the lady demons and their work.

'A power in any community'

At a meeting of the National Commercial Gas Association of America in 1909, Mrs Helen Armstrong presented a paper on the work of lady demons in America. So impressed was Walter King with Armstrong's advice to lady demons and gas companies that he reprinted her entire article with his full endorsement.[5] Armstrong's comments echoed many of the statements articulated by her English counterparts in their 1905 letters. Like her British contemporaries, she emphasised the importance of cookery demonstrations in the company showrooms. She stressed the promotional value of supplying stoves, and regular instruction on how to use them, to local domestic science colleges, school boards and hospitals. She attested to the value of speaking engagements with women's clubs and church groups.

Like Helen Edden and her staff, Armstrong appreciated the effect of the home visit, as female inspectors 'prove a strong practical advertisement for gas as fuel'. She cited the recent successful experiments with home visits by lady demons undertaken by gas companies in New York and St Louis. According to Armstrong, the lady demons provided several important functions. Their regular visits after the installation of a range ensured that the customer was satisfied with the stove and knew how to use it properly. By responding to any additional enquiries and calls, they secured new business via their prompt and personable assistance. Finally, they apprised the business manager of conditions in customers' homes, thus acting as intermediaries between female consumers and male producers. As Edden and her lady demons well knew, it was a job that demanded training in domestic economy and public speaking. It also required a large measure of patience, tact and the ability to offer an opinion without appearing officious or condescending.

Armstrong listed a number of appropriate physical characteristics. The ideal demonstrator was of 'middle age', as she inspired confidence more quickly than a young girl. She should not live locally, as a stranger was less likely to inspire criticism or waste time with gossip. Most of all, Armstrong demanded that the lady demon be neat and tidy, as 'a young woman who is careless in her dress, or has untidy hair has no business in a kitchen'.[6] The chief disadvantage to this list of suitable qualities was that the lady demons were likely to soon marry company men, a possibility our disgruntled demon alluded to in her 1911 poem.

In fact, despite the poem's suggestion that disenchanted lady demons might retreat from the working world into the domestic sphere, it appears that married, widowed and single women alike engaged in this type of work. Among the 11 correspondents to the *Journal of Gas Lighting* debate, five women were single and six were either married or widowed. Some demonstrators, such as Mrs Charles Marshall of Enfield and Mrs Godfrey Sutcliffe of East Croydon, were wives of gas engineers and chose to work for the same companies as their husbands. Their active participation in gas promotion reflected a pattern of cooperation within the family, initiated by women like Marie Sugg and Lillie Richmond, who both worked alongside their respective entrepreneurial male relatives. Moreover, it was possible that married or widowed women appeared to be more reliable culinary experts, the assumption being that they were more experienced cooks or domestic managers than single women. Some of the most celebrated lecturers in the gas industry were, or had been, married. For example, Mrs E. Middletown, a 'favourite with the cooking classes', spoke twice daily for two weeks to packed audiences at the 1907 Manchester Gas Exhibition. Her excellent presentations and oratory skills allegedly drew the crowds like a magnet.[7] Even after the establishment of permanent ladies advisory departments in the gas industry, married or widowed women continued to work alongside single women. Two of the 11 demonstrators pictured in a GLCC group portrait were either married or widowed (see Fig. 3.1). None of the women resembled the delicate 'little maidens' described in this chapter's opening literary tribute, 'Twelve Little Demons'. Judging from the praise lavished on popular demonstrators, like Mrs Charles Marshall, Mrs Middletown and Miss Edden, competence, initiative and a good voice were more relevant to their employers than the women's marital status.

Helen Armstrong suggested that, given the 'nerve-wearing' nature of this career, few women could continue in the job for long. She recommended short hours and good pay, but gave no specifics. Despite the difficulties of the job, Armstrong maintained that the chief service of the lady demon was to 'secure and maintain the goodwill of the housewife' on behalf of the gas company. In this respect, and working through local schools and women's organisations, the lady demons had the potential to become 'a power in any community', given the impact they might have on their customers. By noting that lady demons made excellent wives, Armstrong easily reconciled middle-class womanhood with the demanding world of business and sales; however, she also reaffirmed the typical stereotype of the business girl as a husband hunter.[8]

Nevertheless, Helen Armstrong's assertions about the power of women coincided with contemporary debates concerning the woman question in general, and, particularly, female suffrage. By the Edwardian era, the problem of 'surplus women' was familiar to educators, social

3.1 Women's Advisory Staff, Gas, Light and Coke Company, London

commentators and philanthropists. Late Victorian authors had lamented the growing numbers of single women, particularly among the educated middle classes, an imbalance attributed to a male predilection to delay marriage until they achieved financial success, which often meant foregoing wedlock entirely. Male emigration and colonial service also accounted for the apparent growth in the number of unwed ladies.[9] Proposed solutions included emigration or respectable employment for 'redundant' gentlewomen.[10] While Meta Zimmeck notes that some recommended female occupations were little more than frivolous time-wasters, a growing body of literature on women's professional and artisan work described valid options for those seeking a career.[11] Among social commentators, however, there existed the growing fear that educated women intentionally chose spinsterhood over marriage and motherhood, preferring personal freedom and a dangerous independence. The 'New Woman' was lampooned in the popular press as the undesirable and shocking stereotype of female emancipation.[12] While the rebellious 'Girton Girls' constituted a small minority, according to Judith Walkowitz, women's fiction disseminated the image of the new female; she appeared in the feminist works of Olive Schreiner and the more traditional pages of the *Girl's Own Paper*. Even women who deprecated female emancipation, notably Mrs Eliza Lynn Linton, positioned themselves in relation to the New Woman by asserting the importance of women's maternal role to the continued vitality of the nation.[13]

Fears of the deterioration of the British race and imperial decline added a sense of political urgency to the debates over women's work, marriage and motherhood. Eugenicist Karl Pearson blamed the falling middle-class birthrate on the increased number of educated women who refused to have children. Sexologist Havelock Ellis accepted an 'equal but different' interpretation of gender roles but maintained that women had essential nurturing qualities that made them fit for motherhood in their 'proper sphere'.[14] Women responded to this exaltation of motherhood in a variety of ways. The outspoken feminist writer, Cicely Hamilton, condemned the legal and social restrictions of marriage and demanded female equality in terms of work, education and political rights.[15] Socialist feminist Ethel Snowden used maternal arguments as ample justification for female enfranchisement.[16] She argued that if women's contributions as mothers were critical to the racial health of the nation, then surely they deserved representation. Other female suffragist groups adopted a similarly maternalist position. For example, the National Union of Women's Suffrage Societies cited women's superior morality as reason enough for granting them the vote.[17] Finally, female suffragists asserted that, given the Liberal government's (1906–14) increased concern for standards of public health, rates of infant mortality and the availability of decent housing, women deserved consultation in political matters that affected them directly.

Amid these ongoing social and political debates, the lady demons undertook the public mission of the gas companies: to educate and instruct women customers in their homes. However, in Edwardian London, lady demons of another sort occupied the streets. Both the NUWSS (1897), led by Millicent Fawcett, and the militant Women's Social and Political Union (1903), led by Emmeline Pankhurst, staged elaborate political processions.[18] The marches involved thousands of women from numerous professional groups, including teachers, nurses, artists and writers, gardeners, homemakers and businesswomen.[19] The appearance of Helen Armstrong's article, promoting the educational power of lady demons in their communities, coincided with the political assertions of the female suffragists. The rallies and processions highlighted the economic and social contributions of female workers, whether as professionals, manual workers or homemakers, as ample reason for their enfranchisement.[20] Although the women's suffrage movement fragmented over tactical issues, polarising constitutional moderates and militants, the spectacular processions successfully focused public attention on the rights of working women and their demands for dignity and decent wages.[21]

Unfortunately, it is not known whether any lady demons participated in the suffrage campaign, attended its processions, or even supported the cause. The Association of Teachers of Domestic Science avoided an official stand on the franchise question, although its members were free to participate in rallies if they chose. Despite the suffragists' respect for the important, but often unrecognised, contributions of housewives, the question of teaching domestic subjects to schoolgirls remained a contentious issue.[22] However, not all suffragists saw a conflict between domestic instruction and female emancipation. Miss Baddeley, principal of the Gloucester School of Domestic Science, considered domestic studies a 'potent weapon' for combating social problems such as unsanitary housing and inadequate nutrition. Further, the application of rational scientific methods to home management was just one aspect of a wider reform movement that included female political emancipation.[23]

As noted earlier, the lady demons drew professional inspiration from three different occupational groups: social workers, teachers of the domestic subjects, and, to a lesser extent, shop assistants. Given the similarities in education, professional status and social responsibilities, the lady demons identified most directly with the first two. Nonetheless, the demonstrators were essentially sales personnel, however much they or their employers chose to extend and redefine that role through customer visitations and public lectures. The lady demons combined the duties of all three of the above female occupations and shared similar preoccupations with job status. Amid Edwardian debates over women's suffrage and female employment, all three occupational groups sought professional recognition in much the same manner as the gas industry's lady demons.

Women in the public sphere

Traditionally, charity work ranked among the genteel occupations open to respectable Victorian women. Evangelical authors and activists maintained that women were ideally suited to charity work, because of their apparent natural benevolence, compassion, patience and self-sacrificing tact. The overlap between the duties in working for a religion or charity further enhanced their suitability for feminine involvement, as did the moralistic character of the causes themselves, as they involved temperance, female rescue work and parish visitations.[24] Historians studying female participation in Victorian charities have noted the charities' improved organisation, specialised committee work and cooperation with local government.[25] Although women's involvement in charity work stemmed from their traditional role of aiding the unfortunate, it was also related to the 'increasing bureaucratization' of government and social services. Martha Vicinus claims that by 1893 an estimated 20,000 women in England maintained themselves as paid philanthropic workers, while around half a million worked on a semi-professional or voluntary basis.[26] For example, the development of Octavia Hill's COS, after 1869, exemplifies this movement from voluntary to professional social work.[27] By 1870, the COS had become an umbrella organisation for a number of charities, offering training in visitation methods and social investigation techniques (an early form of sociology). The COS encouraged its members to enter a number of social reform occupations, and their trainees acted as rent collectors, sanitary inspectors and district visitors; however, they were known for their moralistic simplicity, as they often ignored the hard realities of working-class life.[28] Still, historians give credit to the COS for elevating the occupational status of charity work by introducing the case system of formally interviewing clients, a data-gathering method enlisted by Charles Booth's researchers for his studies of London in the 1880s.[29]

Like the COS, the university settlement movement offered charitable employment for middle-class women, and furthered the development of social work as a paid occupation. In 1887, the first women's settlement, Lady Margaret Hall in Southwark, offered female students from Oxford's Lady Margaret College the chance to live among the poor in London.[30] Female settlers involved themselves in their new communities in a variety of ways, by organising clubs for working girls, holding classes on infant health and conducting home visits in the COS style.[31] The leaders of the women's settlement movement were determined to make philanthropy a paid profession; they therefore offered training courses connected to the COS School of Sociology, although only a dozen women graduated per year.[32] Whether volunteers or professionals, settlers increasingly saw themselves as experts on working-class life and poverty. Edwardian female health inspectors were another occupational group that asserted the professional status of social work. Usually affiliated with local medical health offices, these middle-class women often held certificates from the Royal Sanitary Institute. They worked with school authorities by conducting home visits, advising on contagious diseases, offering instruction in nutrition and childcare, and, in some cases, also working as health inspectors in factories and workshops which employed women and children. Like other female social workers, they also engaged in debates – specifically, with local councils and the medical community – regarding their duties, pay and occupational status.[33]

There are some important similarities between these early social workers and the lady demons. Like the COS and the settlement workers, the lady demons made routine calls to working-class homes, keeping official records of their visits (thereby adopting the formal case system). In addition, like the social workers, the lady demons enjoyed a great deal of independence and freedom of movement, the hallmarks of the modern urban woman and the confident female *flâneur*.[34] Both social workers and gas demonstrators were visible public women in communities where they, in many cases, did not reside. The demonstrators' preoccupation with personal appearance and genteel behaviour suggests their desire to provide role models for women in working-class communities and also to ensure that they received a certain degree of respect, much like the female public health inspectors. One demonstrator, Miss M.E. Betts, suggested that gas company lady demons should wear official uniforms or badges to literally clothe them with 'real authority' and also make them easily identifiable to passers-by and their customers.[35] It was well known that uniformed district nurses were free to walk unmolested, even in the roughest neighbourhoods.[36] Possibly, Miss Betts sought similar protection for female gas representatives. Finally, like the professional social workers, the lady demons saw themselves as trained experts surrounded by amateurs. Salaried full-time social workers felt superior to occasional volunteers and donors.[37] Similarly, the 'gas cookery lecturers' considered themselves a highly select group, distinguished from other 'teachers of cookery' by their superior oratory and technical skills.[38] Outspoken lady demons, notably Helen Edden and Mrs Charles

Marshall, feared that their professional image was being undermined by ineffective presenters whose cookery talks reflected 'the principle of the "bus driver's holiday": an inarticulate jumble'. Other lectures, they claimed, deteriorated into a chaotic 'scramble for cakes' as audiences rushed to sample the end results of the demonstrations.[39] Strict standards of presentation, etiquette and appearance became fundamental aspects of the lady demons' evolving sense of professional decorum.

Obviously, despite the demonstrators' attempts to distinguish themselves from the average cookery teachers, the two groups had a great deal in common in terms of class background, training, work objectives and the occupational associations to which they belonged. Yet, compared with the thousands of cookery teachers, the lady demons constituted an elite minority, employed by governors of private corporations and not the ratepayers and councillors of local school boards. As a result, the demonstrators were better paid and enjoyed more professional autonomy than their colleagues in the public sector. Moreover, despite some gas directors' reservations about hiring women permanently, the lady demons were not subject to the same constant academic challenges encountered by the teachers of domestic subjects.

Ignoring the objections raised by feminist educators, domestic subjects were increasingly integrated into the curriculums of elementary and secondary schools after the 1902 Balfour Education Act. Amid this climate of criticism, domestic economy teachers felt compelled to defend both their occupational status and the scientific merit of their discipline. Historians of the English education system have analysed the incorporation of domestic subjects for girls in school curriculums.[40] All of these historians address the impact of the domestic science movement on educational policy and the creation of a gendered curriculum that stipulated that girls in board schools received mandatory domestic subjects training. The extension of domestic training stemmed from concerns over the degeneration of the British race and the notion that better-educated mothers might recoup the 'physical and moral health of the people'.[41] Eugenicists and educators argued that girls, particularly working-class girls, required practical lessons that were applicable in the domestic sphere to prepare them for motherhood.

How best to achieve the desired improvements in public health remained open to debate. For example, in 1906, HM chief woman inspector Maud Lawrence outlined proposals to standardise the domestic subjects curriculums with the focus on simple inexpensive dishes and the elimination of technical information in favour of practical instruction.[42] However, shortly after, the implementation of the school meals programme meant another redirection of domestic instruction. The task of preparing and serving the midday meal often fell to the cookery teacher with class-time devoted to its preparation. Thus, schoolgirls went from preparing bath buns sold for a halfpenny to peeling endless vegetables for the daily soup pot. In neither instance did they learn to prepare affordable family meals. Moreover, astute social critics, such as the Labour activist and feminist Charlotte Despard, noted that the LCC pawned the school meals programme off on to the domestic subjects teachers to save money on food and facilities, with little consideration for the needs of either the children or the teachers.[43]

There was little consensus regarding domestic subjects for female students in the upper grades or in private schools. Some headmistresses resisted the introduction of domestic subjects, not because girls had no need of domestic science, but because it took valuable time away from more important academic subjects. Critics like Ida Freund, a lecturer at Newnham College, Cambridge, feared that the vogue for domestic science would lower teaching standards in girls' schools. Against these objections, supporters like Margaret Gilliland, Lilian Faithfull and Sara Burstall stressed the scientific foundations of cookery, laundry and hygiene.[44] Botany, physics and chemistry provided a solid foundation for the study of domestic subjects as applied sciences, and Faithfull anticipated the day when the distinction 'between the professional woman and woman whose education includes a study of home science will tend to disappear'.[45] As a graduate of Somerville College and former vice principal of the women's department at King's College London, Faithfull realised the connection between education and professional legitimacy. She publicly defended King's College's decision to confer degrees in household and social science, despite the feminist rage vented in the pages of the radical feminist journal, *The Freewoman*.[46] In support of the programme, Hilda D. Oakley, warden of King's College for Women, stressed the scientific components of their domestic science degree, the emphasis on ethics and economics and the university's high academic standards.[47] As Nancy Blakestad's study reveals, King's College offered more than just 'degrees for housewives'; graduates found employment in a range of developing fields including social work and business, although, initially, teaching at the college absorbed the greatest numbers of graduates.[48]

Even before Edwardian educators debated the King's College degree, cookery teachers defended themselves and their discipline against critics within the teaching profession. The prestige of the cookery teacher suffered for a number of reasons. Because they were not on the staff of any particular school, they had no real authority, and headmistresses often resented the disruption of sending girls to the local cookery centres.[49] Inspectors complained that cookery teachers had no control over their classes, were inadequately trained and mismanaged school funds. Teachers of domestic subjects responded in the pages of their professional journal, asserting the 'dignity and greatness' of their work given its impact on national health.[50] They lamented how poorly the nation showed its sympathy for the overworked and underpaid cookery mistresses who supervised saucepans and schoolchildren for a paltry wage of between £40 and £60 per annum. The *Cookery Teacher* condemned the notion that domestic economy teachers were not committed to their professions and were more likely to marry after a few years than other teachers because suppositions of this sort justified the women's continued low salaries. Despite the alleged importance of domestic training to national health, cookery teachers were constantly defending their professional credibility to feminist reformers and also to their fellow educators.

In contrast to the board cookery teachers, the lady demons enjoyed higher wages and possibly greater job satisfaction. Unlike the teacher in the classroom, the lady demon had more individual autonomy and was exposed to far less public criticism. Ironically, the demons' greatest critics were the domestic economy teachers themselves who feared that the entertaining gas demonstrations trivialised their scientific cookery classes.[51] One board school cookery teacher suspected that students who attended these 'peripatetic lessons get ideas of pretty dishes and sweets, and become disgusted with class methods of drill in plain household cookery'. Although the editor defended the lady demons as potentially promoting the 'science' of cooking, a degree of professional differentiation occurred within the domestic economy movement. As the lady demons adapted their training in domestic subjects to the commercial requirements of the public platform and the customers' homes, classroom cookery teachers worried that the image of the corporate domestic expert might somehow damage their own assertions of academic professionalism. After all, the lady demons were really just saleswomen for the gas companies, and as the *Cookery Teacher* reminded its readers, even domestic service was a more 'respectable occupation' than shop work.[52]

As Lady Jeune noted in her 1895 article on shopping, the introduction of the female shop assistant must be attributed to late Victorian changes in retail distribution. The advent of department stores and multiple shops meant that there were fewer shopkeepers and more assistants. It also meant that shopping became recreational, and even addictive.[53] By the last decade of the nineteenth century, female shop assistants dominated certain areas of retail stores where the customers were generally women, including shops for drapery, millinery, hosiery, stationery, flowers, underwear, baked goods and dairy products. Despite the long hours, low pay and physical restrictions of the living-in system, shop work possessed the pretensions of middle-class respectability.[54] This was especially important for working-class girls with social aspirations. However, as Lee Holcombe points out, the public seldom shared the shop girl's self-image of upward mobility. The middle and upper classes considered shop workers to be on the same level as domestics, while the labouring classes mocked their pretensions concerning respectability.[55] Peter Bailey maintains that, for men of all classes, shop girls and other women in the service sector occupied a 'middle' ground of sexuality. Neither 'prostitutes' nor 'ladies', their questionable status offered an invitation to erotic fantasy: if they sold things, did they also sell themselves?[56] Erika Rappaport notes that the popular musical comedies of the 1890s sustained this image by presenting the shop girl as a heroine who used her fashionable style and feminine charms to secure a husband above her social class.[57] But, as Bill Lancaster has shown, there were strict social divisions within the occupation. He notes the development of a 'labour aristocracy' among retail salesgirls, particularly within the large and expensive department stores, such as the Bon Marché, Harrods and Whiteleys. These establishments favoured grammar school graduates and offered special training courses to their girls, who then prided themselves on their manner and appearance both inside and outside the shop.[58]

Perhaps because of the ambiguities of the salesgirl image, the lady demons carefully distinguished themselves from this group of working women. Hence their assertions of the importance of lady-like behaviour, education and class status. Similarly, the lady demons protested against the practice of canvassing door-to-door, a common sales technique utilised by itinerant salesmen and commercial travellers. While the bagman's class and occupational status remained somewhat dubious, except to the members of his own fraternity, commercial travelling was certainly not considered a respectable form of employment for women.[59] In the 1890s, the Richmond appliance company had used female canvassers, or 'lady interviewers', to solicit customers at home in conjunction with local gas exhibitions.[60] Lady demons were divided over the overall effectiveness of this practice and the social ramifications of women going door-to-door. Mrs Charles Marshall dismissed the canvassing suggestion altogether on the basis that 'the visits would place the instructress in an equivocal position; no woman of education would undertake the mission'.[61] Marshall implies that no lady demon of the required class and educational background would ever agree to put herself in such a socially questionable situation, even for her work.

Despite these assertions, however, the lady demons were technically saleswomen, representatives of the gas companies like the male outdoor canvassers. Although the women cultivated a professional self-image and closely identified with social workers and teachers of domestic subjects, at times they were painfully aware of their lowly commercial status. A poem signed 'M.M., another Demon' entitled 'How to Grill a Demonstrator' provides another glimpse into the demonstrator's daily routine. The author, probably Miss M Menhinick, described the discomfort of working outdoors during the hot summer of 1911 when London recorded temperatures over 90 degrees. She related her experience of the tiresome wait on blistering-hot doorsteps until someone answered her persistent knocks, and the agony of trying to maintain a lady-like composure despite heat exhaustion.[62] Standing under the boiling-hot sun until the mistress or her servants opened the door, the demonstrator was reminded that she was, despite her professional pretensions, a servant of the gas company and its customers. Like the shop assistant, the demonstrator's social position remained unfixed and might change with each household that she visited. Evidently, given the realities of the workplace, it was difficult to maintain the image of the placid lady-like professional, no matter how hard one tried.

As combined sales assistants, domestic science teachers and social workers, the lady demons were the first women employed by the English gas companies. They were also the gas industry's first trained sales representatives who possessed a clear sense of occupational identity and unity of purpose. In contrast to the lady demons, and their concerns with business etiquette, the role of the gas salesmen – that is, male showroom attendants – received considerably less attention. A possible reason for this may be the overall indifference of gas managers to the demands of selling, as opposed to making, coal gas. As I mentioned in the previous chapter, until the early twentieth century, public trade exhibitions remained the preferred means of generating business and were staffed by teams of gas engineers, company managers and lady demons. In contrast to these organised itinerant exhibitions, company showrooms were often makeshift commercial spaces: 'a corridor stocked with cookers' or a 'corner of an over-crowded office'.[63] Although the large gas companies, including the GLCC and the SMGC, had appreciated the commercial value of well-appointed showrooms in their business premises since the 1890s – even commissioning professional architects to advise them on retail space design – the smaller gas undertakings responded slowly to this view.[64] More frequently, gas managers set up their showrooms, such as they were, near the accounts desk so that regular customers who came in to settle their quarterly payments were forced to pass the new appliances. Perceptive though this layout may have been, it also enabled the accounts clerks to monitor simultaneously the showroom from their books, thereby pushing them to assume two separate jobs. According to one report, within the smaller gas undertakings, the general managers themselves often had more than one role: running the works as the senior engineer while also managing the showroom, albeit in a 'half-hearted way'.[65] Some company managers assumed that selling gas was the responsibility of all gas employees, especially those who came into contact with the public, such as meter inspectors and gas fitters. Prior to the establishment of a regular sales staff, George Livesey offered cash incentives of up to five shillings to anyone who obtained new business for the SMGC.[66] While Livesey's commissions were an attempt to involve every employee in gas promotions, it likewise suggested that selling gas appliances and services could be successfully undertaken by anyone in the industry.

Among gas engineers, the prevailing image of the male showroom attendant was that of a young man who was too inexperienced or too incompetent to perform any genuinely useful task within the industry. In 1903, in a possible attempt to change this perception, and also to meet the growing demand for gas appliances, the GLCC created new positions for male showroom attendants. At annual salaries of £100, the salesmen's salaries compared favourably with the £40 per annum offered to junior clerks.[67] Presumably, young men were expected to aspire to this job from entry-level clerk positions; however, it remains unclear if any special qualifications were required for a showroom job, aside from being polite and well groomed. Unfortunately, the voices of the showroom salesmen themselves are missing from these early discussions, but it appears that Edwardian gas salesmen continued to occupy a marginal place within their own industry. This should not be surprising considering the overall low social and economic status of male retail workers during this same period.[68] Moreover, in terms of their occupational identity, the showroom men lacked the training and official status of the lady demons, who had constructed their own particular code of business conduct aided by their highly developed sense of professional duty and their middle-class domestic science education. In contrast, the male showroom attendants were ranked alongside the legions of company clerks; however, the latter at least performed recognised administrative functions within the industry. While the GLCC's addition of male showroom attendants in 1903 preceded their decision to employ lady demons on a permanent basis, it remains uncertain what level of cooperation, if any, the company expected from these two sales groups. Despite the gradual movement to hire male showroom attendants, the lady demons were still the acknowledged sales professionals in the Edwardian gas industry, which was a position they held as much by default as by design.

As a new occupational group, the lady demons reflected the social tensions of the pre-war years within the context of the suffrage movement, maternal feminism and Edwardian debates over women's work and education. However, their emerging profession was equally representative of changing commercial strategies within the gas industry itself that had become evident by the Edwardian years. These included an increased attention to customer service and advertising, and the growing diversification of promotional techniques. Women, as buyers and sellers, were an important component in this trend. Because of the similar promotional material of the electricity industry, gas managers admitted they had to market more than gas lighting, cookers and heaters: they had to construct and promote an image of domestic harmony made possible by gas utility services, and they had to sell this message to female consumers.

Progressive advertising and the 'ultimate consumer'

Francis W. Goodenough, chief inspector for the GLCC, was a staunch believer in the advantages of mass advertising. An ambitious young man from Devon, Goodenough joined the GLCC in 1891, but rose quickly through the company's ranks. Beginning as a junior clerk, he moved up to assistant superintendent of the rental department, chief inspector and, finally, to chief controller of the newly created sales department by 1910.[69] For this last position he received a generous annual salary of £1,000, successfully transcending the dream of most aspiring Edwardian clerks.[70] Over the span of his 45-year career within the gas industry, Goodenough worked with numerous business organisations concerned with international trade, advertising and salesmanship. In addition, he was a member of the British Institute of Social Service (BISS), a group dedicated to 'the dissemination of information relating to social service and industrial betterment, in order to improve and elevate national life'.[71]

Through his connections to the BISS, Goodenough joined the Rainbow Circle (1894–1931), a descendant of the earlier London Ethical Society. The Rainbow Circle was a debating and discussion group for 'Progressivism'. According to Michael Freeden, its general aim was the formulation of a 'rational and comprehensive view of political and social progress', with the intention of developing a political and economic doctrine that might ultimately become the basis of a programme for social reform. The Rainbow Circle attracted Liberals and moderate Socialists, with a methodological and ideological focus that Freeden describes as 'new liberal'.[72] Its membership included men of the professional class, politicians, journalists, academics and civil servants, notably the Fabian essayist Graham Wallas, the journalist and liberal theorist J.A. Hobson, Herbert Burrows, co-founder of the Social Democratic Federation, and the nation's first Labour prime minister, Ramsay MacDonald. Of the group's total, 65 members, only five were businessmen with direct connections to industry and trade. This under-representation of the business community highlights the continued class tensions between the professional and the business classes, although, as Harold Perkin has shown, the increasing affluence of the latter group and the corresponding establishment of professional associations to serve them was slowly closing the traditional gap.[73] Francis Goodenough was an active participant of the Rainbow Circle from the time that he joined, in 1911, until its dissolution, in 1931. On four occasions he formally addressed the group; three of his four talks considered industrial and labour matters, such as trade unionism and copartnership, the labour problems during World War I and the future of British industry after the war.[74] Judging by the recorded transcripts of Goodenough's talks, his political views were not entirely in accord with those of his fellow members. For example, following his paper entitled 'Co-Partnership', which he delivered in January 1913, Goodenough received criticism for being 'unduly severe' in his discussion of trade unions and the character of the working man. Although the membership agreed that profit-sharing between labour and capital offered some possible solutions to the current labour difficulties, the success of a copartnership plan depended on industrial stability and the employer's ability to predict profits. Because of the 'steadiness' of the gas industry, it was ideally suited to profit-sharing; however, two-thirds of all workmen went from 'job to job' within a fluctuating and unpredictable labour market. Despite this reaction, Goodenough remained within the Rainbow Circle, evidently committed to the notion that industrial questions could not be settled through direct government intervention, but required greater cooperation between management and labour. According to Goodenough, from both a moral and a practical perspective, employers were responsible for the welfare and safety of their workers, as well as the development of strategies that enabled workers to share in the profits of industry.[75]

Described by his colleagues as a 'great salesman', Goodenough represented a new generation of gas employees whose expertise lay in the areas of marketing and management and not necessarily technical engineering.[76] Sales, not production or distribution, constituted Goodenough's principal concern and he conducted business by the maxim: 'it is no use manufacturing cheaply what you cannot sell'.[77] Although Goodenough appreciated the promotional value of regular trade exhibitions and cookery demonstrations, he claimed that the circulation of effective printed literature and newspaper advertisements remained underdeveloped. One of Goodenough's earlier attempts to address this oversight had been flatly rejected by the GLCC governors when they dismissed his request for £1,500 to print and mount posters for cooking stoves on company property throughout central London.[78] Goodenough persisted in selling the idea of advertising to the gas industry. In 1907, before a meeting of the Institution of British Gas

Engineers, held in Dublin, Goodenough stressed the importance of developing gas sales through cooperative advertising. Advertising offered a sure means of attracting business and educating the public, and, as the official supervisor of the GLCC's lady demons, Goodenough advocated publicity directed at female customers.[79] While he wholeheartedly endorsed the work of the lady demons, Goodenough maintained that sophisticated promotional material could enhance the educational service already performed by the company women in consumers' homes, in the showrooms and at regular trade exhibitions. In addition, comprehensive advertising could appeal more effectively to influential professional groups, including architects, teachers, sanitary inspectors and government officials whose conversion to gas would generate sales indirectly.[80] To this end, Goodenough proposed the creation of an industry-wide publicity organisation jointly funded by subscriptions from participant gas and appliance companies. Such a venture would reduce promotional costs and ensure that even the smaller gas undertakings had access to 'first-class advertising experts'.[81] The proposal also ensured that the small gas companies made some proportional monetary contribution to the costs of promoting the industry instead of reaping the benefits of advertising that was sponsored by the larger companies.

Despite Goodenough's enthusiasm, there was no immediate support for such an organisation. As we have seen, appliance manufacturers and gas companies collaborated at trade shows and these temporary cooperative ventures may have fostered Goodenough's plan for a permanent publicity group.[82] However, despite these occasional joint efforts, individual companies commissioned their own visual propaganda, usually in the form of trade cards and pamphlets, which they distributed to customers in their showrooms and at public exhibitions.[83] At the time of Goodenough's Dublin speech, the majority of gas manufacturers did not advertise beyond a few local campaigns or the industry trade journals.

Eventually, it was the misleading advertisements of the Electricity Publicity Committee,[84] under the direction of H.B. Fenwick, which stimulated the gas industry to take collective action. Not surprisingly, the electricity advertisements contained damaging statements about its competitor, emphasising the 'poisonous compounds' of 'destructive gases'. Electricians warned housewives that 'unless electric light is used in the house, the good effects of spring cleaning will be lost in a few months' with the reappearance of the 'usual disheartening dinginess'. Although Walter King retorted that inverted incandescent lamps left no dirt, smoke or discolouration, he urged 'lethargic' and 'short-sighted' gas managers to retaliate against the competition's propaganda by supporting a gas publicity committee based on Goodenough's earlier proposals.[85] Soon after King's 1911 article, a group of gas managers founded the British Commercial Gas Association. From the beginning, the BCGA was heavily influenced by the GLCC. The GLCCs governor, Corbett Woodall, became the first president of the BCGA and Francis Goodenough served as executive chairman of the new advertising organisation, a post he held for over 20 years.[86] Goodenough's efforts finally paid off, and he dominated the organisation he helped to create for several decades.

Beyond the desire to curry customer favour, the propaganda efforts of the electrical industry fostered the need for a more aggressive advertising strategy on the part of the gas industry. Although the use of electricity for domestic purposes was still not widespread even by the Edwardian era, it had revolutionised urban transportation and communications and was increasingly used in public buildings and theatres.[87] By the end of World War I, electric lighting was still confined to the rich households, and only 6 per cent of all British homes had electrical wiring.[88] Yet, despite the low demand for expensive, inefficient electrical appliances, their manufacturers promised that the 'magic influence' of 'trained lightning' offered consumers a 'future of mechanized ease'.[89] Part of the modern image of electricity derived from continental and American accomplishments, particularly the lighting spectacles of the Paris Exhibitions in 1889 and 1900.[90] At the turn of the century, the electrical industry had emerged as the driving force of the second industrial revolution and the epitome of technological modernity, even though its domestic applications remained insignificant.[91]

One cannot dispute Leslie Hannah's point that, in terms of actual domestic sales, electricity remained a struggling industry until the interwar period. Nevertheless, the impact of the early electrical industry cannot be measured simply in terms of actual sales. The electrification of urban transportation offered consumers a glimmer of what was yet to come; and even though electricity serviced a tiny minority of British homes, the dream of total electrification inspired the public's imagination.[92] Master chef Herman Senn, a founding member of the Universal Cookery and Food Association, raved about the electrical stoves, kettles and grills presented at the association's 1893 exhibition. Senn, an instructor at the National Training College of Domestic Subjects, informed his readers that 'electricity promises a revolution in our kitchen arrangements, an absence of dirt, and danger from fire and explosion'.[93] The undeniably negative aspects of gas, the toxicity and the soot, were highlighted by Senn's unrestrained enthusiasm for electricity.

The advantages of electricity extended beyond the actual comparison of relevant merits. Electricity promised a utopian future in which humanity commanded the forces of nature. Articles and advertisements alluded to the mystery of 'captured lightning' harnessed to the service of modern man and woman. An 1895 article in *Black and White* marvelled at Margaret Fairclough's demonstrations of electrical appliances in her Gloucester Road Cookery School. Possibly having the first electrically fitted kitchen, Miss Fairclough's novel 'school with trained lightning' was supplied by R. Crompton & Company, the exhibitors at the Cookery and Food Association.[94] An endorsement for electrical kettles assured ladies of the safety of placing their new appliance on the table. On the surface it appeared to be an ordinary kettle 'while the water within boiled by electricity, the only visible sign of any unusual connection with the kettle being a delicate line of silken cord, through which the magic influence is conveyed to water'. Described as both 'the most beautiful' and the 'most mysterious of all the forces of Nature', electricity had the power to transform the material world.[95] This was the popular image of the electrical industry at the turn of the century, and, although electricity failed to light many homes and streets, it still fuelled the public's imagination in much the same manner as it had during Edison's 1882 Crystal Palace Exhibition.

If enthusiastic Edwardians hailed electricity as the light of the future, what was the popular perception of the gas industry? Was it viewed as the public servant of the people, as George Livesey confidently stated, or as a monopolistic business whose interests were protected by government legislation, as its critics asserted? The popular image of electricity as a mysterious but natural force, of course, offered a negative comparison with coal gas. The increasing domestic applications of coal gas reminded Victorian householders that the industrial age had permeated every aspect of their lives, while arbiters of Victorian taste described gas lighting as garish and gasoliers as tawdry and cheap.[96] Consumers objected to discrepancies in price between neighbouring companies; however, price changes resulted from technological improvements to gas burners, the reduction of gas waste from leaky mains and the decline of capital costs (because of the steadily declining price of coal after the fluctuations of the early 1870s).[97] Competition between rival gas companies also affected prices. Although 'districting' and amalgamation curtailed this to some extent, London consumers, represented by the Gas Consumer's Protection League, protested about price differences between the GLCC and the SMGC.[98] Mounting consumer and local government agitation concerning the GLCC's higher prices led one company director to defend his industry against accusations of waste and price inflation. To withstand the constant and ill-founded criticisms, he contended that the gas man must 'have a very broad back' because 'he is fairer game than almost any other'.[99] Consumer protests finally resulted in an 1899 House of Commons inquiry under Sir James Rankin. As a witness before the commission, George Livesey accused the GLCC of 'over-capitalization, inadequate engineering skill and incompetent coal buying'. Livesey proposed that his company take over the GLCC interests south of the Thames, even though, as Everard notes, the GLCC customers in those districts already paid the same price as SMGC users, a policy designed to offset customer discontent.[100] While none of the committee recommendations was implemented, the Rankin inquiry damaged the GLCC's public image and strained relations between London's two largest gas companies.

The Rankin inquiry coincided with a series of GLCC retirements and the appointment of managers dedicated to restructuring the company. With the death of Col. William Makins in 1906, Corbett Woodall, the chief engineer at the Imperial Gas Co., assumed the company directorship. After receiving heavy criticism from the Rankin committee, general manager, John William Field, retired to a seat on the board of governors. His successor, David Milne-Watson, provides another example of the new management class. A young barrister from Edinburgh, he joined the GLCC in 1897 as Field's assistant, assuming the general manager position in 1903. With the death of GLCC director Corbett Woodall in 1916, Milne-Watson filled the position of managing director. He was appointed GLCC director in 1919 and knighted in 1927.[101] Like Goodenough, his expertise lay in the fields of management and not engineering. The new leadership favoured increased departmentalisation, and specialist officers at the company headquarters took control of designated operations. Goodenough's promotion to chief inspector of the new sales department reflects this management trend.[102]

The improvement of customer service to restore public confidence was an important aspect of the modernisation plan, particularly after the negative publicity of the Rankin inquiry. Although gas sales and cooker rentals appeared unaffected by the controversy, Goodenough remained convinced that the gas industry could not afford to take its customers for granted. To provide an improved service to the consumer the industry needed to develop a highly trained sales staff, which included lady demons. But, equally, it must rely on 'judicious' advertising, the benefits of which could not be underestimated in terms of generating sales, fostering public goodwill and assuming the offensive against any misinformation disseminated by the electrical industry.[103] Like Goodenough, company director Corbett Woodall appreciated the strategic importance of advertising. Against those critics who saw 'publicity work as beneath the dignity of the gas profession', he claimed that advertising was no longer simply 'a blatant business principally associated with quack medicines'. Advertising was an 'honourable and scientific profession based on truth and honesty' and therefore 'an integral part of every progressive business'. To prove itself as truly 'progressive', the gas industry needed a publicity organisation.[104]

In Britain, publicity was critical to the electrical industry because it had a well-established and less expensive rival, but one whose faults were evident to most consumers. If it is true that the 'second industrial revolution' was powered not by electricity and oil, as some late nineteenth-century writers asserted, but by advertising and credit, then the electrical industry still retained prominence within the retailing revolution of the 1890s.[105] By the same token, the gas industry was an equally important force, because of its developing marketing strategies which included advertising at trade exhibitions, permanent displays at company showrooms, the use of lady demons and pre-payment schemes to serve customers of varying economic backgrounds. All of the above techniques were widely practised by 'progressive' late Victorian businessmen and placed the gas industry at the forefront of the retail revolution.

Corbett Woodall's claim that the advertising business was 'honourable and scientific' reflected the exact sentiments of those involved in the developing field of advertising. Edwardian businessmen increasingly agreed that advertising was a sure means of securing business and educating the public.[106] Although the medium was still exploited by confidence men and the purveyors of questionable products and pornography, by the turn of the century the advertising industry faced increasing regulation, both from public authorities and the industry itself.[107] The 1907 Advertisements Regulation Act empowered local councils to enact by-laws to control excessive 'bill posting', while the industry's Advertisers Protection Society expressed concern over the publication of fraudulent or misleading advertisements in newspapers. This latter concern arose not from the desire to protect consumers, but to safeguard the interests of its members who believed that cheap 'catchpenny announcements' undermined the credibility of all advertising material. While the publishers responded that they could not verify every commercial claim, the philosophy of a paper was still measured by the quality and character of its advertisements. *The Times* exemplified a superior moral standard in 1909 declaring that:

> ...[the] best modern advertising has the publication of facts for its basis. The day of successful claptrap and vulgarity, still more the day of exaggerated and deceptive misrepresentation, is quickly passing away. So far from these being fostered by advertising agents, the whole tendency of the best and most successful agents is to repress them. Even apparently trivial inaccuracies are, on principle, excluded from advertising.[108]

Such was the ideal of socially responsible advertisers. But, beyond simply telling the truth about a product, some advertisers assigned themselves a higher objective. In the words of psychologist and business author Walter Dill Scott, advertisers were no longer 'bamboozlers', like P.T. Barnum, simply out to make a profit by whatever fraudulent means:

> The advertiser in the past may have been the exploiter of the public, but the new generation of advertisers are becoming more and more the protectors of society. In the past they may have in all too many instances misled the unwary, but the successful advertisers of today are becoming the trusted guides of the ultimate consumer.[109]

Scott emphasised the importance of winning the confidence of the public not only through advertising, but also by good reliable service, the most valuable asset that any businessman could possess. No longer the Victorian charlatan or master of puffery, the modern advertiser was a scientific expert with a moral duty to inform, protect and educate consumers, many of whom were women. This emphasis on the dual importance of advertising and cultivating consumer confidence suggests that early advertising spokesmen, like Walter Dill Scott, did not make a clear distinction between product marketing and public relations. Persuasive advertising and the presentation of a positive corporate image were directed towards the common goal of increasing market sales.

Astute Edwardian advertisers recognised that shopping, for both necessities and luxury goods, was a predominantly feminine activity. Clarence Moran stated bluntly that advertising 'must appeal to women' because 'it is their peculiar function to spend'.[110] Even the cleverest advertisements were unsuccessful if they failed to attract women's attention. To win the notice and the custom of female shoppers, Moran suggested that advertisements should be 'wholesome, simple and attractive'. Like Walter Dill Scott, Moran assumed that responsible advertisers were educators and advisors able to counsel consumers. Although women were the principal buyers of household goods, business writers assumed that they were more vulnerable than men to the misleading statements of false and sensational advertising. As self-appointed consumer guides, Edwardian businessmen stressed the need to ensure that female consumers made rational choices. Moreover, while shopping potentially extended the range of women's influence beyond the private domestic sphere, businessmen and male advertising agents attempted to control and direct that consumer activity. If truthful advertising served women's best interests in the marketplace, it also confirmed the continued importance of male breadwinners, as the advertisers generally maintained that women were responsible for spending their husbands' hard-earned wages.

From its inception, the BCGA embodied a lofty set of business ideals. The organisers, Goodenough and Woodall, proclaimed their group's dedication to serving the public interest by educating consumers through reputable advertising. In addition, they also aspired to create and maintain a positive public image for the gas industry. Without noting the possible contradiction of these two aims, BCGA executives and staff integrated advertising and public relations into a single organisation. Reflecting on the original aims of the BCGA, general manager William Mason noted that the organisation combined advertising and public relations to increase sales through publicity, but also to overcome lingering prejudices against the industry.[111] This dual mandate was particularly significant for the embattled gas companies responding to criticisms from customers and challenges from the electrical companies.

Theoretically, the BCGA executive dedicated their organisation to meeting the needs of all consumers of every class with the goal of improving the welfare and health of the nation by raising the quality of domestic life. In particular, Woodall spoke of the need to address their advertising towards the working classes: the wives of 'labourers and poorer mechanics', who still had a limited appreciation for gas technology or were unable to gain access to gas services because of the intransigence of negligent landlords. He wryly noted that all the domestic science training the school system had to offer 'would be of small value in the absence of appliances'.[112] Even when addressing the competition, the BCGA organisers retained their lofty idealism. Despite the provocative assertions of the Electricity Publicity Committee, the BCGA executive claimed that their intent was not to damage electric interests. They were 'rivals' but not 'foes' and Woodall urged the electricians to abandon 'cheap journalism'. Alluding to the gentleman's sense of honour and fair play, even in matters of business, Woodall urged his competitors 'to respect British traditions and compete like sportsmen'. While competitiveness and aggression were acknowledged masculine virtues in the corporate arena, adopted as much from Herbert Spencer's 'social Darwinism' as the Victorian sports field, Edwardian business writers proclaimed that self-control and honourable behaviour were equally fundamental to occupational success.[113] By their apparent use of slanderous advertising, Woodall accused the electricians of resorting to the lowest form of business conduct. In the interests of the consumers, who needed accurate information, he exhorted his competitors to act like gentlemen by exercising honesty, integrity and good taste in all their advertising material. Besides, he concluded, not all electricians were direct competitors, as some BCGA members were municipal corporations with combined gas and electric services.[114]

The BCGA

In early 1912, the BCGA launched its first advertising campaign from a suite of offices in central London. The association executive, led by Goodenough and Woodall, hired a small staff of journalists and illustrators under the direction of general manager William Mason.[115]

They also hired journalist Maud Adeline Cloudesley Brereton as staff writer and editor-in-chief of the association's two monthly magazines. The first of these publications, *Bulletin*, was an industry paper aimed at gas employees, such as gas fitters, general workmen, salesmen and clerks, who probably did not read the more technical gas journals. The second publication, *A Thousand and One Uses for Gas*, addressed professionals whose conversion to gas would directly benefit the industry. While each issue targeted a specific audience – including architects, builders and health officials – the underlying message remained the same: gas as a public service could improve the health of England's citizens, whether in schools, factories, hospitals or homes. In addition to the monthlies, the BCGA produced its own advertisements that appeared in a variety of magazines from *Queen*, *Home Chat*, *Country Life* and *Punch* to the *Journal of British Architects*, *Building World*, *Lancet* and *Baby*.[116] In total, they advertised in 32 journals and six London dailies – including *The Times*, the *Daily Express* and the *Daily Mail* – to ensure that their message received the widest possible circulation among consumers of different classes. The organisation was funded by subscriptions from members who were usually fitments manufacturers and gas corporations. In 1912, these included 250 gas companies and the major appliance makers – Richmond, Sugg, Davis, Main and Parkinson – indicating that the publicity group was genuinely an industry-wide collaborative effort. Contribution rates were based on the annual revenues of companies that were members and therefore ranged from several guineas to thousands of pounds.[117] Goodenough and William Mason anticipated that by sharing expenses and staff the volume of advertising could be increased to the advantage of the entire industry, with the BCGA undertaking national and district advertising, but leaving local publicity up to individual companies.

The formation of the BCGA in 1911 exemplifies the movement towards specialisation in business management and the trend towards greater departmentalisation. This pattern of organisation was evident in the GLCC's pre-war management restructuring which resulted, for example, in the establishment of a special sales department and the recruitment of business professionals who were not gas engineers. From its inception and up until nationalisation in 1949, the driving forces within the BCGA were executives of the GLCC, from Woodall and Goodenough to Milne-Watson. Goodenough held the chairmanship of the executive committee until 1936, and, as a 'pioneer' in advertising and marketing, he was credited with the success of the venture. In terms of the advertising profession as it existed in 1912, the BCGA was a distinctively modern organisation. Although the London Post Office Directories listed 339 advertisement agents and contractors in 1906, according to T.R. Nevett's study, the vast majority of these firms were brokers of advertising space who worked for a publisher's commission in addition to their business fee. In contrast to the space brokerage agent, advertising consultants wrote and designed advertisements and might either work independently or on contract with an agency. Unlike the brokers, consultants did not accept commissions from the media but represented the business interests of their clients.[118] By asserting control over the production and distribution of the industry's advertising and arranging contracts directly with publishers, the BCGA avoided commercial dealings with intermediary agencies or consultants. In this respect, they were able to strive for, and maintain, the high degree of professional integrity so important to Woodall and Goodenough. And, on the practical level, it made good business sense because it was cheaper and more efficient. The gas industry apparently organised the BCGA in response to a similar but short-lived initiative by the electrical engineers. Nevertheless, the formation of an advertising and public relations agency that was controlled, funded and dedicated to one industry must still be considered ambitious, particularly considering the size of the gas industry and the general conservatism of many of its managers. Similarly progressive was the decision to appoint a female public health specialist as head of its editorial staff.

'A woman's point of view'

An outspoken advocate of public health, modern housing and 'the science of housekeeping', Maud Adeline Cloudesley Brereton combined social activism with business. After 20 years' training in public health and authorship, she confessed that she chose to represent the gas industry because it addressed issues of health and sanitation from a practical perspective.[119] Although, it remains unclear exactly how Mrs Cloudesley Brereton became interested in health matters, we do know that she began her career as a teacher. Born in London and educated at Hockerill College, Bishop's Stortford, Maud Adeline Ford held several prestigious teaching posts within the first few years of her teaching career. In 1893 she served as the first headmistress of St Andrew's Girls' Secondary School, Willesden, and, in the following year, she became headmistress of the Baroness Burdett-Coutts School, St Anne's in South Highgate. She joined the staff of Homerton Training College, Cambridge, in 1895 as a junior lecturer and, two years later, at Easter, she married John Charles Horobin, the college principal, whose first wife had died several months earlier. With plans to retire as principal at Homerton, Horobin died suddenly in 1902 at the age of 52, leaving Maud and an infant son. Upon the death of Horobin, Maud Adeline Horobin was appointed acting principal of Homerton, and was favoured by the board of governors as her husband's successor. Apparently, however, the young widow began to 'neglect her duties', and, in 1904, the board asked Maud Adeline to resign 'as she had not kept true economy' with the school's finances.[120] Her dismissal from Homerton may have been the inevitable outcome of her remarriage that year to Cloudesley Brereton, and what appears to have been a new preoccupation with public sanitation.

Maud Adeline's interest in public health seems to have increased after her marriage to educator Cloudesley Brereton in 1904. A graduate of St John's College, Cambridge and the University of Paris, Mr Cloudesley Brereton taught for a short time at the London School of Economics and worked as an inspector of modern languages with the London County Council. He lectured in America, Germany and France, wrote special reports for the Board of Education, conducted government studies on secondary schools and published translations of the works of the French philosopher, Henri Bergson. Significantly, Mr Cloudesley Brereton was a member of the Rainbow Circle from 1906 until 1931, and his contributions to that society focused on topics related to educational reform.[121] It is entirely possible that Francis Goodenough met Mrs Cloudesley Brereton through her husband's connection to this intellectual association.

Probably due to Mr Cloudesley Brereton's influence, Maud Adeline was one of only two women ever to address the Rainbow Circle. This accomplishment appears all the more remarkable since, according to Jose Harris, this debating society of 'London's progressive intelligentsia never included a single woman among its members, and often listened to lectures that emphasized women's moral and physical incapacity for civic affairs'.[122] Maud Adeline's chosen topic, 'Motherhood, the Home and Medical Inspection in Schools', dealt with the health of elementary schoolchildren. To develop a better awareness of health issues in the home, she urged greater cooperation between school medical inspectors and parents, beginning with public clinics for infants and pre-school children. During the discussion that followed Mrs Cloudesley Brereton's talk, the audience expressed their overall support for her proposals, although, apparently, there was some divergence of opinion regarding the attitude of the poor towards school doctors.[123] It is probable that Maud Adeline shared her second husband's conviction that education and health were inseparable in terms of the nation's well-being. In 1907, she was honoured by the French government 'for services to International Public Health'.[124] By 1912, Mrs Cloudesley Brereton belonged to several health organisations, including the Royal Sanitary Institute and the Institute of Hygiene. While her early contributions to the field of public health remain unclear, in a letter to *The Times* she expressed her disgust at Southampton's vermin-infested customs houses. Complaining of the body lice found on her belongings, 'of such a size as I have never seen in all my experience working amongst filthy slums', Mrs Cloudesley Brereton questioned why, in the interests of sanitation and national prestige, the port authorities were unable to disinfect the inspection sheds.[125] Given her expressed familiarity with both urban poverty and lice, it is possible that Maud Adeline had worked as some sort of district visitor, the type of work often undertaken by female associates of the Royal Sanitary Institute.

Although married twice and the mother of three sons (two with Cloudesley Brereton), domestic life apparently did not impinge on Maud Adeline's career, if one might judge from her long list of achievements. For example, she joined the Institute of Journalists, was vice president of the Society of Women Journalists and was a consultant with the Ministry of Food during World War I. In 1926, Maud Adeline became the first female honorary fellow of the Institute of Sanitary Engineers. She also shared her second husband's dedication to educational reform. Given her association with the gas industry, she focused her attention on technical education and cooperative schemes between employers and schools; from 1923–24, she chaired the Association for Education in Industry and Commerce.[126] In his retirement speech to the BCGA after serving 25 years as its executive chairman, Goodenough paid tribute to Maud Adeline Cloudesley Brereton, describing her as 'that very able woman' who 'did yeoman service in the foundation days of the Association'.[127] Mrs Cloudesley Brereton retired from the BCGA in 1932, several years before Goodenough; however, she continued to serve as a consultant for the organisation that she had helped to build.

Maud Adeline Cloudesley Brereton's professional contributions appear all the more impressive because she ran a household in Hampstead and raised three children. Fabian feminists admitted the difficulties of combining a career and a family, regretting that 'for most women, the two are alternatives'.[128] But in Mrs Cloudesley Brereton's view, while marriage and motherhood were important, domestic duties should not prevent women from engaging in remunerative work or public service if they so desired. Her positive description of modern companionate marriage, found in her book *The Mother's Companion* (1909), emphasised the importance of intellectual compatibility between husband and wife; this model for married couples did not receive serious attention by social theorists until the 1920s. Her early advocacy of companionate marriage, and indeed her use of that still unfamiliar term before World War I, suggests that Mrs Cloudesley Brereton enjoyed a degree of personal autonomy that few married middle-class women shared.[129] So firm was her belief in the positive value of such 'true partnerships' that Maud Adeline Cloudesley Brereton believed that the settlement of the 'many outstanding women's questions' depended on the arbitration of married couples who possessed a 'full and comprehensive view of the two-sided problem'. The resolution of gender-related issues, she argued, should not be left to 'those who boast that in celibacy is their strength' no matter how 'invaluable and indeed essential their assistance'.[130] For Mrs Cloudesley Brereton, contemporary debates over women's work, education and motherhood extended far beyond the immediate suffrage goal. The woman question encompassed a broad range of social problems that could only be resolved by cooperation between both men and women as, ultimately, it affected the 'prospects and interests of the future generation'.

As an Edwardian advice book on how to succeed in marriage, *The Mother's Companion* spanned a number of topics of relevance to the married woman. Throughout the text, Mrs Cloudesley Brereton assumed a frank and practical tone, even when dealing with taboo subjects, like menstruation, sex education and pregnancy. She wrote that marriage and motherhood were the most challenging and difficult occupations that women undertake because they combine the duties of many professions: business partner, book-keeper, doctor, nurse, teacher and wage-earner. Although she apologised to her readers for her 'horribly business-like and unromantic tone', Maud Adeline asserted that only by 'facing the business side of marriage fairly and squarely, and mastering it, that there is any room left for romance'.[131] Quoting Fabian feminists Lady McLaren and B.L. Hutchins, Brereton defended women's natural right to work as free citizens, whether out of genuine economic necessity, personal satisfaction or altruism. Noting the economic importance of female domestic labour, she stated: '... in so far as the woman is a skilled housekeeper, she is a wage-earner even in her own home, and some money value should be put upon the work she does'.[132]

From a maternal feminist perspective, Mrs Cloudesley Brereton defended motherhood as a source of feminine power, in both the public and private sphere. By asserting the principle of 'universal motherhood' as ample justification for women's civic, social and economic importance, she employed the same maternal arguments as Millicent Fawcett's suffragists in their demands for political representation.[133] Mrs Cloudesley Brereton's feminism rested on the ideals of progressive liberalism; she was confident that women would soon receive the vote as a recognised right within a modern enlightened society. However, she maintained that suffrage was but a single victory in a much larger struggle. Consequently, she urged women to consider their civic duties as mothers, to look beyond the confines of their homes, and through public work, to help raise women's status, both inside and outside of the home.

Like a number of middle-class reformers, including Fabian socialists and Liberal imperialists, Brereton expressed concern over the declining birthrate and the apparent degeneration of the urban population. Dr Launder Brunton, a witness for the Inter-Departmental Committee of Physical Deterioration (1904) wrote the preface to *The Mother's Companion*. Maud Adeline was sceptical of the interventionist solutions offered by eugenicists like Francis Galton and Karl Pearson. With a note of sarcasm, she wrote that the apparent racial deterioration among the British population could be reversed if all of the 'learned people' active in the eugenics movement would simply have children themselves. But she was equally wary of benevolent proposals that encouraged 'wastrals and criminals' to 'breed like rabbits'.[134] Brereton supported environmental solutions to the problem of racial decline, including proposed improvements to urban sanitation and education; these reforms were aimed at developing, to quote Michael Freeden, an 'equality of opportunity' for all classes.[135] Her pronouncements regarding social undesirables suggest that she favoured some means of controlling the growth of the lowest strata of the population. While responsible parenthood and civic duty were inseparable, she still held definite reservations: she asserted that some individuals would benefit the race more if they did not have children, a view she shared with a number of liberal progressives, including L.T. Hobhouse and J.A. Hobson.[136]

Throughout her lifetime, Maud Adeline remained a supporter of the birth control movement in England. She was a personal friend of one of its leading advocates, Dr Marie Stopes. Correspondence between the two women indicates that Stopes sent her manuscript for *Radiant Motherhood* to Maud Adeline, requesting that she read and comment on the book prior to its publication in 1920.[137] Maud Adeline also used her official position within the gas industry to further the birth control cause, even when the support of her male associates for the movement was less than clear. For example, she placed advertisements for the BCGA in Dr Stopes's *Birth Control News*, thus putting the gas industry among the few businesses that visibly supported the controversial paper. Letters between herself and Stopes suggest that Maud Adeline personally decided to advertise in the *Birth Control News* without the official approval of her male colleagues at the BCGA because she instructed Stopes to bill her directly for the advertisements and not to send the invoices to her office. Although Maud Adeline admitted that she was unable 'to commit the committee of the British Commercial Gas Association to birth control', she still did not hesitate to use its name for a cause that she personally supported.[138]

As a treatise on motherhood and women's rights, Maud Adeline's book, despite its apparent conservatism, reveals her genuine commitment to women's social and economic emancipation. Her personal associations with outspoken women reformers corroborate her written opinions. In addition to Dr Stopes, guests Maud Adeline entertained in her Hampstead home included Hilda Oakley and Alice Zimmern.[139] Dr Stopes, herself a Hampstead resident, lived near the Cloudesley Breretons; in early July 1914, the time of Maud Adeline's reported 'at home', Stopes was recently estranged from her first husband, Reginald Ruggles Gates, and was staying with another group of Hampstead friends. Given Maud Adeline's earlier criticism of British divorce laws, it is likely that she sympathised with Stopes's situation. Similarly, the two women agreed on married women's right to work, however much they exalted the state of motherhood; this theme was soon to be taken up by Stopes in both *Married Love* and *Radiant Motherhood*.[140] Among Mrs Cloudesley Brereton's other female acquaintances, Hilda Oakley was an outspoken advocate of higher education for women. As the warden of King's College for Women, University of London, she was instrumental in establishing the first degree-granting programme in domestic science, for which she drew both criticism and praise from feminist educators. Oakley defended the programme and insisted that King's College provided young women with scientific training not simply for the domestic sphere, but as preparation for careers in social work, nursing, teaching and business.[141] Mrs Cloudesley Brereton's friend, Alice Zimmern, was also an educationalist and an avowed feminist. A graduate of London's Bedford College and Girton College, Cambridge, Zimmern ardently supported women's higher education and female suffrage; she was a member of the International Women's Franchise Association and the Ladies University Club. Given her awareness of the social and economic barriers that hindered single, professional women, Zimmern expressed her concern over the need for low-cost accommodation for the thousands of unmarried working women living independently in urban areas.[142] Maud Adeline moved in a social circle that included professional women who were articulate and committed, in various ways, to the women's movement.

Clearly, Maud Adeline Cloudesley Brereton advanced the objectives of the Edwardian domestic subjects movement by emphasising the relationship between scientific housewifery, public health and female emancipation. Like other women professionals in the domestic subjects movement, including cookery teachers and lady demons, she occupied the paradoxical position of reinforcing 'women's home-centredness' while expanding their own personal horizons outside the household. In Annmarie Turnbull's words, they increased their own autonomy 'without fundamentally challenging male power within the accepted sexual division of labour'.[143] Nevertheless, as Turnbull rightly concedes, by advancing the needs of women in the home, instructors of the domestic subjects aimed to help women conquer 'their own sphere'; within this context, their contributions to the pre-war women's movement cannot be readily dismissed. By looking to the gas industry to relieve women 'from the treadmill of scouring and scrubbing and scrubbing and scouring, day in and day out, from year end to year end', Maud Adeline forcefully asserted the 'women's point of view'. She thereby enlisted the BCGA to help free women from the domestic sphere.[144]

'The silent servant'

Despite the contradictions between profit margins and social welfare, the image of modern capitalism serving the needs of ordinary men and women was impossible for gas managers to resist. The National Gas Exhibition in 1913 at Shepherd's Bush, London, celebrated this idea with elaborate displays, entertainment, lectures and conferences intended to illustrate to the public the continued advantages of this underrated service.[145] The exhibition marked the centenary of the gas industry in Britain and was advertised extensively throughout the country. Mrs Cloudesley Brereton, a member of the organising committee, hosted an exhibition preview which hundreds of reporters attended. *The Times* commented favourably on the collective exhibition, praising the industry's tenacity and adaptability against electrical competition, with the result that the gas industry had retained 'a far larger share of private and public lighting than at one time appeared possible'.[146] This seemed remarkable considering earlier predictions of the industry's rapid decline following Edison's fabulous displays in the Crystal Palace in 1882.[147]

But the National Gas Exhibition of 1913 was more than self-congratulatory grandstanding. The accompanying lectures and conferences revealed a genuine interest in the connections between public health and domestic economy. For example, invited lecturers at a conference on food and cookery included Herman Senn of the Universal Cookery and Food Association and Dr Joseph Priestley, medical officer of health for the borough of Lambeth. Prominent Labour Party member Dr Marion Phillips was a conference vice president.[148] Speaking on cookery in its relation to health, Dr Priestley reinforced his advice on nutrition through the preparation of an enormous four-course meal cooked in two hours at the cost of a penny for gas.[149] On the same platform, domestic economy teacher Mildred Atkinson, of Battersea Polytechnic, upheld the value of domestic instruction to working-class children as a means of improving national health. Margaret McKillop of

Somerville College, in a paper entitled 'The Penny-in-the-Slot Meter: the Working Man's Friend', illustrated how to achieve the greatest cooking value for a penny. She recommended the economy of the double-boiler pot and urged the companies to sell pots and pans at cost price to their customers to ensure that all classes of gas consumers had access to proper, efficient kitchenware.[150] While this may appear as a small and rather patronising point, it illustrates that, in many cases, working-class housewives lacked even basic housekeeping tools.

Whether the gas cooker was in reality the 'working man's friend' remained questionable. According to the Fabian social investigator, Maud Pember Reeves, the penny-in-the-slot system often resulted in undercooked food, as a penny of gas gave insufficient time to cook a proper meal.[151] Apparently, when the gas meter expired half-way through one family's dinner, in a 'flash of genius' a Lambeth customer saved the meal from being underdone by extracting a penny from the baby's umbilical dressing. A district health visitor had bandaged the coin in place to put pressure on the infant's abdomen.[152] Recipes for scotch oatmeal and recommendations for double-boiler pots appear equally misguided in light of Reeves's descriptions of working-class budgets and ill-equipped sculleries. Nevertheless, the list of conference participants, which included scientists, industrialists, teachers, architects and doctors, suggests a professional concern for the health of all classes that extended beyond conventional advertising. And even Reeves admitted that an inefficient gas cooker was still far better than a smoky coal grate.

The most obvious beneficiaries of improved gas technology were middle-class groups of urban dwellers, including professionals, shopkeepers and skilled artisans, and members of the working class with steady employment. Despite Woodall's earlier determination to appeal to the working classes, advertising for the 1913 National Gas Exhibition loosely targeted the middle classes; this was probably because the middle classes were deemed most likely to attend the exhibition. Six weeks prior to the opening, the *Daily Express* carried a full-page illustrated advertisement depicting middle-class household settings where gas-fuelled heating stoves were in use. The vignettes included that of shirt-sleeved gentlemen enjoying a game of billiards, a little girl at her piano lesson and a young woman in a dressing gown brushing her hair before a gas fire.[153] Each scene suggested domestic comfort and convenience while the notice promised that 'half the work of the house is saved where there are no grates to clean, no coal to carry, no fires to lay, and where fire irons and scuttles do not exist'. For middle-class householders, particularly those of the lower-middle class struggling to maintain social appearances on slim salaries, gas utilities presented a possible solution. Middle-class families with annual incomes of less than £300 were increasingly unable to afford permanent domestic assistance, and, even if they were able, it was still difficult to attract and keep reliable help when better positions in more affluent households were plentiful.[154] The elimination of servant problems for the middle classes remained a popular theme in gas advertising.

To coincide with the 1913 National Gas Exhibition, the BCGA compiled a documentary film entitled *The Manufacture and Applications of Gas*. While the first part of this advertisement visually outlined the process by which gas was made, purified and transported, the second half depicted the use of gas in the home.[155] Among the images of domestic gas use was that of a female servant, opening a gas cooker to view the contents (see Fig. 3.2). The woman's neat appearance and crisp white apron suggest that gas technology has simplified the running of this home, while the new medium of the film links the gas industry with modernity. Customers might expect more favourable culinary results from their kitchens since a gas cooker potentially improved the cook's temper and lightened her workload. The message that gas offered an easy solution to domestic problems also appeared in an illustrated book about domestic gas use that was co-edited by Mrs Cloudesley Brereton and Edden. In one illustration, a young housewife, with her sleeves rolled up, in the midst of her baking, appears to be interrupted

3.2 The idea that gas technology could ease the chronic servant problem was a recurring promotional theme, and, here, a uniformed domestic servant opens a gas cooker to check the contents

by the arrival of unexpected guests in their new motor car. However, if the new technology of the automobile enabled guests to 'drop in' on short notice, the gas stove offered a modern solution. Thanks to her gas cooker, the housewife could accommodate extra meal preparations in her stride.

Domestic technology was particularly helpful to those housewives of the lower-middle classes who did not have servants, and whose preoccupations with gentility and appearances often demanded that their households appear more prosperous than they actually were.[156] As a passage from Arnold Bennett's novel *Buried Alive* suggests, women of the lower-middle classes used the gas stove to replace daily domestic help: 'With the charwoman, a pair of gloves for coarser work, and the gas stove, you made naught of domestic labour.'[157] In a popular lecture delivered at the 1913 National Gas Exhibition, entitled 'The Silent Servant', Miss Constance Williams assured her listeners that gas would eliminate the need for domestic help altogether. With apparent seriousness, she added that 'a gas cooker in the house is quite as important to the ordinary housewife as a good, kind husband'.[158] Although Miss Williams may not have had actual grounds for comparison, her point was well taken.

But was the gas cooker the answer to women's domestic problems as Miss Williams so confidently stated? Probably not. A gas cooker would not prevent a restless servant from departing or keep home the husband who liked to drink. In some respects, the gas stove introduced new domestic problems. As noted earlier, the penny-in-the-slot meter and its money box tempted customers to rob the utility companies. While most meter thieves were male, women also resorted to this crime if family circumstances demanded it; for example, a young Stratford woman stole one shilling and eight pence from the gas meter because her husband had been out of work for two months.[159] If the meter presented a temptation to distressed customers, the gas stove itself offered a permanent release from domestic problems, as it came to be used as an instrument of suicide. According to Olive Anderson, suicides by gas poisoning were unusual in the 1880s and 1890s; however, in 1909 the registrar general reported 130 cases. Women resorted to this method more frequently than men, possibly because it was 'painless, accessible and reassuringly familiar'.[160] Ironically, on the same day that the London dailies reported the opening of the National Gas Exhibition in Shepherd's Bush in 1913, they also described in considerable detail the suicide of a woman, 30 years of age and the wife of a Fulham clerk, who was found with her head in the gas cooker.[161]

Gas and water feminism in the Edwardian home

Despite such obvious negative publicity, Maud Adeline Cloudesley Brereton's remarks to the BCGA revealed her continued faith in the emancipatory potential of domestic technology; she forcefully made this point at the BCGA's second annual conference, held in conjunction with the 1913 National Gas Exhibition. Maud Adeline was one of several key speakers and the only woman on the roster. Presumably, these official gatherings with their solitary female delegate posed a social conundrum for the male participants. How does one seat an unescorted, but married, woman among 250 men? At the first annual conference, held the previous October in Manchester, the BCGA executive overcame the problem by arranging a separate reception and dinner for Maud Adeline and some local female dignitaries. While the official association dinner was in progress, in a separate suite, Maud Adeline entertained the mayor's wife; also among the women present were Sara Burstall, head of Manchester's Education Committee and Jane Redford, chair of the Ladies' Public Health Committee.[162] Maud Adeline's ladies-only reception may have avoided an awkward situation at the men's banquet, but it also enabled her, as BCGA editor, to meet informally with other women interested in matters of public health and education. By all reports, the separate gendered events were successful. Although Maud Adeline briefly addressed the BCGA during a question period, her presence at the first conference was low-key (see Fig. 3.3).

MRS. M. A. CLOUDESLEY BRERETON,
Officier d'Académie,
Editor.

3.3 Maud Adeline Cloudesley Brereton at the British Commercial Gas Association's ladies-only reception

In contrast to the 1912 meeting, Maud Adeline Cloudesley Brereton occupied a more prominent place at the second BCGA conference. It might have alarmed some of the more conservative members that this beautiful and impeccably dressed lady, who the previous year had hosted a quiet, tasteful separate reception, was chastising them from the podium at their annual meeting. From her opening remarks, Mrs Cloudesley Brereton's address to the BCGA reveals the same maternal feminist position found in her earlier writings. She called upon the gas industry to hasten 'the social and economic emancipation of women' by providing efficient and labour-saving appliances. Speaking before the predominantly male audience, she reminded her listeners of the ameliorative potential of their industry since their commercial interests were 'so directly bound up with personal and public health and preventive medicine'.[163] Because they directly controlled the distribution of this modern technology, gas engineers thereby had the power to 'revolutionize our home life'. Maud Adeline cautiously avoided a suffrage stance and denied the intention of 'dragging in any question of votes or no votes for women'. Yet, she forcefully decried the 'certain type of man [who] talks the utter rubbish about women staying home and minding her business or her babies'. Instead, she argued that women's emancipation depended on their release from wasteful domestic labour, and that their time might be directed to more generous and useful purposes, such as paid employment or voluntary service. In this manner, the modern woman might use her 'broader and deeper education' to extend her sphere of influence beyond her own private domain and thereby affect the 'Home-life' of the community in general that others 'may profit by her knowledge and experience'.[164] She insisted that the gas industry consider the needs of its female customers of all classes. According to Maud Adeline, working women living in poorly serviced homes experienced the greatest material deprivations. For example, she asserted that the sanitary standards imposed on working women by school inspectors were entirely unrealistic in the absence of hot water.

Although Maud Adeline skirted the suffrage question, her broad definition of the 'Home' and her defence of women's social contributions echoed the maternal feminist arguments of English female suffragists and the militant suffragettes. Similarly, in her demands that women be released from 'repetitive and soul-killing drudgery' through labour-saving technology and efficient home management, Maud Adeline returned to a theme already outlined in her 1909 book, *The Mother's Companion*. Her arguments reflected the current debates over domestic organisation.[165] The servant shortage of the early twentieth century led middle-class women to explore methods of simplifying housework as they themselves undertook more of the actual work. Plans for cooperative housekeeping with modern flats serviced by communal kitchens appeared in Fabian tracts, the radical feminist press and even the pages of the popular press, including *T.P.'s Weekly*.[166] The letters in *T.P.'s Weekly* explored practical alternatives to housekeeping difficulties and problems with servants which included proposals for flats with communal dining rooms and housekeeping staffs, not unlike the suggestions put forward by Alice Zimmern in her earlier article. In the modern low-maintenance flats, sculleries and kitchens were eliminated in favour of the common dining room; water closets and bathing facilities were provided, operating on penny-in-the-slot meters. However, as Alison Ravetz observes, plans to centralise cooking areas were not necessarily revolutionary. The Victorian kitchen, that 'distant cavern from which dinners emerged' was not the focus of family life, and, therefore, its removal did little to challenge gender roles or architectural conventions.[167]

Edwardian socialists and feminists assessed housekeeping from a more ideological perspective. Various American utopian socialist communities, including the Shakers and the Owenites, had practised communal housekeeping with limited success; however, according to Delores Hayden, Charlotte Perkins Gilman offered the most practical critique of the home and domestic ideology. Described as a material feminist, because she considered women's struggle for economic independence of greater importance than the suffrage question, Gilman drew inspiration from a variety of sources, including the American home economics movement and English Fabian socialism.[168] In her written work, Gilman forwarded her vision of the feminist housing complex where domestic services, including cooking, laundry and even childcare, were provided for residents by salaried, professional domestic workers.[169] She noted that while technology blurred the distinction between male and female spheres, sentiment still dictated that women should remain exclusively in the home. For example, although gas technology in the form of 'the gilded pipes of the steam heater and the flickering evanescence of the gas range' released women from the physical necessity of tending the hearth, social convention still restricted women to the home. As Gilman put it, 'the matchbox has freed the housewife from that incessant service, but the feeling that women should stay at home is with us yet'.[170] Admittedly, Mrs Cloudesley Brereton's proposals never radically challenged the organisation of the home in the manner of Gilman's vision of communal housekeeping; however, her remarks to the BCGA protested against traditional definitions of women's sphere. While she asked gas managers to support the domestic revolution by extending the use of technology in women's homes, she also defended her own right to leave the hearth untended.

To attain the type of domestic transformation that she envisioned, Mrs Cloudesley Brereton advanced recommendations for improving customer service and sales. Too often, she lamented, the gas managers underestimated the power of their salesmen and left this important job to inexperienced, albeit charming, young men. Mrs Cloudesley Brereton insisted that 'the commercial side should be as much respected and as well paid as the purely professional side', reminding her listeners that, in terms of the quality and quantity of work, employers never received more than they paid for. Effective gas salesmen should be educated technical experts, equally able to advise the 'woman doctor' or the poor housewife. She recommended the appointment of well-paid regional sales consultants who understood the economic and social side of the gas campaign and could effectively cooperate with existing sales staff – including the lady demons, showroom attendants, outdoor canvassers and gas inspectors – thereby creating a multi-level sales team. Mrs Cloudesley Brereton also urged gas managers to become more involved in public health organisations and to subscribe to various periodicals for their sales staff, in an effort to familiarise them with current debates concerning domestic economy from the perspective of teachers, school committees, women and homemakers.[171]

Her comments elicited a generally favourable acceptance. Although the respondents appreciated the sentiment of her suggestions, they were, nonetheless, hesitant as expert salesmen and reading rooms were costly experiments. Conceding Mrs Cloudesley Brereton's point regarding their youthful salesmen, the managers agreed that showroom attendants were inadequately trained and often quite unfamiliar with their customers' needs. On the 'woman question', however, the gas men were less supportive and dismissed Mrs Cloudesley Brereton's anticipation of gender equality as being beyond present-day expectations, and even utopian. W. Armitage, an alderman from Doncaster, quipped that some of the suggestions he 'read with fear and trembling', and he wondered how Mr Cloudesley Brereton felt about his wife's apparent aversion to domestic labour.[172] Against these comments, Maud Adeline replied that, utopian though her vision of the modern society might be, 'unless we do long for something that is much beyond our grasp, we cannot get anything'. She spelled this out very clearly in her reply to the BCGA:

> ... the domestic problem does not begin and end with the individual home, but is at the bottom as home life is at the bottom of all life: and that you who have in your keeping, as doctors and other professional men, such enormous responsibilities and enormous power for good or evil over the lives of the people whom you serve, should look to the broader aspects of the work you have undertaken.[173]

Maud Adeline Cloudesley Brereton's remarks to the BCGA considered the social and moral responsibility the gas companies owed to their customers, and women in particular. It was an invitation to cooperate with housing authorities and local governments to ensure that workman's homes were properly fitted with basic gas appliances like cooking stoves and water boilers. It was a cogent reminder of the need to assure middle-class consumers that, despite the promises of electricity, gas was their 'silent servant', the answer to domestic labour problems and the means of freeing women from useless toil. Maud Adeline concluded that comfortably heated, cleaner homes and better food equalled a healthier population, and hence a stronger nation. By offering the 'woman's point of view', she was speaking not only on behalf of female consumers, but also for an increasing number of Edwardian women who anticipated that women brought to the business community a very different sensibility than their male co-workers. According to Kwolek-Folland, some business commentators noted the influence, by the 1910s, of the activist perspective of the progressive 'New Woman' on the role of women in business. Armed with their moral strength and reformist convictions, some observers asserted, women could improve the ethical climate of business by their selfless perspective.[174] By asking gas managers to look beyond immediate profits and to consider their role as agents of urban and social reform, Maud Adeline clearly represents the progressive woman of business, appealing to a higher goal than simply making money.

In some respects, Maud Adeline's message was both familiar and problematic. Her emphasis on women's needs as consumers adopted a uniquely feminist perspective; however, her references to the regenerative potential of gas technology rephrased a well-worn argument. The 1904 Inter-Departmental Report on Physical Deterioration, which cited nutrition as the critical factor in determining the size and general health of urban dwellers, had already given cooker sales a social dimension which the industry eagerly exploited. Gas managers supported the teaching of domestic subjects in board schools by supplying free classroom cookers for girls' instruction and for the preparation of school meals for destitute pupils.[175] But even the most socially conscious businessmen were essentially capitalists and not philanthropists. While training schoolgirls to use gas stoves could be seen as an investment in future sales, the return on other types of promotional spending, including advertising, salesmen's schools and lady demons, proved more difficult to measure. R.W. Edwards of Aldershot cautioned the BCGA membership that even benevolent capitalists must consider their responsibility to pay shareholders the expected dividends, and that customer service was an expensive practice.[176]

A thousand and one ways to domestic bliss

Despite ongoing debates among gas men concerning the merits of propaganda,[177] the campaigns of the BCGA and the influence of Maud Adeline Cloudesley Brereton underscore the gas industry's underlying commitment to advertising and customer education. Mrs Cloudesley Brereton continuously promoted the interests of women as consumers in the home and the workplace, and, although she was a corporate publicist, she wrote and spoke as a sanitation reformer. In this respect, her work continued an earlier Victorian feminist campaign for improved domestic architecture. According to Annmarie Adams, between 1870 and 1900, medical doctors and women involved in the sanitation movement identified the Victorian house as a potential cause of disease. Reformers had recognised long ago the health risks associated with urban slums; by the late nineteenth century, the middle-class home had become suspect as well. In developing their theories of contagion, Victorian doctors attributed some illnesses, such as typhoid and consumption, to 'sick houses' or those with poor ventilation and faulty plumbing; they blamed architects and builders for placing aesthetics or cheapness above the health concerns of the buildings' occupants. House manuals written by female designers criticised the traditional layout of the narrow four-storey middle-class house, with its basement kitchen, garret nursery, multiple staircases and narrow passages. Their labour-saving house plans eliminated the subterranean scullery and provided space for children near the main floor. Both of these changes contravened Victorian notions of propriety which dictated that the commotion of meal preparation and child-rearing be removed as far as possible from the centre of the home to preserve the tranquillity of the house, particularly for the men. In the interests of women's health, however, feminist decorators insisted that houses be designed to minimise fatigue for mistresses and their servants. By the turn of the century, with the advent of town planning and the garden city movement, professional architects assumed the lead in housing reform, while, according to Adams, women took a secondary role in the movement they helped initiate.[178]

It is undeniable that male architects, including Raymond Unwin and Ashton Webb, dominated Edwardian housing debates. Their reports and pamphlets made recommendations for suburban land use, housing density, minimum building standards and government subsidies for housing development. Principally concerned with working-class housing, they advanced designs for two-storey terraced or semi-detached cottages in the quasi-rural setting of the model suburbs, like Hampstead and Golders Green.[179] However, like the feminist interior designers, the reformist architects agreed that the middle-class multi-storey townhouse was flawed in design. Architect Percival M. Fraser listed the attributes of the 'well-planned house' as 'weather, damp and vermin-proof, and at once substantial, comfortable, noiseless, convenient, healthy, homelike, economical in cost and service, beautiful and as pleasing to live in as look upon'. He described the basement as a 'great evil', criticising labour-intensive fixtures, like stone steps and polished metal fittings. To ensure the occupants' health and comfort, he recommended well-ventilated ceilings at least 8 foot, 6 inches in height, and self-cleaning sanitary fittings for water closets. Nevertheless, despite his appreciation of labour-saving designs, Fraser still insisted on the value of the coal range and the open fire. And, because he disliked cooking odours in the house, he advocated distancing the kitchen from the dining area.[180] Evidently, male architects did not share the same priorities as female housing critics.

Despite the predominance of male architects, women were not excluded entirely from Edwardian housing debates. Through her work with the BCGA and from the perspective of a sanitation reformer, Maud Adeline Cloudesley Brereton continued her feminist campaign to redesign homes in the interests of women, children and domestic workers. Her principal interests were the provision of sanitary and labour-saving fixtures for cooking, washing and heating for all classes. At a meeting of the Institute of Sanitary Engineers, she presented a paper on domestic lighting, cooking and heating. Introducing her topic, she criticised the faulty design of Victorian townhouses 'with their steep stairs, stupid and meaningless corners and cornices' whose inefficient layouts resulted in a 'wicked waste of precious human health and energy'. In particular, Mrs Cloudesley Brereton condemned the basement of the Victorian home as being inconvenient and unsanitary. Unfortunately, it was also the site of many routine domestic chores, including laundering and meal preparation. Because of recent developments in 'modern sanitary science', Mrs Cloudesley Brereton granted that architects no longer favoured basements, noting that model houses combined kitchens and sculleries on the first floor. The problem of modernising existing basements remained, however, and, with this objective in mind, she presented to the Institute of Sanitary Engineers five designs for homes ranging in annual rents from £30 to £250, prepared in consultation with architect E.C.P. Monson. Speaking to the assembled architects, housing authorities and sanitary engineers, she defended women's rights to efficient, well-ventilated homes in the same manner that factory employees had the right to safe work premises.[181]

All of the designs illustrated how the 'infernal regions' could be eliminated, either by moving the kitchen and scullery to the first floor, or by rationalising the layout of the basement kitchen-scullery and installing a dumbwaiter between the kitchen and the first-floor dining area. Maud Adeline contended that by adapting or abolishing the basement and using gas appliances throughout the house, domestic labour could be reduced substantially, either for the servant, housewife or the mistress with a small staff. Labour-saving fixtures, like gas fireplaces, stoves, water heaters and even a gas-powered incinerator for household refuse, eliminated physically demanding jobs for all classes of women, such as carrying water for baths, fetching coal and cleaning coal grates (tasks that were often relegated to young female domestics).[182] Although the Victorians attributed the spread of disease to 'miasmas' of polluted air and water, by 1914, the connection between germs and illness was an accepted piece of household wisdom.[183] Given this recognised association between sanitation and disease prevention, Maud Adeline's provisions for hot water and rubbish disposal went beyond labour-saving advice. Modern gas appliances, particularly hot water boilers, helped ensure a healthier population because they eliminated germs in the home. Addressing the cost factor, she advised her listeners that the gas industry provided free consultations, bore the capital costs of installation and hired out appliances to both tenants and landlords. Her audience was less convinced, not so much on the need to remodel basements, but on gas technology as the means of providing domestic bliss. In response to audience protests that gas fires blackened ceilings as much as coal did, emitted noxious fumes and lacked the comfort of the coal stove, Mrs Cloudesley Brereton offered personal testimony as to the satisfactory nature of gas installations. Her own domestic help objected to the seven coal fires in her Hampstead house and 'struck', leaving her with two adolescent kitchen maids, a large house and five children. She installed gas fires to alleviate the servant problem, thereby managing on a small staff.[184] To illustrate her point, Mrs Cloudesley Brereton included a plan of her own altered house on Lyndhurst Road.

Similarly, Mrs Cloudesley Brereton recognised the beneficial effects of gas technology on the working lives of female teachers, nurses and factory workers.[185] Reporting on the numerous applications for gas technology, she argued that the achievement of domestic bliss depended not on women's relegation to the private sphere, but on mechanised efficiency in both the home and the workplace. Mrs Cloudesley Brereton used the example of gas-operated laundries to illustrate her point. She contended that the arduous task of doing the weekly wash, 'which makes most homes no home', should be assigned to efficient commercial laundries. The time and energy saved by sending the washing out might be directed towards less physically demanding remunerative work, like dressmaking, which in turn could offset the laundry charges.[186] She insisted that laundries, predominantly staffed by women, install gas-operated steam washers, mechanised mangles, rollers and gas irons to lighten the work performed by female employees. Like Fabian author Clementina Black, an outspoken critic of female sweated labour, Mrs Cloudesley Brereton supported legislation to limit laundry workers' hours and to dictate improvements for commercial facilities.[187] In addition, Mrs Cloudesley Brereton praised the efforts of the government inspectors responsible for enforcing the new standards of ventilation, hygiene and temperature for laundry workrooms.[188] According to Mrs Cloudesley Brereton, gas technology could potentially eliminate the drudgery associated with laundry work, whether it was done at home or on commercial premises. In either case, the benefits to female workers were significant. Nevertheless, despite the occupational improvements made possible by legislation and technology, and the increased demand for skilled female operatives to use specialised ironing equipment, Edwardian investigators still considered laundry work a difficult trade 'suited to strong girls only'.[189]

Still, Mrs Cloudesley Brereton's conviction that technology could improve the lives of working women reflected her commitment to female workers and public health. Like contemporary groups of feminist activists, including the Women's Industrial Council and the Women's Labour League, Mrs Cloudesley Brereton defended the need for improved working-class housing, for clean and well-lit work premises and for sanitary rest areas for employees. In common with these organisations, she advanced the importance of motherhood, infant health and nutrition.[190] Similarly, she contested the perceived differences between separate male and female spheres, and used maternal feminist arguments to defend women's participation in public life. But, most importantly, she sought these objectives not through government lobby groups or overt political activism, but by working from within the male-centred corporate world.

Conclusion

The establishment of the BCGA illustrates the gas industry's dedication to increasing sales and improving customer service. Despite the conservative image of the gas industry, the creation of an industry-wide publicity organisation that combined public relations and advertising must be seen as a modern business strategy. In addition, the growing numbers of lady demons and their continued participation in gas exhibitions and cookery lectures illustrate women's importance in the industry's evolving public relations strategy. While the lady demons remained central to customer service and education, the advertising material produced by the new publicity organisation provided them with powerful tools to assist them in their work. The lady demons were the first sales workers in the gas industry who upheld a sense of occupational identity based on their shared experiences as travelling lecturers, their concern for business procedures and their sense of social purpose. In the bid for occupational status, the lady demons also distinguished themselves from their masculine counterparts in gas sales. Given their higher education and class status, they distinguished themselves from the youthful untrained male canvassers and showroom attendants. For young men, a promotion to a sales job could be read as a marginal advancement from the clerking ranks or confirmation that they were ill-suited to any useful position in the industry. The lady demons provide an example of a group of female workers who constructed a collective professional identity within the masculine sphere of industry and commerce. Nonetheless, it should be remembered that selling was still not a high priority among the majority of Edwardian gas managers, despite the crusading work of Francis Goodenough and the BCGA. This alone may account for the reason that it was left to cookery teachers and young clerks.

Maud Adeline Cloudesley Brereton became an important ally in advancing the professional mandate of the lady demons. She urged gas managers to consult on a regular basis with their lady demons, who maintained close contacts with female customers in their homes. She also argued in favour of training for salesmen and greater cooperation between showroom attendants, demonstrators and gas inspectors, to create efficient multi-level sales teams. Similarly, she articulated the needs of female consumers in both the private and the public sphere, not only to gas managers, but to architects, builders and employers. By advocating labour-saving technology to assist women, she challenged existing assumptions about the low value of domestic work and female labour in general. Her feminist assertion that women could combine home and public life, could work both inside and outside the house, and still be excellent and efficient mothers, must be considered a progressive suggestion, coming as it did from a married middle-class Edwardian working mother. Her belief that 'home life was at the bottom of all life' asserted the importance of women's contributions beyond domestic matters; in this respect she contested the perceived differences between separate male and female spheres of influence. While her concern for infant and maternal health, public sanitation and housing was shared by other groups of Edwardian female investigators, Mrs Cloudesley Brereton remains distinct because she attempted to effect material and social change via the business sector.

Notes

1 'Twelve Little Demons', *Co-Partners' Magazine* (September 1911), 130.
2 In August 1911, England experienced an intense heatwave with recorded temperatures of over 90 degrees Fahrenheit.
3 The conservative image of the gas industry is upheld by business historians and historians of design. Business historians suggest that the gas industry's technical innovations were motivated by the belief that electricity threatened to undermine their industry. See Everard, *History of the GLCC*, 262–263, 277; Barty-King, *New Flame*, 138, 144–145. Design historian Adrian Forty focuses on the innovative nature of electrical designs but overlooks the gas industry's comparable accomplishments. Adrian Forty, *Objects of Desire: Design and Society since 1750*, 2nd edition (London, 1995).
4 Bowden and Offer, 'The Technological Revolution that Never Was', in de Grazia with Furlough (eds), *The Sex of Things*, 267.
5 'Demonstration work by the gas companies', *JGL*, 20 April 1909, 158.
6 Ibid., 159.
7 'Manchester gas exhibition', *JGL*, 29 October 1907, 347; 'Manchester gas exhibition', *JGL*, 5 November 1907, 415; *JGL*, 'Manchester gas exhibition', 12 November, 1908, 489.
8 Kwolek-Folland, *Engendering Business*, 65–67.
9 Jane Lewis, *Women in England 1870–1950: Sexual Divisions and Social Change* (Sussex and Bloomington, Ind.), 1984, 75.
10 Elizabeth L. Banks, 'New paid occupations for women', *Cassell's Family Magazine* (1893–94), 586–587.
11 Zimmeck, 'Jobs for the Girls', in John (ed.), *Unequal Opportunities*, 157. Guidebooks for women's employment include Margaret Bateson, *Professional Women Upon Their Professions* (London, 1895); Phillips et al. (eds), *A Dictionary of Employments Open to Women*; M.K. Lyttleton, *Women and Their Work* (London, 1901); Apprenticeship and Skilled Employment Association (ed.), *Trades for London Girls and How to Enter Them* (London, 1909); Bird, *Women At Work*.
12 Walkowitz, *City of Dreadful Delight*, 72. Also see George Romanes, 'The capacity of women', *Nineteenth Century* (May 1887), 654–667.
13 Walkowitz, *City of Dreadful Delight*, 72; Lewis, *Women in England*, 99. Eliza Lynn Linton condemned the independent New Woman and accused unwed middle-class women of shirking their 'racial duty' to the state.
14 Lewis, *Women in England*, 99.
15 Cicely Hamilton, *Just to Get Married* (London, 1909).
16 Ethel Snowden, *The Feminist Movement* (London, 1912).
17 Lisa Tickner, *The Spectacle of Women: Imagery of the Suffrage Campaign 1907–1914* (Chicago, 1988), 216–219. Tickner notes that 'motherhood' legitimised women's participation in a wide range of reform movements from rescue work to municipal politics.
18 Norman McCord, *British History, 1815–1906*, 3rd edition (Oxford, 1995), 453–454. The National Union for Women's Suffrage was formed in 1869. The amalgamation of several suffrage groups, including the NUWS, created the NUWSS in 1897.
19 Tickner, *Spectacle of Women*, 80–86, 90–98.
20 Ibid., 100.

21 Ibid., 111–119, 122–131. Tickner describes suffrage processions on 18 and 23 June 1910, organised respectively by the militant WSPU and the constitutionalist NUWSS. The Women's Coronation Procession, 17 June 1911, was the largest demonstration with contingents from throughout the empire, marching female workers and historical pageantry.
22 St John, 'Educate or domesticate?', 194–199.
23 King's College/University of London/KWA/GPF 11, Miss Baddeley, unpublished speech at the opening of the extension of the Gloucester School of Domestic Science, 7 October 1911, 4–6.
24 F.K. Prochaska, *Women and Philanthropy in Nineteenth-Century England* (Oxford, 1980), 5–13.
25 Patricia Hollis, *Ladies Elect: Women in English Local Government, 1865–1914* (Oxford, 1987); Martha Vicinus, *Independent Women: Work and Community for Single Women 1850–1920*, (London, 1985), chaps 2 and 6; Walkowitz, *City of Dreadful Delight*, chap. 2.
26 Vicinus, *Independent Women*, 211–212.
27 Henrietta O. Barnett, 'Women as Philanthropists', in Theodore Stanton (ed.), *The Woman Question in Europe*, (London, 1884), 114–115; for a discussion of Victorian classifications of 'the poor', see Rachael Vorspan, 'Vagrancy and the New Poor Law in Late-Victorian and Edwardian England', *English Historical Review*, **92**, 362 (January 1977), 65.
28 Hollis, *Ladies Elect*, 202–203. Hollis is understandably critical of the COS hard line; however, some caseworkers, notably Beatrice Webb and Eleanor Rathbone, understood the systemic nature of poverty and redirected their attentions to broader social and economic issues.
29 Many of Charles Booth's investigators were COS rent collectors and visitors, including Beatrice Potter, Ella Pycroft and Clara Collett.
30 Ross, *Love and Toil*, 15–17; Vicinus, *Independent Women*, 221; Walkowitz, *City of Dreadful Delight*, 52–53.
31 Vicinus, *Independent Women*, 214–220.
32 Ibid., 242.
33 Evans, 'The essential qualifications of a lady health visitor', 306–318.
34 Walkowitz, *City of Dreadful Delight*, 68; Rappaport, *Shopping for Pleasure*, 6–7, 111.
35 'The lady lecturer and teacher', *JGL*, 31 January 1905, 282.
36 Walkowitz, *City of Dreadful Delight*, 58.
37 Edith J. Morley (ed.), *Women Workers in Seven Professions: A Survey of Their Economic Conditions and Prospects* (London, 1914), 234. Morley made a distinction between the professional women serving the public municipal authorities, calling them 'health inspectors' as opposed to the 'health visitors' who were appointed by private charities. According to Morley, the latter usually had 'no universally recognized standard of attainments' while the municipal inspectors had attended university and held government certificates. Female inspectors received £65–100 per annum. Also see Vicinus, *Independent Women*, 227. Full-time social workers often received no salary, and paid £30–35 per annum for their room and board in a settlement house. Department heads sometimes received a small stipend.
38 'The lady lecturer and teacher, letter from Miss M.E. Betts', *JGL*, 31 January 1905, 283.
39 'The lady lecturer and teacher, letters from Miss Helen Edden and Mrs. Charles Marshall', *JGL*, 31 January 1905, 283.

40 Attar, *Wasting Girls' Time*, 99–104; Carol Dyhouse, *Girls Growing Up in Late Victorian and Edwardian England* (London, 1981), 84–98; Turnbull, 'Learning Her Womanly Work', in Hunt (ed.), *Lessons for Life*, 83–100; Felicity Hunt, 'Divided Aims: the Educational Implications of Opposing Ideologies in Girls' Secondary Schooling, 1850–1940', in Felicity Hunt (ed.), *Lessons for Life: The Schooling of Girls and Women, 1850–1950* (Oxford, 1987), 3–21; Purvis, *A History of Women's Education in England*, 26–29.

41 J.C. Buckmaster, 'Cooking as a Branch of Domestic Education', in C. Herman Senn (ed.), *The Universal Cookery and Food Association Cookery Annual* (London, 1906), 52.

42 Dyhouse, *Girls Growing Up*, 95.

43 Charlotte Despard, 'The feeding of school children', *The Times Educational Supplement*, 7 April 1914, 66.

44 Dyhouse, *Girls Growing Up*, 162–169; see also Margaret Gilliand, 'Home Arts' (1911), in Spender (ed.), *The Education Papers*, 316–324; Lilian Faithfull, 'Home Science' (1911), in *The Education Papers*, 325–327.

45 Faithfull, 'Home Science', in *The Education Papers*, 327.

46 Dyhouse, *Girls Growing Up*, 168.

47 Hilda D. Oakley, 'The teaching of home science and economics', *The Times Educational Supplement*, 1 August 1911, 112.

48 Nancy L. Blakestad, '"What's the use of learning this?" Careers of KCHSS graduates, 1910–1949', *Envoy: Queen Elizabeth Old Student's Paper* (Autumn, 1992), 7–14. Blakestad notes that between 1910 and 1919, the majority of King's College's KCHSS graduates took jobs in teaching (45.7 per cent). Social work absorbed only 8.7 per cent, as did jobs in industry and business. Blakestad comments on the lower than expected numbers of graduates that went into social work, an area the programme founders had anticipated their programme would serve almost exclusively. Equally curious were the numbers of women who entered business and industry, a job market that the programme founders had not anticipated. Blakestad attributes the entry of KCHSS graduates into business as the result of 'new opportunities in the gas and electrical industries as home service advisors, demonstrators and in product development'.

49 Minutes of evidence of Mr J. Lewis, West Green Board, North London, *Interdepartmental Committee of Physical Deterioration* (1904), vol. 1, 321.

50 'Domestic economy teaching in schools', *Cookery Teacher*, **1**, 3 (May 1904), 34.

51 'Letter to the editor: cookery demonstrations', *Cookery Teacher*, **1**, 1 (March 1904), 19.

52 'Domestic service', *Cookery Teacher*, **1**, 4 (June 1904), 60–61.

53 Jeune, 'The ethics of shopping', 120–121.

54 Margaret G. Bondfield, 'The effects on health of women's employment in shops', *Journal of the Royal Sanitary Institute*, transactions, 25, 1904 (London, 1905), 738–744.

55 Holcombe, *Victorian Ladies At Work*, 106–107, chap. 5; Lancaster, *The Department Store*, chap. 8.

56 Peter Bailey, 'Parasexuality and glamour: the Victorian barmaid as cultural prototype', *Gender and History*, **2**, 2 (Summer 1990), 1–2, 35, 32.

57 Rappaport, *Shopping for Pleasure*, 196–99.

58 Lancaster, *The Department Store*, 137–38.

59 Christopher P. Hosgood, '"The knights of the road": commercial travellers and the culture of the commercial room in late-Victorian and Edwardian England', *Victorian Studies*, 37, 4 (Summer 1994), 524, 534, fn. 9. For a contemporary account of commercial travelling, see 'On the road', *Chamber's Journal* (1902), 715–718.
60 Goodall, *Burning to Serve*, 142–43.
61 'The lady lecturer and teacher', *JGL*, 31 January 1905, 283.
62 'How to grill a demonstrator', *Co-Partners' Magazine* (September 1911), 132. The poem was probably written by Miss M. Menhinick, the only demonstrator with those initials at that time. The middle stanzas convey the discomfort of the job which those who have worked long hours outdoors will understand. 'Then place her on a doorstep, 'Neath the suns hot rays: Never mind her knocks and rings, She must stand that blaze. As the front door blisters, Let her do so too; The sun with scorching heat, Will grill her through and through. Degrees 90 in the shade, 120 in the sun: Oh! the Demonstrator's lot, Is surely awful fun!'
63 George Clarry, 'Showrooms in Relation to Gas Supply', in M.A. Cloudesley Brereton (ed.), *First Annual Report and Conference Proceedings of the British Commercial Gas Association*, 8–9 October 1912 (1913), 172.
64 LMA/B/GLCC/*Minutes of the Board of Directors*, 18 October 1889, 376. The GLCC set up its first showroom at Goswell Road at a cost of £630. Five years later, the GLCC invested £2,500 in its Kensington showrooms, with an additional £1,500 annual rent for the property. See GLCC, *Minutes of the Board of Directors*, 28 September 1894, 80; GLCC, *Minutes of the Board of Directors*, 22 March 1912, 32. The company engaged architect W.J. Tapper to submit suggestions for the 'artistic arrangements of exhibits in company showrooms'.
65 Clarry, 'Showrooms in Relation to Gas Supply', 172–173, 183.
66 LMA/B/SMetG/III 11/1, 2, *Minutes of the Board of Directors*, 4 May 1892, 132.
67 LMA/B/GLCC/42/ *Minutes of the Board of Directors*, 26 June 1903, 71. The board approved the creation of the position of showroom attendant at a starting annual salary £100, rising to £150 maximum by £5 annual increments. Junior clerks could expect pay rises in £5 increments up to a maximum of £100 per annum.
68 Hosgood, 'Mercantile monasteries', 322–352.
69 'Twenty-five years' service as executive chairman: presentation to Sir Francis Goodenough', *JGL*, 14 October 1936, 121; Everard, *History of the GLCC*, 288. Goodenough held the position of chief controller of gas sales until his retirement from the GLCC in 1931. He was knighted in 1930.
70 LMA/B/GLCC 45/*Minutes of the Board of Directors*, 4 February 1910, 171; Spurr, 'Those who are obliged to pretend that they are gentlefolk', 102.
71 Freeden *Rainbow Circle*, 5–10, appendix II.
72 Ibid., 1–7.
73 Perkin, *Rise of Professional Society*, 83–90.
74 Freeden, *Rainbow Circle*, 226, 258, 282, 322, 354.
75 Ibid., 226, 323.
76 Everard, *History of the GLCC*, 284, 289, 308–310.
77 'Annual report: remarks by Corbett Woodall and F.W. Goodenough', in M.A. Cloudesley Brereton (ed.), *First Annual Report and Conference*

Proceedings of the British Commercial Gas Association, Manchester, 8–9 October 1912 (1913), 33.

78 LMA/B/GLCC 44/*Minutes of Board of Directors*, 16 March 1906, 43.

79 Everard, *History of the GLCC*, 288, 295. As controller of gas sales, Goodenough was responsible for all gas inspectors, canvassers, showroom attendants and lady demons.

80 'Remarks by Woodall and Goodenough', in Cloudesley Brereton (ed.), *First Annual Report and Conference Proceedings of the BCGA*, 24.

81 W.M. Mason, 'Co-operative advertising of a public utility', *GJ*, 23 July 1924, 391.

82 'Earl's Court Exhibition: list of exhibitors', *JGL*, 4 October 1904, 19.

83 Bodleian Library/John Johnson Collection/Gas and Gas Appliances, Box 1 & 2. Examples of gas advertisements in this collection include: John Wright Co. 'Wright's Gas Stoves', May 1901; GLCC, 'Modern Gas Lighting and Heating', 1910; GLCC, 'A Talk on Hot Water Comfort', 1910.

84 This organisation appears to have been short-lived, and historians of the electrical industry described the 1919 British Electrical Development Association as the first publicity group. Hannah, *Electricity Before Nationalisation*, 77; R.A.S. Hennessey, *The Electric Revolution* (Newcastle-upon-Tyne, 1971), 134.

85 'The publicity campaign', *JGL*, 21 February 1911, 511–512.

86 'Remarks by Woodall and Goodenough', in Cloudesley Brereton (ed.), *First Annual Report and Conference Proceedings of the BCGA*, 19, 33.

87 Hannah, *Electricity Before Nationalisation*, 14–22; Paul Thompson, *The Edwardians: The Remaking of British Society* (London, 1975), 294.

88 Hannah, *Electricity before Nationalisation*, 187.

89 Bodleian Library/John Johnson Collection, Box 1. H J. Dowsing (ed.), *Crystal Palace Electrical Exhibition Catalogue* (London, 1892), section F.

90 Rosalind H. Williams, *Dream Worlds: Mass Consumption in Late Nineteenth-Century France* (Berkeley, 1982), 84–90.

91 C.F. Brackett and F.L. Pope (eds), *Electricity in Daily Life* (London, 1891). This collection of essays described the existing and anticipated applications of electricity to industry, transportation and the home.

92 *Illustrated London News*, 17 May 1884, 472–473; Dowsing (ed.) *Crystal Palace Electrical Exhibition Catalogue*, section F.

93 C. Herman Senn, 'Cooking by electricity', in C. Herman Senn (ed.), *Cookery Souvenir of the Universal Cookery and Food Association* (London, 1894), 123.

94 Quoted in Davidson, *Woman's Work*, 37.

95 Bodleian Library/John Johnson Collection/Box 1, *Electricity: The Light of Today* (London, 1893), preface.

96 Milan, 'Refracting the Gasolier', 88–89.

97 Denek Matthews, 'Laissez-faire and the London gas industry in the nineteenth century: another look', *EHR*, 39, 2 (February 1986), 256–260. Nowhere in Matthews's detailed analysis of gas pricing does he mention electrical competition as a factor. According to Matthews, price changes depended on coal prices, capital expenditures, such as the price of iron, and the reduction in wasted gas because of improved gas mains.

98 Ibid., 257.

99 'Gas works management and consumer's interests', *Chamber's Journal* (February 1898), 190.

100 Everard, *History of the GLCC*, 284–285. In 1898, the GLCC north of the Thames charged three shillings per 1,000 cubic feet of gas; the SMGC charged two shillings and three pence for the same measure and, when the gap widened, the government undertook an investigation. Everard notes that Livesey's charges were accurate and that the GLCC suffered, despite its 'admirable organization on paper', from 'inexpert execution, uninspired technical advice and a pedestrian engineering policy'.
101 Ibid, 289, 293.
102 Ibid., 295–296, chap. 19.
103 Bodleian Library/John Johnson Collection, Box 2. Institution of Gas Engineers, *Gas Versus Electricity – Which is the Better Light?* (London, 1902). A comparative assessment that warns consumers 'You will be told that electricity is the cheaper light – NOT TRUE'. The consulting chemist for the Institution of Gas Engineers, Vivian B. Lewes, Royal Naval College, Greenwich, was also a spokesman for the Smoke Abatement Society.
104 'Remarks by Woodall and Goodenough', in Cloudesley Brereton (ed.), *First Annual Report and Conference Proceedings of the BCGA*, 19–20, 33.
105 Coffin, *Politics of Women's Work*, 95.
106 Clarence Moran, *The Business of Advertising* (London, 1905), chap. 1.
107 T.R. Nevett, *Advertising in Britain*, chap, 6.
108 Quoted in Nevett, *Advertising in Britain: A History* (London, 1982), 125.
109 Walter Dill Scott, *The Theory of Advertising: A Simple Exposition of the Principles of Psychology in Their Relation to Successful Advertising* (Boston, 1905), 395, 407–408.
110 Moran, *Business of Advertising*, 15–16.
111 Mason, 'Cooperative advertising of a public utility', 23 July 1924, 391.
112 'Remarks by Woodall and Goodenough', in Cloudesley Brereton (ed.), *First Annual Report and Conference Proceedings of the BCGA*, 24.
113 Kwolek-Folland, *Engendering Business*, 48–49.
114 'The British Commercial Gas Association', *JGL*, 12 November 1912, 492; H. James Yates, 'Aims and claims of the BCGA', in Cloudesley Brereton, *First Annual Report and Conference Proceedings of the BCGA*, 135. Yates, a Birmingham gas engineer, noted that municipal concerns managed both gas and electric utilities. He suggested that the rival industries needed 'to recognize the legitimate sphere of each and their respective advantages', meaning that the 'legitimate sphere' of electricity was transportation and communications, while gas maintained its lead in the areas of lighting and fuel.
115 Mason, 'Cooperative advertising of a public utility', 391.
116 'Remarks of Woodall and Goodenough', in Cloudesley Brereton (ed.), *First Annual Report and Conference Proceedings of the BCGA*, 40.
117 Mason, 'Cooperative advertising of a public utility', 391. W.M. Mason does not give the exact percentage but indicates that the largest subscriber, presumably the GLCC paid over £6,000 in annual dues. See also LMA/B GLCC, *Minutes of the Board of Directors*, 10 January 1913, 139. This subscription method was used to raise money for other joint ventures. In 1913, the GLCC pledged £2,708 to the National Gas Exhibition based on a rate of one guinea per 10 million cubic feet of gas sold in the previous sales year.
118 Nevett, *Advertising in Britain*, 108.

119 'Mrs. M.A. Cloudesley Brereton, Response to H. James Yates', in Cloudesley Brereton (ed.), *First Annual Report and Conference Proceedings of the BCGA*, 149–151; see also 'The British Commercial Gas Association: the first conference', *JGL*, 12 November 1912, 492.

120 *Who Was Who* vol. IV, 1941–1950 (London, 1952), 137; Maud Adeline had two sons with John Horobin, but only one survived. T.H. Simms, *Homerton College, 1695–1978* (Cambridge, 1979), 45. My thanks to Peter M. Warner, senior tutor at Homerton College, for shedding some light on the Homerton career of Maud Adeline Horobin. Because the time when the board of governors declined support for Maud Adeline coincided with her remarriage in 1904 to Mr Cloudesley Brereton, it is conceivable that it was this event that divided her interests.

121 *Who Was Who*, vol. III, 1929–1940 (London, 1941), 157; Freeden, *Rainbow Circle*, 15, 184–85, 270–71, appendix II.

122 Harris, *Private Lives, Public Spirit*, 31.

123 Freeden, *Rainbow Circle*, 11, 204.

124 'Mrs. M.A. Cloudesley Brereton', *The Woman Engineer*, 3, 12 (September–October 1932), 180.

125 Mrs M.A. Cloudesley Brereton, 'Southampton docks and public health', *The Times*, 14 September 1912, 18.

126 Mr Cloudesley Brereton died in 1937; Maud Adeline died in 1946. 'Mrs. Maud Adeline Cloudesley Brereton', *The Woman Engineer*, 3, 12 (September–October 1932), 180.

127 'Twenty-five years service', *JGL*, 14 October 1936, 120.

128 Fabian Women's Group, *Women's Fabian Tracts*, ed. Sally Alexander (London, 1988), 274.

129 M.A. Cloudesley Brereton, *The Mother's Companion* (London, 1909), chap. 1; for a discussion of companionate marriage during the 1920s, see Christina Simmons, 'Modern Sexuality and the Myth of Victorian Repression', in Christina Simmons and Kathy Peiss (eds), *Passion and Power: Sexuality in History* (Philadelphia, 1989), 160–162.

130 Cloudesley Brereton, *Mother's Companion*, 12.

131 Ibid., 133.

132 Ibid., 90.

133 Sandra Holton, *Feminism and Democracy: Women's Suffrage and Reform Politics in Britain* (Cambridge, 1986), 26.

134 Cloudesley Brereton, *Mother's Companion*, 16.

135 Michael Freeden, 'Eugenics and progressive thought: a study in ideological affinity', *Historical Journal*, 22, 3 (1979), 652.

136 Ibid., 645–671.

137 British Library Manuscripts Collection, Marie Stopes Papers, Add. 58687 ff. 6–8. Letter to Stopes from M.A. Cloudesley Brereton, 11 June 1920 and reply to M.A. Cloudesley Brereton from Stopes, 12 June 1920.

138 British Library, Manuscripts Collection, Marie Stopes Papers, Add. 58692, f. 23. Letter to Marie Stopes, from M.A. Cloudesley Brereton, 10 August 1922. See *Birth Control News*, 1, 1 (May 1922), 1, for an example of the advertisement. The BCGA advertisement appeared regularly in the *Birth Control News* in 1922 and 1923.

139 'Events of the week', *The Times*, 6 July 1914, 11.

140 June Rose, *Marie Stopes and the Sexual Revolution* (Boston and London, 1992), 82.

141 Dyhouse, *Girls Growing Up*, 167–169.
142 Alice Zimmern, *Women's Suffrage in Many Lands* (London, 1909); Alice Zimmern, 'Ladies dwellings', *Contemporary Review* , 77 (January, 1900), 97–104.
143 Turnbull, 'An isolated missionary', 82.
144 M.A. Cloudesley Brereton, 'The Study of the Problems of Domestic Economy as a Factor in Salesmanship', in M.A. Cloudesley Brereton (ed.), *Papers Delivered to the Second Annual General Meeting and Conference of the British Commercial Gas Association, London, 21–22 October 1913* (London, 1914), 112.
145 'The National Gas Exhibition', *Co-Partners' Magazine* (October 1913), 161; 'The National Gas Exhibition', *JGL*, 7 October 1913, 19–32.
146 'Coal gas and its applications: exhibits at Shepherd's Bush', *The Times, Engineering Supplement*, 1 October 1913, 21.
146 For example, see: 'Electric light in New York', *Chamber's Journal*, 29 October 1881, 696; 'Two recent improvements in lighting', *Chamber's Journal*, 11 June 1887, 374; 'The modern wonder worker: electricity', *Chamber's Journal*, 19 March 1892, 177. All of the above articles predicted the imminent and steady demise of the gas industry.
148 British Commercial Gas Association, *Gas in the Home, Office and Factory: Hints on Health, Comfort and Economy: A Series of Popular Lectures at the National Gas Congress and Exhibition* (London, 1913), ii.
149 'Food and health: the value of proper cooking', *The Times*, 6 October 1913, 3.
150 'Conference on food and cookery', *JGL*, 7 October 1913, 33.
151 Reeves, '*Round About a Pound a Week*, 59. Reeves stated that gas cooking 'ought to be an excellent way of cooking, but under the penny-in-the-slot system, it is a way which tends to underdone food', suggesting that the gas available for a penny was insufficient to cook a meal.
152 Ibid., 52.
153 'Advertisement: the house comfortable', *Daily Express*, 1 September 1913, 1.
154 Perkin, *Rise of Professional Society*, 92–96.
155 'The BCGA cinematograph film', *JGL*, 28 October 1913, 359; 'The Manufacture and Applications of Gas', *JGL*, 4 November 1913, 424.
156 Crossick, 'The Emergence of the Lower Middle Class in Britain', in Crossick (ed.), *The Lower Middle Class in Britain*; Bailey, 'White collars, gray lives?', 273–290.
157 Arnold Bennett, *Buried Alive*, reprint of 1912 edition (London, 1949), 78–79.
158 Miss Constance Williams, '"The Silent Servant": How the Use of Gas Can Minimize Life's Little Worries', in BCGA (ed.), *Gas in the Home, Office and Factory: Hints on Health, Comfort and Economy: A Series of Popular Lectures at the National Gas Congress and Exhibition* (London, 1913), 72.
159 'Temptation of the slot meter', *The Times*, 24 March 1913, 24.
160 Olive Anderson, *Suicide in Victorian and Edwardian England* (Oxford, 1987), 367, fn. 95.
161 'A wife's last words', *Daily Telegraph*, 1 October 1913, 5; 'Woman smoker's suicide', *Daily Express*, 1 October 1913, 3.
162 'Reception of ladies by the editor,' in Cloudesley Brereton (ed.), *First Annual Report and Conference Proceedings of the BCGA*, 122.

163 Cloudesley Brereton, 'The study of the problems of domestic economy', in Cloudesley Brereton (ed.), *Papers Delivered to the Second Annual General Meeting and Conference of the BCGA*, 108.

164 Cloudesley Brereton, 'The study of the problems of domestic economy', in Cloudesley Brereton (ed.), *Papers Delivered to the Second Annual General Meeting and Conference of the BCGA*, 112.

165 Dyhouse, *Feminism and the Family*, chap. 3. For a discussion of housekeeping alternatives in America, see Hayden, *The Grand Domestic Revolution*, chaps. 8 and 10.

166 See Dyhouse, *Feminism and the Family*, 111–123 for a discussion among the Fabian women and the proposals in the *Freewoman*. Regarding the popular interest in cooperative housekeeping, see a series of letters in *T.P.'s Weekly*: 'Cooperative house-keeping in villadom', *T.P.'s Weekly*, 26 December 1902, 210; 'Duty and drudgery', *T.P.'s Weekly*, 2 January 1903, 247; 'Cooperative housekeeping in villadom', *T.P.'s Weekly*, 9 January 1903, 278; 'Cooperative housekeeping: a practical experiment', *T.P.'s Weekly*, 16 January 1903, 314; 'Cooperative housekeeping in villadom', *T.P.'s Weekly*, 30 January 1903, 342; 'Cooperative Housekeeping', *T.P.'s Weekly*, 6 February 1903, 406–407; 'Cooperative housekeeping', *T.P.'s Weekly*, 13 February 1903, 438–439; 'Cooperative housekeeping', *T.P.'s Weekly*, 20 February 1903, 474; 'Cooperative housekeeping', *T.P.'s Weekly*, 6 March 1903, 438.

167 Alison Ravetz, 'A View from the Interior', in Judy Attfield and Pat Kirkham (eds), *A View from the Interior: Feminism, Women and Design* (London, 1989), 192.

168 Hayden, *The Grand Domestic Revolution*, 4–11, 186–188.

169 Charlotte Perkins Gilman, *Women and Economics: The Economic Factor Between Men and Women as a Factor in Social Evolution*, reprint of 1898 edition, with foreword by Carl Degler (New York, 1966), 249–269.

170 Charlotte Perkins Gilman quoted in Dyhouse, *Feminism and the Family*, 114.

171 Cloudesley Brereton, 'The study of the problems of domestic economy', in Cloudesley Brereton (ed.), *Papers Delivered to the Second Annual General Meeting and Conference of the BCGA*, 117–127. To facilitate the use of lady demons by the gas companies, the BCGA maintained a registry of qualified women who were available for lecture work at gas exhibitions. See 'British Commercial Gas Association and lady demonstrators', *JGL*, 11 March 1913, 659; 'Lady Demonstrators', *JGL*, 10 June 1913, 748.

172 Cloudesley Brereton, 'The study of the problems of domestic economy', in Cloudesley Brereton (ed.), *Papers Delivered to the Second Annual General Meeting and Conference of the BCGA*, 128–139. Respondents to Mrs Cloudesley Brereton included David Milne-Watson, GLCC; Douglas Helps, Reading; Alderman F.S. Phillips of Salford; R.W. Edwards of Aldershot; and Alderman W. Armitage, Doncaster, 128–139; see 135 for Armitage's remarks.

173 Ibid., 135.

174 Kwolek-Folland, *Engendering Business*, 55–56.

175 LMA/B/GLCC 44, *Minutes of the Board of Directors*, 6 November 1908, 448. Isidore Salmon of the LCC asked the GLCC for the 'free loan of gas apparatus' to prepare school meals for underfed children; the company agreed to 'gladly co-operate with the LCC'.

176 Cloudesley Brereton, 'The study of the problems of domestic economy', in Cloudesley Brereton (ed.), *Papers Delivered to the Second Annual*

General Meeting and Conference of the BCGA, 132.

177 W.M. Mason, 'Notes on salesmanship and advertising', *JGL*, 14 October 1913, 150–151.

178 Adams, *Architecture in the Family Way*, chaps. 2, 3 and 5.

179 Raymond Unwin, 'Housing and town planning', *Journal of the Royal Institute of British Architects*, 9 December 1911, 110–11; also see Mark Swenarton, *Homes Fit For Heroes: The Politics and Architecture of Early State Housing in Britain* (London, 1981), 27–47.

180 Percival. Fraser, 'Suggestions for a well-planned house', *Journal of the Royal Institute of British Architects*, 23 March 1912, 390.

181 'Domestic lighting, heating and cooking: problems considered from the women's point of view', *JGL*, 7 April 1914, 40–43. Before a meeting of the Institute of Sanitary Engineers, Mrs Cloudesley Brereton presented layouts for remodelling five houses, each one representative of different economic levels, with the objective of eliminating the unhealthy, damp basement kitchen.

182 M.A. Cloudesley Brereton and Helen Edden, *Cooking by GAS* (London, 1913), preface.

183 Horsfield, *Biting the Dust*, 92–100.

184 'Domestic lighting, heating and cooking', *JGL*, 7 April 1914, 40.

185 'The twentieth century fuel', *ATAOUFG*, July, 1913, 3–4; 'Laundries and their Equipment', *ATAOUFG* (December 1913), 3–7; 'Domestic hot water from the cottage to the mansion', *ATAOUFG* (September 1915), 3–14.

186 Cloudesley Brereton, *Mother's Companion*, 95–96.

187 Clementina Black, *Sweated Industry and the Minimum Wage* (London, 1907), 31–32. Black reports on the long hours in commercial laundries and quotes a factory inspector, Miss Vines. Vines prosecuted an employer for keeping women for 28 consecutive hours, with a 2.5-hour rest on the floor during the night; another firm employed an ironer for 37 consecutive hours, a 'calendar worker' put in 32 straight hours including meal times and short breaks. In 1905, the Laundry Act extended the terms of the factory acts to cover commercial laundries because they used mechanised equipment; however, inspectors had difficulty enforcing the specified 12-hour day.

188 'Laundries and their equipment', *ATAOUFG* (December 1913), 3–4.

189 Apprenticeship and Skilled Employment Association, *Trades for London Girls and How to Enter Them*, 105–107. Wages in commercial laundries ranged from two to three shillings per day for operating the washing machines or 'preparing' articles for ironing. Skilled shirt and collar machinists and 'fine ironers' earned between 18 and 25 shillings per week.

190 Ellen Mappen, *Helping Women at Work: The Women's Industrial Council 1889–1914* (London and Dover, 1985), 23–24; Christine Colette, *For Labour and For Women: The Women's Labour League, 1906–1918* (Manchester, 1989), 110–128. See also Dr Marion Phillips, *The School Doctor and the Home*, (London, n.d.); Women's Labour League, *The Needs of Little Children: Report of a Conference on the Care of Babies and Young Children* (London, 1912).

CHAPTER 4

'Keep the Gas Fires Burning!' War and Reconstruction

> It will come as a surprise to the housewife and a triumph to the cook to learn that it is more patriotic to boil the kettle on the gas stove than on the kitchen fire; more patriotic to sit in a room with an unromantic gas fire than in one heated with glowing coal; and that every shovelful of coal means the loss up the chimney of valuable gases that ought to be used in killing Germans.[1]

The British government's declaration of war against Germany in August 1914 eventually drew the nation into the largest continental military conflict it had ever experienced, exceeding even the Napoleonic Wars. By the autumn of 1914, however, the anticipated short encounter had evolved into a stalemate between opposing trenches on the western front from the Swiss border to the English Channel. Come February 1916, the war against Germany, that was to last only a few months, was dragging through its second winter. Dwindling reserves of wheat and sugar were compounded by the decline of commercial shipping, creating food shortages throughout Britain. Female workers replaced the thousands of unmarried men recently conscripted under the provisions of the Military Service Act implemented in January 1916. Those numbers would soon grow as a second act, issued in May of that year, recruited all men between the ages of 18 and 41 for military service.[2] As the British nation hunkered down for a third year of war and uncertainty, the propaganda on behalf of the Home Office accelerated, in an effort to maintain popular support for the war and bolster morale on the home front. According to some historians, when the war in Europe became one of attrition, 'the organization of material resources counted more heavily than individual bravery on the battlefront'. Consequently, the home front was strategically important, as the war could be won or lost according to how successfully a nation utilised its economic resources and rallied its citizens behind the military effort.[3]

Being the relatives of enlisted men and also replacement workers, women were of critical importance to the war effort at home. Women's daily routines were imbued with patriotic symbolism as everything from knitting socks to voluntary food rationing and even lighting the gas cooker could provide support for the soldiers fighting abroad. And although it may have been a surprise to some housewives that the chemicals extracted during coal-gas production provided the raw materials for munitions manufacture, the opening quote demonstrates how gas established a direct connection between the English home and the battlefields of Europe. Although readers of the *Daily Mail* might have seen the gas fire piece when it originally appeared in that paper on 17 January, no doubt the editors of *Co-Partners' Magazine* were sufficiently impressed by the positive and patriotic references to their industry to reprint the article in their February issue. At any rate, it seemed to summarise the wartime activities of the gas industry: continued service to the domestic consumer and housewife on the home front, and aid to the munitions industry through the supply of raw materials. Moreover, the editorial also appeared next to an article that outlined the jobs currently being undertaken by women in the gas industry, the first of several *Co-Partners' Magazine* reports of this nature.[4]

War depressed the manufacture and sale of gas appliances such as cookers and water heaters because the use of metal was restricted to military purposes. In light of this diminution of domestic sales, the gas industry's advertising agency, the BCGA, called a truce with the electrical industry, concentrating instead on wartime propaganda. More than ever, they confidently claimed that gas offered middle-class housewives a solution to chronic servant shortages, exacerbated by the growing wartime demand for female workers to replace enlisted men. Equally, gas technology assisted the thousands of women in the workforce, who had less time for domestic chores and childcare. After 1916, anticipating the need for post-war housing, the BCGA promoted its services to builders, architects and municipal councils on the premise that the new 'homes fit for heroes' required gas technology. The gas industry assisted the war effort at several levels: by providing primary materials for arms manufacture, feeding workers and soldiers at canteens and supporting housewives on the domestic front. In addition, the industry looked forward to peacetime prosperity, fully prepared to work in cooperation with public housing authorities and private builders to resolve the housing shortage.

Historians have noted that, by early 1915, the drain in manpower from the factories to the battlefields, coupled with the need for manpower in munitions, gradually created an unprecedented demand for female workers in numerous occupations.[5] With assurances that it was a temporary measure, trade union representatives agreed to support the government's 'Charter for Labour' by relaxing trade practices. The terms of the Treasury Agreement (March 1915) and the Munitions of War Act (July 1915) enabled munitions manufacturers and industrialists to maintain production in essential war work by substituting unskilled male and female workers for skilled men.[6] Given its connection to arms production via chemical manufacture, the gas industry was entitled to meet its labour requirements under the terms of the above agreements.

As of 1915, the lady demons were no longer the only female gas employees; however, women's wartime participation in the gas industry remains largely undocumented, which was possibly due to their relatively small numbers compared to the thousands of women working in other fields, including munitions.[7]

This chapter examines three related areas. First, it considers the experiences of female replacement workers at the gasworks, and their contributions to the war effort. As industrial workers, the female 'gassies' challenged previous definitions of male and female work. Although few women stayed on as outdoor workers after 1918, their success as industrial workers, drivers, fitters and inspectors called into question the gendered notion that women were somehow less mechanically inclined than their male co-workers. Secondly, while the war generated industrial opportunities for women, it also presented new challenges for the lady demons, the second theme of this chapter. As they cooperated with local food committees and government officials on the 'domestic front', the demonstrators further legitimised their professional status within the gas industry, ensuring their public role as spokeswomen for their customers. Thirdly, this chapter explores the gas industry's participation in the reconstruction debates over post-war housing. Maud Adeline Cloudesley Brereton continued her previous campaign for well-serviced, rationally designed houses in conjunction with the BCGA's newly appointed housing consultant, H.H. Creasey. Both Mrs Cloudesley Brereton and Creasey articulated the industry's position in terms of class and gender, championing the cause of working-class women who asked for affordable homes fitted with modern conveniences. Serving the needs of female consumers became more than a matter of good business policy, as customer service was politicised by housing legislation and competition with government-funded electrical plants. This struggle over the domestic market, acted out in the post-war reconstruction debates, set the stage for the interwar rivalry between gas and electricity. Women, as buyers and sellers of domestic technology, were at the very centre of this commercial contest. However, before considering the above subjects it is necessary to outline the general impact of the war and subsequent government policy on the gas industry.

An industry at war

At the outset of the conflict in August 1914, British politicians were optimistic that the fighting would be over in a few months. Businessmen like H.E. Morgan of W.H. Smith booksellers supported this view. His wartime maxim, 'business as usual', implied that the war would soon be won, while politicians and businessmen alike adopted W.H. Smith's slogan in agreement that government interference in the economy was unnecessary. Winston Churchill used this phrase to defend non-intervention, as he remained convinced that the profits of normal enterprise were required to win the war.[8] However, trench warfare was costly in terms of troops, and Lord Kitchener called for an army of 600,000 men. Understaffed and ill-equipped, the War Office struggled with the practical and administrative problems of outfitting and supplying the thousands of new recruits. Amid reports of profiteering and unfilled army contracts, the political leadership realised that, despite the revered principles of laissez-faire liberalism, the current situation necessitated some form of government intervention to organise industry, thus ensuring the availability of military provisions and equipment, including munitions.[9] The mobilisation of labour and industry intensified with the creation of the Ministry of Munitions in late May 1915 by the Asquith coalition government. Moreover, the implementation of conscription, under the Military Service Acts passed in January and May of 1916, served as further proof of the growing subordination of private interests to the national cause.[10]

Despite the initial 'business as usual' maxim and the government's early reluctance to intervene in private enterprise, the gas industry experienced government controls from the very beginning of the conflict. Chemical by-products of coal gas production, particularly toluene and benzene, were essential raw materials in the manufacture of high explosives, notably TNT (trinitrotoluene), the new explosive adopted by the British Army just before the outbreak of the war. Although parliamentary regulations, in place since the late 1860s, already set gas prices and dividend rates payable to stockholders, immediately after the declaration of war with Germany, the Asquith government imposed wider restrictions on the industry. Orders-in-Council sanctioned chemical exports and requisitioned stocks of essential chemicals; later sets of regulations affected gas prices, production operations and labour relations at the gasworks. These regulations, primarily designed to maintain the required chemical supplies for ammunition manufacture, also meant that domestic consumers purchased lower-quality gas at higher prices, produced with the assistance of female replacement workers. With the coming of war, the gas industry became the focus of government attention and one of its partners in arms production.

Although the processes of chemical recovery from coal gas had been first developed in Britain, for several reasons the chemical industry was not well established for commercial purposes at the outset of the war. The curriculum of Britain's elite public schools and universities, notably Oxford and Cambridge, which traditionally favoured classical education over the natural sciences, failed to train and graduate sufficient numbers of qualified chemists and engineers. Government policies, including restrictive patent laws and free trade on imported chemicals further hampered the development of new industries at home. In contrast, the German chemical industry flourished in connection with the artificial dye industry, supported by industrial entrepreneurs. Ironically, although the scientific discoveries may have originated in British laboratories, the German chemists successfully applied these breakthroughs to manufacturing.[11] For this reason, significant amounts of the chemicals extracted from Britain's coal gas were exported to Germany as raw materials for the dye industry and for the production of drugs and photographic chemicals. After 1914, the War Office not only wanted to conserve chemical materials for domestic arms production, it was also anxious to deprive Germany of access to British war materials.[12] The government therefore lost no time in assuming control over essential chemical stocks, thereby building up reserves. Moreover, the export sanctions imposed on the gas industry helped develop later strategies for the control of private enterprise. The prohibition of exports, stock requisitions and compulsory changes in production processes were all strategies initially applied to the gas industry that were later extended to other sectors as the war continued.

The first step in this process of government intervention consisted of export restrictions imposed on essential chemicals. On 5 August 1914, the War Office prohibited the export of munitions materials, including benzene, coal-tar toluene and phenol, under the auspices of the Army (Supply of Food, Forage and Stores) Act.[13] Despite these measures, the manufacturers of explosives complained to the government that speculators withheld chemical supplies, especially toluene. The growing crisis in the supply of raw materials was exacerbated by the realisation that there would be no short and easy victory for the Allies. The munitions industry required increased government supervision to eliminate profiteering and increase the ammunition supplies available for the armed forces. To alleviate the chemical supply problem, in November and December 1914, Lord Moulton's Committee on the Supply of High Explosives issued requisition orders to all gas companies, tar-distillers and coke oven operators. The order was authorised by a clause in the Defence of the Realm (Consolidation) Act (DORA), inserted in late November 1914.[14] W.F. Wintour, the director of Army Contracts, and R.H. Brade of the War Office, informed all gas undertakings distilling their own tar that, for the duration of the war, the government was to assume control of their 'whole output' of toluene stocks and substances containing in excess of 3 per cent toluene. Gas undertakings which distilled their own tar, such as the GLCC and the SMGC, were required to submit returns estimating their current chemical output, and all future dealings were subject to permits issued by Lord Moulton's committee. To stimulate production, the government offered a generous fixed price at three shillings, four shillings, one halfpence per gallon for pure toluene and three shillings for lower-grade commercial toluene because it required an additional distillation process to remove the benzene residues.[15] As a result of these measures, the War Office requisitioned practically all of the gas industry's toluene output for TNT manufacture, leaving a small percentage for the dye industry.[16]

Government controls over chemical stocks and prices were soon followed by recommendations for actual chemical production processes with the object of increasing the levels of pure toluene extracted from coal tar. Commercially, benzene was the most important by-product extracted from coal gas since it was used as motor fuel in England and France. The distilled benzene contained a significant amount of toluene, as much as 10 per cent; its presence improved the quality of the benzene because toluene rendered the liquid fuel less likely to freeze. Given the large volume of domestic benzene production, government officials estimated that nearly one half of the nation's supply of toluene was being lost through the sale of toluene-rich Commercial benzene. In December 1914, Lord Moulton's Committee on the Supply of High Explosives ordered all distillers, including gas companies, to recover this toluene, by placing restrictions on the sale of benzene that contained in excess of 3 per cent toluene. Producers could substitute the toluene with naphtha to lower the freezing point.[17]

Despite these attempts at increasing chemical stocks and regulating supplies, shortages of British toluene continued because domestic production simply could not fulfil the demands of the munitions industry. Munitions factories required enormous quantities of toluene: two tons of toluene went into every three tons of TNT.[18] Purchasing agents for British munitions makers competed for essential chemicals on the American market, forcing up prices and production costs on munitions contracts. In early 1915, the War Office intervened, albeit reluctantly, by purchasing and distributing chemicals to inexperienced contractors at the new shell factories. The shortage of toluene was so critical that, in February 1915, the War Office purchased all toluene available on the American market at the 'ridiculous price' of two dollars per gallon.[19] Throughout the spring of 1915, the continued shortages of essential chemicals were reflected in the shell crisis on the western front; according to military commanders, low ammunition stocks contributed to Britain's dismal showing at the second battle of Ypres and the failed advance at Festubert. Public accusations of incompetence appeared in *The Times* and the *Daily Mail*, seriously undermining Asquith's leadership. The controversy resulted in the Asquith–Bonar law coalition government in late May 1915, followed by the formation of the Ministry of Munitions, in June 1915, headed by David Lloyd George.[20]

Historians have attributed the 'shell crisis' of mid-1915 to inefficient armaments production facilities, manpower shortages and distribution delays. There can be little doubt, however, that the shortage of essential chemicals was another important contributing factor. In the manufacturing process for coal gas, benzene and toluene passed into the gas, as both chemicals improved illuminating power and the calorific value of the final product. Reviewing the returns from Lord Moulton's gas industry surveys, the government estimated that 5,000 tons of toluene per annum escaped with coal gas; as a result, government officials appealed to the gas companies to alter production processes to achieve maximum toluene recovery. Charles Carpenter, director and chief engineer at the SMGC, responded to the emergency by developing a process for synthesising phenol from benzene to produce toluene.[21] Adopting Carpenter's discoveries, the larger gasworks installed special plants for 'scrubbing' the gas through exposure to creosote oil. The process was initiated at 12 of the largest gasworks in March 1915, and, by November, about 600 gas undertakings had converted to the process.[22] The government took on the cost of erecting and operating the special distillation plants, which were run by the gas companies but supervised by inspectors from the government's explosives supply department.[23] After June 1915, the Ministry of Munitions assumed responsibility for monitoring the chemical distillation plants at the gasworks.

Because of the new distillation process, the gasworks doubled their toluene production; however, this was at a cost to the domestic consumer. By late 1915, the chemical shortage was less critical, as a result of the economic incentives to the gas companies, increased domestic production of toluene and organised government supervision, thus remedying the ammunition crisis of the previous spring. The 'scrubbing' process successfully extracted most of the benzene and toluene from coal gas, but the process reduced its candle power by 50 per cent and the calorific value by 4–5 per cent. To maintain a standard of quality for consumers, the explosives supply department temporarily granted the gas companies permission to replace the benzene extracted by the 'scrubbing' process; however, this concession was rescinded when the demand for benzene increased after August 1915. No enrichment substitute could be found in adequate supply, with the result that the quality of gas sold to British consumers fell well below the official standards laid down by the Board of Trade.[24] In order to protect the gas companies, should they be unable to meet pre-existing quality levels, the government issued a circular in November 1915 under the provisions of the DORA, making gas 'scrubbing' mandatory and the final product exempt from previous calorific standards.[25] Moreover, local authorities were requested to overlook the deterioration of gas quality in the interests of 'meeting national needs'. To prevent consumer protests, the government asked municipal officials to stop testing gas for heat and illuminating power, or, at least, to withhold the results of quality control tests from the public.[26] Combined, these regulations effectively extended government control over the production, pricing and marketing of coal-gas chemical by-products, thereby bringing this aspect of the gas industry under direct government supervision.[27]

Despite the usual reluctance on the part of businessmen to accept government intervention in private enterprise, the gas industry conceded to the new regulations. From the outset, the government paid generous fixed prices for requisitioned stocks of essential chemicals, meaning that the gas industry was well compensated and guaranteed a market. As the demand for chemicals increased, government funding subsidised the installation and maintenance costs for the new gas 'scrubbing' facilities. The gas companies appreciated that public intervention enabled them to expand their previously neglected chemical production works and gas managers anticipated the benefits of post-war diversification in the chemical industry (an area that had remained underdeveloped in comparison to Germany). Finally, gas managers were no strangers to government regulations, as gas quality and prices were already regulated before the war by the Board of Trade.[28]

Gas companies readily adapted their gasworks to meet the increased demand for chemical coal-gas by-products. The GLCC directors boasted that they were 'absolutely the first in rendering this national service'.[29] As early as September 1914, their chief research chemist, W. Gordon Adam, issued instructions for expanding the production of 'essential chemicals' to aid munitions manufacture.[30] The shortage of gas chemists or engineers with chemical training put further pressure on the gas companies attempting to cope with new production processes and regulations. It was only in February 1918 that the *Gas Journal* continued to report shortages of experienced chemical workers. Although the chemical laboratories of the major gas undertakings were sufficiently staffed, the smaller companies still desperately needed trained chemists. Under the Munitions of War Amending Act (May 1917) and its Schedule of Protected Occupations, skilled munitions workers were exempt from overseas military service;[31] the list included chemical engineers required in the gas industry. For some months, female replacement workers had made up the shortfall in the chemical testing laboratories. For example, the SMGC journal reported the excellent work being done by 'lady' chemical technicians, many of whom were science graduates.[32] In June 1916, the industry press praised the work of the female technicians, noting their apparent competence at 'doing research and testing work'.[33] Nonetheless, while the government financially assisted the gas companies to expand chemical facilities in order to meet increased demands, politicians could do little about the chronic shortage of experienced chemical technicians required to supervise the chemical works.

Although the war created a ready market for gas by-products by boosting the chemical industry, the overall impact of the war on the gas industry remains difficult to assess. According to some studies, the conflict had a decidedly negative affect on the gas industry. German U-boats targeted gas company colliers, endangering critical coal supplies, destroying company property and killing gas employees. At the end of the war, the GLCC's fleet of 18 colliers, purchased or built since 1912, had been reduced to six. In addition, because of the shortage of materials and labour, works maintenance fell behind, leaving many gasworks producing well below their capacity.[34] To conserve gas and minimise the risk of air raid attacks at night, the government instructed local councils to reduce public lighting, thereby decreasing revenues from municipal gas consumption. Although gas companies received guaranteed prices for their chemical by-products, this revenue was offset by increased production costs, which was then reflected in the prices of coal, oil and labour.[35] Faltering equipment exacerbated rising production costs, which eventually forced the gas industry to increase prices for domestic consumers.[36] In August 1918, the SMGC director, Charles Carpenter, estimated that the cost of gas had risen by approximately 100 per cent from a pre-war price of two shillings and two pence per 1,000 cubic feet, to the 1918 price of four shillings.[37]

Since the 1870s, government legislation had determined the gas companies' prices and profits using a sliding scale set by the Board of Trade. Under this system, the government set a 'standard price' charged to consumers and a corresponding 'standard dividend' paid to shareholders. Any increase in the price of gas, above the standard rate, reduced the allowable dividend; conversely, if gas companies reduced the cost of gas, they might increase their dividend payments. The system ensured that consumers paid reasonable prices for gas services, rewarding companies for careful management while creating incentives to keep prices low.[38] During World War I, given the high production costs and controlled prices for residual products, the gas companies raised their prices above the allowable standard, thereby reducing their dividends to below the standard rate determined by the sliding scale. To address this situation, the gas companies appealed to the government to abandon, temporarily, the sliding scale rates. The resulting Statutory Undertakings (Temporary Increase of Charges) Act, passed in June 1918, authorised companies to increase gas prices to maintain their shareholders' dividends at three-quarters of the pre-war rate.[39]

Consumers complained. For example, in a letter to the *Gas Journal*, J.G. Hawkins of Wisbech accused the gas companies of profiteering, because of their apparent reluctance to reduce shareholders' dividends to offset rising production costs. Instead, the companies applied to the government for permission to raise their prices for domestic gas.[40] Hawkins had legitimate grounds for complaint. At a general meeting of company stockholders in early 1918, the governors of the GLCC reported an available surplus of over £500,000, despite the rising costs of labour, shipping and coal. According to company chairman, John Miles, domestic sales were up by 6 per cent over the previous year. Revenues from the rental of stoves, meters and fittings brought in £36,000; gas sales yielded an additional £541,000. While Miles was obviously pleased by the GLCC's wartime prosperity, he described the situation 'in these times' as 'almost embarrassing', considering the increased charge the company made to domestic consumers that year.[41]

Complaints about rising costs were combined with public concerns over the deterioration of the quality of the gas. According to Mrs C.S. Peel, of the Ministry of Food, some consumers experienced nausea and faintness from 'bad gas'.[42] Still, the Ministry of Munitions exhorted the public to choose gas over coal. Public authorities were asked 'not to test too searchingly the quality of their gas supply', while private customers were advised 'to tolerate small inconveniences', presumably ill-health and expensive fuel, in view of the national emergency.[43] For most of the war, the debased quality of gas remained unchallenged, until February 1918 when the Portsmouth Corporation prosecuted the Portsea Island Gaslight Company for not maintaining its required calorific standard. Fearing additional litigation against other gas companies, the Ministry of Munitions intervened, asking the treasury solicitor to represent the Portsea Island Gaslight Company in its appeal against the court's decision. The Ministry of Munitions wanted to reassure the gas companies that they were not liable for reducing gas standards if engaged in supplying raw materials for armaments.[44] With the government's assistance, the court reversed its ruling against the Portsea Island Gaslight Company, sending a clear message to municipalities and consumers that the gas companies were exempt from pre-war quality controls, leaving customers no recourse. Individual irate consumers, like Mr Hawkins, were branded 'un-English' while gas men dismissed public criticism as another indication that the 'war was getting to everyone's nerves'.[45] In light of the gas industry's sustained profits, despite cost increases, and the addition of new government-subsidised facilities for chemical production, it appears that the large gas companies, like the GLCC and the SMGC, did not fare all that badly during the war.

Gas managers realised that, once the war was over, they faced the enormous task of restoring normal customer service and rebuilding consumer confidence. Perhaps, in an attempt to deflect the mounting criticisms of war-weary consumers, the gas industry drew attention to its most positive contributions. Throughout the war, the industry heavily publicised the role of its female replacement workers. It made extensive contributions to public service work in cooperation with various government departments. In addition, the industry's advertising service, the BCGA, initiated active preparations for the anticipated housing boom when reconstruction finally came, by promising gas stoves and hot water boilers for every British hero and heroine.

Women at the works

In the initial stages of the war, female unemployment increased, particularly in the traditionally female 'sweated trades', like dressmaking and millinery, since the patriotism of wealthy women abstaining from luxury purchases put labouring women out of work.[46] The situation soon reversed. By mid-1915, constant military enlistment resulted in a general labour shortage, which was felt equally in transportation, the civil service and manufacturing, while the demand for supplies created industrial jobs which women increasingly filled.[47] Under the terms of the Treasury Agreement, negotiated in March 1915 between Lloyd George and the participating trade union representatives, the unions agreed to relax restrictive trade practices for the duration of the war, including regulations prohibiting women and unskilled workers from entering some industrial trades.[48] The Munitions of War Act, passed in July 1915, further legitimised the practice of labour 'dilution', substituting women and unskilled men for skilled male workers in essential war industries like munitions.[49]

Because of its relationship to armaments manufacture as a supplier of raw materials, the gas industry easily justified labour dilution. After mid-1915, companies maintained production by replacing enlisted workers with women dilutees until the 'ordinary breadwinners' were no longer 'required by their country'. Initially, the women undertook the 'work for which the special faculties of the sterner sex are not needed', such as the jobs of office clerks, meter readers and coin collectors, inspectors, chemical laboratory technicians, appliance repairmen, delivery van drivers and lamp-lighters.[50] As conscription further depleted the male workforce, women assumed even the heaviest and dirtiest jobs at the gasworks, including loading and hauling coke and coal, and emptying and recharging spent oxide from the gas purification tanks (see Fig. 4.1).

Women were also hired as general outdoor labourers, working as mates to bricklayers, carpenters and pipe fitters.[51] Company general manager, David Milne-Watson, reported that the GLCC relied on approximately 1,600 female workers, distributed evenly among the clerical staff and the industrial labourers at the various gasworks.[52] By late 1917, the SMGC employed 1,915 women; over one-third were retort house and chemical production workers, as these were jobs which director Charles Carpenter considered to be the most physically taxing in the business.[53] These estimates compare favourably with the findings of the War Cabinet's Committee on Women in Industry, whose report states that, in July 1918, some 5,000 women worked in the gas industry, most of whom replaced male workers.[54]

In comparison to the men at the gasworks, female 'gassies' worked the same hours, but usually for only two-thirds of the pay. Like the men, women did regular eight-hour shifts, with an hour break for dinner. They put in the same 48-hour week, doing six continuous shifts, including night work that the women apparently accepted with no complaint. Gas managers claimed that, in terms of actual physical work performed in an eight-hour shift, a female worker accomplished two-thirds the output of a man, thereby entitling her to two-thirds of the pay.[55] Pay scales for piece-rate jobs, like filling and loading sacks of coke, did not differentiate

4.1 Women gas workers removing spent oxide from a gas purifier. South Metropolitan Gas Company, 1917

between male and female workers. For piecework, managers noted that the output and earnings of female workers approximately equalled that of males, despite the accepted wisdom that women did one-third less work than men. A reporter for the *Gas Journal* noted that women preferred time-rates for good reason. When a thoughtful SMGC foreman supplied a female crew with smaller shovels, the women requested the larger ones used by the men so that they could earn the same amount of money.[56] Female gassies were also entitled to the 'war wages' bonus, but received nearly 50 per cent less than their male co-workers. The GLCC paid its men 15 shillings per week in war bonuses while women received eight shillings and six pence.[57] Where women and men worked for the same piece-rates, they earned the same 12½ per cent war bonus, although, in trades other than the gas industry, only male workers over age 21 qualified for the same bonus.[58]

Apparently, the London gas companies attempted to provide for the comfort, safety and hygiene of both its indoor and outdoor female gas workers. According to Mrs J.S. Lee, the women's matron at the GLCC, although the women clerks worked alongside their male counterparts, the company furnished its new female clerical staff with their own rest room where they might leave outdoor clothing and take their meals. Complete with sofas, upholstered chairs and a small library, the rest room for the indoor female workers looked reassuringly like an Edwardian parlour and thereby offered female clerks a domestic, private space within the unfamiliar public sphere of the corporate office. To enhance their skills, female clerical workers were eligible for courses in languages, accountancy and shorthand. Mrs Lee's chief complaint on behalf of the indoor staff was that the women and girls required 'some open space' for a bit of 'much-needed' exercise despite their generally favourable working conditions.[59]

Forewomen supervised the female outdoor workers and they too enjoyed separate cloakrooms and canteens, but without the libraries, language courses and sitting rooms. The GLCC provided 'rational dress' for its female labourers – overalls, smocks and caps – made by staff seamstresses based at the Horseferry Road offices.[60] Obviously, the need for exercise was less of a problem for outdoor workers engaged in physical labour; however, access to fresh air at the gasworks was another matter. One former gas worker, a Miss J. Tewkesbury recalled: 'many is the time the girls would be affected by the gas, the remedy being to walk them up and down in the fresh air, and then drink a bottle of Guinness'.[61] That women were encouraged to take a drink on the job, even for medicinal purposes, suggests a rather liberal attitude on the part of the women's employers and their supervisors at the gasworks. Wartime discussions concerning public canteens stressed their importance to munitions workers as a means of solving the food, drink and housing problems exacerbated by the war. While men were encouraged to dine in the public canteens, because these facilities did not serve alcoholic refreshments, the advocates of public canteens assumed that women workers would naturally shun the pub and would therefore require somewhere respectable and inexpensive to eat. Canteens were often privately operated and staffed by groups that supported temperance, like the YWCA.[62] Employers who valued temperance among their female workforce even monitored the social behaviour of their women workers out of hours, through random inspections of local pubs conducted by female welfare officers. In terms of industrial health and safety, offering a woman a bottle of stout to relieve the ill-effects of coal gas inhalation was an entirely questionable practice and it remains doubtful that Adelaide Anderson, H.M. chief lady inspector of factories, would have approved. However, Miss Tewkesbury's comment does indicate that the gas industry did not necessarily impose on its women workers the types of social restriction that were evident elsewhere.

The GLCC and the SMGC both appointed lady superintendents to oversee their female workforce. These women assumed important managerial functions within their companies, surpassing anything that women had undertaken in the industry before the war. The lady superintendents were responsible for personnel decisions, in terms of assigning women to particular jobs, and were expected to ensure the safety and health of their outdoor women workers. They also represented their employers in public discussions about women and industrial employment. No doubt it was a challenging and stressful job, considering the numbers of women the lady superintendents oversaw and their apparent lack of formal managerial training. For example, in late 1915, due to the 'circumstances of the war', the SMGC created the position of lady superintendent. Miss Fawcett, already an employee of the company, was appointed to the job at an annual salary of £200; she was also nominated to the company's copartnership committee. After holding this position for two years, Fawcett married and left the SMGC, taking with her the generous gift of £500 in company stocks.[63] Her successor, Miss R. Stubbington, the daughter of a gas employee from the SMGC's Greenwich works, was promoted from her job in the engineer's office at Vauxhall to the position of lady superintendent with an annual salary of £250.[64]

Giving evidence before the War Cabinet's Committee on Women in Industry, Stubbington reported on her company's relationship to its female employees. She noted that every forewoman was trained in first aid in case of accidents and in hopes of avoiding injuries on the job. Stubbington and Fawcett assigned women to particular tasks according to their physical ability. To accommodate the women, where possible, the company replaced unnecessarily heavy tools with lighter ones and adjusted wheelbarrows to take some of the weight off the workers' arms. Stubbington noted that, although her company had found that the output of women workers was roughly two-thirds that of men, the quality of the women's work was equal or better, particularly in tasks requiring manual dexterity. The women equalled the men in terms of 'timekeeping' and absence from sickness or other causes. Employee turnover was fairly low: 60 per cent of those women who had started with the company in July 1915 were still working for the SMGC in November 1918. The company estimated that about 50 per cent of its female workers were married, some of them wives of gas employees engaged in military service.[65] The SMGC preferred to hire the female relatives of its enlisted workers in order to give the jobs, when possible, to the families of its workmen.[66]

The fact that many of the SMGC's replacement workers were married and members of gas workers' families may partially explain the stability of the gas companies' female labour force. According to government studies, the wastage of female labour constituted a serious problem, particularly in the munitions industry. In March 1918, Captain M. Greenwood reported to the Health of Munitions Workers Committee that only 10.81 per cent of female munitions workers remained at any one factory for longer than a 12-month period. Even the Cardonald National Projectile Factory, which government investigators praised as a model factory in terms of its welfare provisions for workers, reported that only 13.6 per cent of its female workers employed between 1916 and 1918 had actually stayed in excess of two years.[67] In comparison, the low turnover rate of female gas workers appears all the more remarkable.

Occasionally, the companies' lady superintendents organised recreational events for their outdoor female employees. In August 1917, 120 women employees at the GLCC's tar and ammonia products works embarked on a 'pipe fitters holiday' at Theydon Bois. The works matron, Mrs Thomson, planned games, races and an outdoor meal; the cost for the outing was supported by subscriptions from the works superintendent, staff and foremen.[68] In another show of company spirit, several Beckton Gasworks women played an exhibition cricket match, on the field of the Beckton cricket club, as the Gasworks team met their sisters from the Products Works. Arranged by the lady superintendents of both groups, the match was an unusual display of female athleticism, outside the company's regular sports leagues, which catered for men.[69] This exhibition game indicates that the women were attempting to fill every aspect of the gas worker's job, at the works and on the company cricket pitch.

As women proved themselves capable of even the most laborious tasks, the wartime female gas worker, in sooty smock, clogs and trousers, symbolised the feminisation of industrial work, as much as the 'munitionettes', despite the former group's smaller numbers and lack of publicity. On 29 July 1918, a deputation of female gassies from the GLCC and the SMGC, dressed in their work uniforms, participated in the procession of 2,540 women war workers who marched from Hyde Park to Buckingham Palace for an official presentation to the king and queen.[70] The GLCC delegation consisted of nine lady supervisors, in broad-brimmed hats and overcoats, and 11 female labourers, wearing the standard issued baggy trousers, smock and cap. Commenting on the women's reaction to the public display, Mrs J.S. Lee, the GLCC women's superintendent, stated that the gas women were 'specially pleased to be allowed to take part in this occasion, as they felt they were now recognized as war workers'.[71] Mrs Lee and a Mrs Gibson, from the SMGC, were among the 25 superintendents actually presented to the monarchs, an honour that was reserved for management. Evidently, the women workers responded positively nonetheless, satisfied that their vital war efforts at home had received the official recognition it deserved.

Apparently, the gas companies took into consideration the needs of its female employees, making every attempt to avoid injury and health risks as a result of the heavy physical labour. According to Stubbington's testimony before the War Cabinet's Committee on Women in Industry, in general, the women gas workers experienced improved health despite the difficulty of the work. The lady superintendent noted that Dr Ethel Bentham regularly examined the SMGC's female gas workers for occupational health problems. A prominent Fabian and Labour Party member, Dr Bentham later supported protective legislation for female workers to exclude them from dangerous trades, such as underground mining, and also to prohibit women from undertaking industrial processes that involved heavy lifting, exposure to excessive heat or the cleaning of dangerous machinery.[72] Given the process by which gas was extracted from coal, many female outdoor gas workers would have performed their jobs under the above conditions. However, despite her later support for protective legislation, Dr Bentham reported that her routine inspections of female gas workers failed to indicate that the women developed any health problems as a result of the routine heavy lifting and heat exposure at the works.[73] Similarly, in her report on female replacement workers in the gas industry, Agnes E. Conway, the honorary secretary of the Women's Work Subcommittee of the Imperial War Museum, opined that some tasks at the works, particularly stoking retort furnaces, 'looked appallingly heavy, hot and unsuitable'. Nevertheless, Miss Stubbington assured Conway that 'the girls' were 'extremely healthy' and actually preferred the heavy work to the lighter, cleaner jobs at the works.[74] A final report written by factory inspectors at the end of the war confirms Stubbington's views on the women's health. Contrary to the expectations of the inspectors, employment in the gasworks 'has been extremely popular among the women, and many of those working in the open air processes have declared their health has improved since they entered this occupation'.[75] What exactly constituted unfit conditions for women gas workers became a highly contested issue throughout the war as some companies assigned growing numbers of women to increasingly hazardous and physically demanding jobs.

By 1917, the London gas companies utilised female replacement workers in almost every capacity to maintain production levels. With appropriate tools, clothing and training, gas managers considered few tasks too arduous or dangerous for female workers. David Milne-Watson, managing director of the GLCC, believed that emptying and recharging the gas purifiers was the heaviest job that his company assigned to women. Workers used picks and shovels to break up the spent oxide in the bottom of the purification tanks; the iron oxide was then removed from the tank by shovelling it into containers that were then emptied into wheelbarrows and transported outside to revivify for later use. Milne-Watson also admitted that jobs in the products works plants involved exposure to chemical by-products of gas production, some of which were extremely unpleasant and poisonous, including prussiate and naphthalene. Although the GLCC was willing to expose women workers to toxic substances, Milne-Watson and many of his fellow managers in consultation with the Reserved Occupations Committee still considered women unsuitable for stoking furnaces in the retort houses, apparently because of the searing heat and the heavy lifting.[76] Faced with the critical shortage of male labour in mid-1917, however, and in preparation for the upcoming winter season, the SMGC challenged this preconception by experimenting with female stokers. That summer, Thomas Curr, the chief engineer of the Vauxhall works, brought in female crews to operate the stoking equipment at three of the four retort houses at this South London works, thereby proving that women were capable of undertaking every task that manufacturing coal gas entailed.[77]

The SMGC stressed the importance of proper training to develop the women's skills and confidence in the retort house. Carbonising foreman, J. Strutt, implemented a three-week programme designed to acclimatise the women to the retort houses while providing instruction in the various tasks, beginning with the least skilled (forking up coke) and ending with the most difficult task (operating the hydraulic coal chargers). Trainees worked partial shifts, observing skilled female workers in action, before permanently joining a crew. A trained forewoman headed every crew. Strutt described the women as 'apt pupils' who acquired the necessary skills with astonishing rapidity.[78] Moreover, the company praised the women for their overall stamina and courage. Strutt reported that the female stokers 'carry on' even during air raids, 'replacing their caps by steel helmets'.[79] After a raid, the crews remained until the next teams arrived, and, under some circumstances, the women worked double shifts.[80] During an influenza outbreak in October 1918, the normally five-person retort crews at the Greenwich works were reduced to groups of three; however, apparently, the women managed remarkably.[81] Although the company recruited its retort women from all occupations, the majority were former 'laundry girls' between the ages of 18 and 28.[82] Possibly, Strutt and Fawcett favoured this occupational group because they were already accustomed to heavy work under hot conditions. Despite the demands of the job, retort women supposedly enjoyed good health while, apparently, the examining physicians, Dr Ethel Bentham and Dr Margaret Stirling, failed to produce physical evidence that the work was detrimental.[83] While the women's improved constitutions may be attributed to the constant exercise, adequate nourishment must also have been a factor. Even with the long hours and hard work, studies indicate that women's overall health improved during the war years because working women had more money to spend on food for themselves, while company canteens offered regular affordable meals.[84] Women gas stokers had access to hot food and drinks, like tea, coffee and mineral water, at a company canteen which catered exclusively to the needs of the SMGC's female gas workers.[85] Although the provision of canteens for workers was a wartime innovation aimed at improving workers' efficiency, some of the larger gasworks, like the GLCC and the SMGC, had provided workers' canteens since the late nineteenth century.[86] However, facilities designed specifically for female gas workers were products of female war work.

As an example to other companies, the SMGC's experiment with female stokers received widespread publicity in the gas trade press. Company directors invited gas managers from other companies to visit the Vauxhall works and witness for themselves their female crews in operation. In preparation for the winter of 1918, and at the request of the dilution department of the Ministry of Labour, the SMGC extended its training programme at its Vauxhall works to accommodate women gassies from other companies. Several gas corporations took advantage of this service, including Sheffield, Newcastle, Richmond, Woking and Birmingham.[87] The SMGC formed a special crew of female stokers, the 'Women Operative Gas Demonstrators', who were available to be sent anywhere in the country to train and assist local retort women. Vauxhall's crew of female stokers created an entirely new type of lady demon. The WOGD worked in close cooperation with the dilution department. The Ministry of Labour paid the salaries for this special team while a Mrs Pennington, a female dilution officer, spent a month herself at the Vauxhall works observing the women retort workers.[88] Pennington then visited gas companies throughout the country, urging them to adopt the use of female stokers, thereby overcoming the critical labour shortages and releasing men for the armed forces. In Figure 4.2, a female retort house worker with the SMGC pauses during a shift. While the young worker's stance indicates that she was posing for the camera, she wears the heavy leather gloves, wooden clogs and woollen sweater that were mandatory protective clothing issued to retort house women. Gas workers preferred the traditional wooden clogs, as they were more durable and heat resistant than leather-soled boots, and the women stokers adopted the same uniform as their male counterparts.

Government committees investigated the possibility of extending the use of female employment in the retort houses. Early in 1918, representatives from the National Gas Council, an organisation of gas managers, met with members of the Reserved Occupations Committee, the Ministry of Munitions and Neville Chamberlain, the director of National Service,

4.2 A female retort house worker with the South Metropolitan Gas Company, 1917. Note the protective work clothes including the heavy leather gloves and wooden clogs, favoured by stokers because of their heat resistance and durability

to determine if more men could be spared from the gas industry for active military duty. As a result of this meeting, the BCGA published a list of guidelines for gas managers, listing suitable industrial and indoor occupations for women at the works. Regarding retort house work, the report suggested that women should not perform the most dangerous and difficult tasks, like operating 'hand charged' retorts, filling and cleaning the coal furnaces, or clearing the ascension pipes through which the gas escaped from the oxidised coal. However, the report concluded that women were entirely capable of working, charging and discharging coal retorts fitted with electrically operated hydraulic equipment.[89] Throughout that winter, 1917–18, the SMGC successfully used female crews on hydraulically fitted retorts. As yet, it was the only company to do so.

Not all government officials were as willing to sanction the use of women for increasingly difficult industrial labour. Giving evidence before the War Cabinet's Committee on Women in Industry, Miss Adelaide Anderson, HM chief inspector for women industrial workers, noted several distinct phases in terms of wartime female labour recruitment. She described the initial growth of female workers in areas that normally employed women, which was followed by women's substitution for men in new occupations to release eligible males for military service. The third stage involved the concentration of women in repetitive engineering and munitions work. Stage four, the most critical in Anderson's view, occurred after 1917 when women in greater numbers were pressured to substitute men in industrial processes. According to Anderson, women were expected to undertake jobs that were too heavy for them, including employment at gasworks and in retort houses. In Anderson's view, as a result of this 'last and striking change', employers had pushed their female workers 'to the utmost limits of what we [factory inspectors] thought it was possible for them to do'.[90] Evidently Anderson upheld the conviction that stoking retorts, by hand or with mechanical pulleys, was still expecting too much from women workers. A 1919 report published by the Home Office similarly condemned the use of female labour in the manufacturing processes at gasworks because the work was laborious and dirty and also involved working during the night and on Sundays. The report conceded that women were admirably suited to less physical jobs in the district workshops: for example, doing meter maintenance and repair, inspection work or employment in the company laboratories.[91] The Home Office investigators were most critical on the issue of women undertaking work in the retort houses, a contentious issue throughout the war and up until the spring of 1919.

Gas managers readily acknowledged the existing success of female labourers at the gasworks, albeit in performing unskilled or repetitious industrial tasks.[92] Stoking retort houses was another matter. It required not only strength and endurance, but also mechanical skill to operate the heavy hydraulic equipment that was used to load and discharge the coal ovens. A retort worker performed several specialised tasks, undertaken in quick succession, as a member of a four-man crew, each responsible for different sets of duties. Experienced stokers were agile and strong, with good manual dexterity and methodical efficiency. The majority of gas engineers and male stokers were convinced that women were inherently unfit for retort work, because it was hot, tiring and dangerous.[93] Although they refused to give the exact reasons for this assumption, possibly the workmen and managers believed that women were unable to master the required mechanical skills to load and discharge the retort ovens, could not work collectively in crews, and would collapse under the strain of the eight-hour shifts.

It was also a matter of gender. Female stokers infringed on the last male preserve in the gas industry. Their presence threatened to erode a traditional hierarchy of job skills that was inherently masculine and political. Since the mid-nineteenth century, stokers were the most militant labour activists in the gas business. As noted in Chapter 2, they had organised illegal strikes and had formed the first gas workers union in August 1889 under the leadership of stoker Will Thorne, while the SMGC stokers had been blacklisted for their union activities.[94] Stokers were among the highest paid labourers at the gasworks, and took great pride in being the physical backbone of the industry. Male stokers were not readily inclined to surrender their jobs, their occupational status or their wages to female dilutees, even during wartime labour shortages. It was no coincidence that the first company to test the militancy of the stokers by hiring female retort workers was the SMGC, a company traditionally known for its antipathy to trade unionism. Although the SMGC's gas stokers raised no objections, the introduction of female stokers drew negative reactions in other quarters.

The members of the National Union of General Workers were 'strongly opposed' to the introduction of women into retort houses because women were 'unsuitable' for the job. In early August 1918, Labour MP Will Thorne, the current secretary of NUGW, publicly denounced the use of female stokers before the House of Commons, despite his patriotism and general support for the terms of the Treasury Agreement. In response, Arthur Kellaway, the Minister of Munitions, defended the gas industry's efforts to overcome the shortage of male workers, thereby maintaining a level of productivity that was vital to the munitions industry. Further, he praised their successful training programmes for female stokers.[95]

The chief dilution officer at the Ministry of Munitions urged gas companies to use the slower summer months to train women stokers and avoid the labour shortages of the previous winter. Having witnessed the success of the SMGC with female retort crews, the Ministry of Munitions circulated a letter among gas companies. It stated that future company requests for the exemption of male gas workers from military service or employers' demands for replacement army reserve munitions workers would be subjected to stringent review. The industry's labour demands must be met 'by the employment of women' unless managers could demonstrate that every effort failed to secure sufficient female replacement workers.[96] Nevertheless, Kellaway wished to avoid confrontation with male gas stokers that might result in possible work stoppages. To alleviate tensions and assure the men that the use of female stokers was an emergency measure, he proposed consultations with workmen's organisations in the various localities where the introduction of women retort house workers was proposed or had already been adopted. Some corporations, like the GLCC, wished to avoid labour confrontations with stoking crews by respecting the trade union's objections to female stokers. Although women were used to clean and maintain retorts, the GLCC did not use female stoking crews, because, according to general manager David Milne-Watson, 'women were not suited for this type of work even in wartime'.[97] Adelaide Anderson, who later expressed her misgivings about female retort crews, would no doubt have approved of Milne-Watson's restraint.

In most cases, the consultation process successfully diffused labour protests. Initially, female stokers met with resistance in both Sheffield and Birmingham. At the Neepsend works, Sheffield, the company superintendent, Mr Johns, and lady superintendent, Mrs Cockshot, introduced 70 women stokers in August 1918. Of these female stokers, 30 received their training at Vauxhall, as did Mrs Cockshot.[98] Protests by male stokers were immediate. To address the situation quickly, the company called a meeting between the workers, company managers and government dilution officers. To convince the men that the company had no ulterior motives but to maintain production until 'the boys came home', three Sheffield stokers accompanied their local NUGW secretary, Mr Blackburn, to London for the purpose of touring the Vauxhall works. Reassured that the measure was temporary, the union representatives returned to Sheffield in support of the women stokers.[99]

Birmingham presented a similar case. When the issue of dilution in retort houses was broached, the municipal gas corporation sent a deputation to London to see the Vauxhall works, resulting in a signed agreement between the gas corporation and the Municipal Workers' Union outlining a policy for experimenting with dilution at the Nechells works retort houses. When the women arrived on site, however, some 25 men downed their tools in protest, despite the earlier agreement of cooperation. The men convened a meeting and passed a resolution that the policy of dilution, which the men had already agreed upon, must be honoured. The striking stokers returned to work with no penalty, provided that they agreed to assist the women as previously negotiated.[100] The implementation of women stokers aroused less resentment at Newcastle, Richmond, Bournemouth, Prescott and Woking. Several explanations may account for this. Cooperation may have been related to the impact of positive industrial relations between men and management. In the example of Woking, whose gasworks were described as 'destitute of men' since the beginning of the war, there were few men present to protest against the use of female stokers while the need for replacement workers was obvious.[101]

After the Armistice was signed in November 1918, women remained at the gasworks throughout the winter of 1918–19. Although the munitions factories were already dismissing their male and female workers, gas managers were reluctant to replace skilled female gassies with unskilled redundant munitions men, many of whom were medically unfit for gas work. Managers considered wounded soldiers and sailors returning from Europe equally unsuitable for the physical demands of the gasworks. Although gas managers requested the immediate release of 'trained stokers from the Colours', they anticipated a continued shortage of skilled gas workers throughout the coming winter, a shortage which could only be met by keeping on female substitutes, and even extending the use of female stokers.[102] However, the gas companies reassured former employees that their jobs were waiting in accordance with the terms of the Treasury Agreement and the Munitions of War Act of 1915. The SMGC issued the statement that 'each man who wishes to come back to his work will find his job, or one at least as good, ready for him'. The female replacement workers who faced unemployment understood the circumstances. Nevertheless, the directors expected hundreds of job vacancies, offering first consideration to those workers who had efficiently 'carried on' during the war. Praising the female stokers for their 'admirable courage and determination', the SMGC's director anticipated optimistically that these workers would find 'softer jobs in the great world of industry'.[103] Despite their successful pioneer efforts, the women stokers were a temporary institution, but one that worried male trade unionists all the same.

Throughout the war, unions and workmen expressed a variety of responses to increased female employment in industry. Some industrial unions, particularly those representing unskilled and poorly organised groups, claimed to support the notion of workers' equality, regardless of sex. In contrast, the skilled workers' unions, especially in the aircraft and engineering trades, voiced the loudest protests against women workers, fearing that female substitution would depress wages, undermine job status, and create post-war unemployment for men.[104] The NUGW, which represented both male and female gas workers, was generally supportive of female workers, provided that they did not threaten male wages or job status. Historically, the NUGW was one of the first general unions to organise and admit unskilled female workers.[105] Giving evidence before the War Cabinet's Committee on Women in Industry, union representative John Jones estimated that, in 1918, the NUGW represented 350,000 workers; 60,000 were women, of whom 40,000 had joined during the war.[106] Jones reported that the union was not hostile to the employment of women in industry. Its membership advocated a policy of equal pay for equal work, disagreeing with the stipulation that an equal output must be produced. According to Jones, there was no such stipulation in the case of men who also differed in individual output. However, Jones criticised the use of female workers in trades 'for which they are unsuited' and he included in this list of unsuitable occupations the use of women in carbonisation works, in chemical products works and in the gas retort houses. He gave no specific reasons for female exclusion from these jobs, but simply stated that the introduction of women must not depress wages.[107] In this respect, Jones and the NUGW offered the same criticisms of female workers as did the more specialised unions that excluded female membership altogether.[108]

Jones made special reference to the gas stokers and the impact of female labour on their jobs. Prior to the war, many NUGW members were naval reservists employed as stokers. With the war, stokers were called up to fight and their positions were gradually taken over by women who worked for two-thirds of the pay at six shillings per eight-hour shift. Jones predicted 'very serious trouble' when the men came back, even though he admitted the existence of a post-war shortage of skilled male labour for retort house work. He worried that gas managers would overcome the labour shortage by retaining their wartime female crews in the retort houses. In a telling remark, which indicates the strategic importance of the equal pay policy, Jones stated that 'equal pay alone will not prevent the employment of women on unsuitable work', such as stoking gas retorts.[109] Jones favoured a 'uniform minimum wage' for unskilled men and women and the implementation of a gender-specific employment code set by joint industrial councils to designate occupations as either male or female work. Although Jones admitted that the women stokers unanimously agreed that returning men should have their old jobs back, he failed to comment on the mounting pressure being put on the gas industry to replace their skilled female stokers with unskilled male workers recently released from munitions work.

Jones's testimony reflected the wartime debates over work, equal pay and gender equality.[110] The NUGW was outwardly supportive of its female members, advocating equal pay for equal work. This ensured that women doing men's work, like the female gassies, obtained adequate reward for their labour without depressing the wages of skilled workmen. However, NUGW leaders anticipated that equal pay for equal work would discourage female industrial employment after the war because women workers would appear less attractive in comparison to men. Moreover, the NUGW's support for designated male and female occupations suggests a definite strategy to keep women out of certain trades. Jones's specific example of the gas stokers indicates that, despite the apparent cooperation between management, male stokers and trade unionists, and the small number of companies that relied on female stokers, the NUGW still feared that this example of women doing 'men's work' permanently undermined their wages and job status. Will Thorne used these same arguments when he protested to Arthur Kellaway, in the House of Commons, against the introduction of female stokers. The continued presence of women at the gasworks after the Armistice presented further cause for concern. Unlike the munitionettes, whose jobs began to disappear in mid-1918 even before the war ended,[111] the female gas workers awaited a more gradual displacement as a result of military demobilisation. Despite the women's relatively small numbers (for example, the SMGC employed approximately 1,800 women in outdoor work and, of these, less than 400 did retort house work) the trade unionists considered women gas workers to be a dangerous precedent.[112] In addition, the increased use of hydraulic equipment further undermined arguments that women were unsuitable for employment in heavy industry.

In reality, Jones and the NUGW had little to fear. In compliance with the law, gas managers reassured the government and the NUGW that men returning from the war had jobs waiting for them. Before the same committee that, heard Jones, Albert Stokes, of the SMGC, explained that, despite its satisfaction with women stokers, his company fully intended to disband the female crews as soon as skilled male replacements could be found. Nevertheless, he offered no definite timetable. Stokes agreed with John Jones that retort houses were not places for women. Restrictions on female employment came under the Restoration of the Pre-war Practices Act, passed in August 1919, which compelled employers to release 'unskilled' replacement workers or face fines levied by tribunals.[113] Although the companies appeared in no immediate haste to replace competent female gassies with untrained men, particularly for the winter season of 1918–19, by mid-1919, the crews of female stokers had been disbanded.

Long after the war, both the SMGC and the GLCC retained some of their female employees, albeit in more appropriate 'peacetime jobs' such as lighter labour, clerical jobs, technical duties in chemical laboratories and meter index reading.[114] Outdoor women workers disappeared from the works, but not the general offices, showrooms or testing laboratories. The displaced workers were not the unionised male gassies, but the male indoor staff who witnessed the post-war feminisation of clerical work and the entrance of women into light industry.[115] As of 1918, the lady demons were no longer the only women employed by the gas industry. Yet, they retained an elevated status, as the first women's department. Throughout the war years, as the gas industry dedicated its services to the war effort at home, the lady demons widened their sphere of influence beyond the showrooms and the customers' kitchens.

Women on the 'domestic front'

In terms of the lady demons' career development, the Edwardian years were characterised by the expansion of professional opportunities within the gas industry. During World War I, the gas industry's domestic experts undertook public service work to assist women in the home, thereby extending their professional authority. In some respects, the staff members of the BCGA, particularly Mrs Maud Adeline Cloudesley Brereton, deserve credit for this last development. Mrs Cloudesley Brereton continued as editor of the BCGA publicity journals throughout the war. Since coming to work for the BCGA in 1911, she had anticipated that women would have a much larger role to play in both the gas industry and public service. Her recommendations were soon put to the test, not necessarily because gas managers intentionally created more opportunities for women, but because of the political events of August 1914.

The lady demons, known after 1912 as home service advisors, were enlisted to fight the war on the home front. They helped women to cope with the daily stresses of food shortages, fuel scarcities, rising prices and servant difficulties. As public relations officers, the home service women unofficially represented the government by disseminating information for the Ministry of Food, the Board of Education, and the Board of Agriculture's Department of Food Production. Officially, they represented their own industry, literally putting a positive public face on the gas industry's war work. As the war dragged on, and consumers grew impatient with high gas prices and poor quality, the home service advisors' task became even more strategic for the companies as they maintained a buffer between consumers and producers in the face of growing public dissatisfaction.

David Lloyd George's coalition government, formed in December 1916, led to the creation of five new departments designed to take control of domestic problems related to the war. Food and coal shortages prompted the organisation of the Ministry of Food in late December 1916, with

Lord Devonport, the ennobled retailer, appointed food controller. Despite severe shortages of butter, sugar, bread and meat, the government did not favour official rationing, apparently because of the anticipated problems of administering the programme. Instead, Lord Devonport's department made recommendations to the public to practise 'voluntary rationing', to grow vegetables in allotments, to limit restaurant dinners to three courses, and to avoid rice-throwing at weddings. However well intentioned, such measures failed to curb profiteering and hoarding or to eliminate shop queues, the serious problems of wartime food distribution. Housewives complained about the chronic shortages and escalating costs of food staples like potatoes, wheat, sugar and margarine. In some cases, the food controller intervened by imposing 'price controls', but, in the absence of systems for enforcement and distribution, the maximum price guides were simply ignored, or circumvented by food retailers and producers. During the potato shortage of early 1917, farmers evaded the Ministry of Food's price controls by overcharging for additional retail services, like 'carting'.[116]

Food shortages were exacerbated by German U-boat attacks on British ships. Public frustration bordered on panic over the feared depletion of food supplies.[117] Liberal journalist G.K. Chesterton worried that an exaggerated preoccupation with food shortages created a famine frenzy on the home front, sending a disturbing message to the Allies and a heartening one to the enemy.[118] Labour unrest and workmen's walkouts, including the nation-wide engineers' strike of May 1917, coincided with the food shortages.[119] Mounting public discontent over dilution and conscription was exacerbated by the government's inability to solve the nation's food problems.[120] Ostensibly, for reasons of poor health, Lord Devonport resigned from his post in May 1917.

The new food controller, Lord Rhondda, a coal magnate who served on the Local Government Board, assumed more direct interventionist policies, expanding his staff to implement the new regulations. Under his control, the Ministry of Food stabilised prices by issuing subsidies to bakeries and by setting and enforcing rigid price controls on staples, including milk, sugar and bread. In addition, under Lord Rhondda's control, the Ministry of Food consulted with consumer groups, notably the War Emergency Workers' National Committee (WEWNC). A workers' advocacy organisation originally formed to concentrate on industrial relations, the WEWNC also focused on workers' living conditions outside the workplace; for example, they monitored the prices of food, rent and coal, and put pressure on the government to assume more responsibility for the pricing and distribution of food supplies.[121] According to Margaret Barnett, Lord Rhondda was anxious to reassure the labour movement that he was a consumer advocate, despite his long-term support for free trade practices. An amendment to the DORA, in June 1917, gave the food controller the power to requisition food stocks from suppliers. This measure was followed by a series of 'maximum prices orders' and attempts to limit the wholesale profits on the sale of meat, butter and cheese.[122] Unfortunately, price controls disrupted systems of distribution. As Barnett notes, once retailers were limited by what they could charge, merchants at a distance from distribution centres could not compete against those with lower transportation costs. Throughout 1917, severe shortages continued in rural areas and munitions centres, renewing consumer protests and labour discontent.[123] Finally, in January 1918, Lord Rhondda's Ministry of Food introduced ration coupons for some items, such as sugar, meat and margarine. Moreover, he empowered food committees to extend local rationing, where necessary, for tea, cheese and jam. By July 1918, the rationing programme overcame the worst defects of the distribution system, eliminated the queues and soothed the labour unrest associated with food shortages.[124]

Several accounts of wartime food rationing bestow credit for its success on Lord Rhondda and his staff. They compare the failed Devonport policies for 'voluntary rationing' and 'meatless days' with those of his successor, asserting that Rhondda's team of civil servants brought the food crisis under control by fixing prices and directly purchasing and requisitioning food from manufacturers, importers and distributors.[125] This interpretation relies heavily on evidence provided by the Ministry of Food itself, particularly the writings of William Beveridge, who acted as undersecretary to the Ministry of Food's permanent secretary, Ulick Wintour.[126] In contrast, other reports of wartime food policy suggest that Lord Rhondda's staff responded to mounting public pressure from labour and consumer groups, like the WEWNC, and was not without its own share of administrative bungling. Lord Rhondda's staff finally introduced official rationing after several municipalities, Birmingham being the largest, implemented local ration programmes to solve their own distribution problems. The creation of a consumers' council, an advisory body to the food controller that reported on the practical impact of food policies, provides further evidence that the Ministry of Food implemented official rationing to satisfy the insistent demands of ordinary consumers and workers.[127]

The wartime preoccupation with food supplies and the rising cost of living cast new importance on the promotional work of the gas industry. BCGA advertising reflected the themes of official propaganda by imparting images of national service, fuel conservation and individual sacrifice for the greater cause. A 1915 BCGA advertisement in *Punch* ran the slogan 'Better Economise in Fuel than in Food' beneath an illustration of two uniformed men sitting down to eat, under the approving gaze of a female domestic (see Fig. 4.3). The implication was that fuel-efficient gas cookers constituted the patriotic alternative to burning expensive coal and that the men in uniform required the proper fuel and nourishing well-cooked food to keep them in good fighting form.[128] Another wartime advertisement connected gas fires to health, hygiene and physical recovery (see Fig. 4.4). Under the heading, 'Gas Appliances as an Aid to Good Nursing', an anxious uniformed nurse watches a solitary inert man, the wounded soldier, recuperating at home or in a private hospital. The text suggests the various ways in which gas appliances add valuable comfort to the sickroom, thereby facilitating the patient's full recovery and the eventual return to the front, or, more optimistically, to normal life.[129] Both advertisements identified the gas industry with the war effort and presented women in active roles, either preparing men for battle or hastening their recovery afterwards. By showing women in domestic support roles, fighting the war at home by aiding men before and after the fighting, gas advertising reflected the official propaganda of women at war, working behind the scenes but in a critical capacity. Following this same theme, the BCGA publications featured articles on women's war contributions. Praise for female patriotism was combined with descriptions of workers' canteens and hospital kitchens serviced with gas.

Not surprisingly, given its pre-war interest in attracting female consumers, the gas industry addressed the wartime difficulties that many housewives experienced. Mrs Cloudesley Brereton advised middle-class housewives how to manage comfortably despite the shortages of coal, food and domestic help. The BCGA endorsed the recommendations of the Coke and Coal Supplies Committee, a government group appointed in January 1916 which sought ways of limiting coal consumption to essential purposes. For obvious reasons, the government's public appeals to substitute gas and coke for coal pleased the gas industry. Gas managers and publicists were less enthusiastic about the Coal Conservation Committee's report that the coal supply, if 'properly utilized' was capable of producing three times the current fuel output.[130] To impress on consumers the national importance of using gas, to save coal and produce munitions, the BCGA and the industry press reiterated the official propaganda of the food controller, warning housewives that 'the war must be won in the homes of our people'.[131]

The shortage of domestic help represented another wartime frustration for some segments of the population. The demand for female workers in factories and offices meant that fewer numbers entered domestic service, because of the low pay, irregular hours and limitations on personal freedom that characterised this increasingly unpopular occupation.[132] In recognition of this, Mrs Cloudesley Brereton published a book on gas and the one-maid house, offering 'many valuable hints

4.3 'Good Cooking for Good Men', British Commercial Gas Association

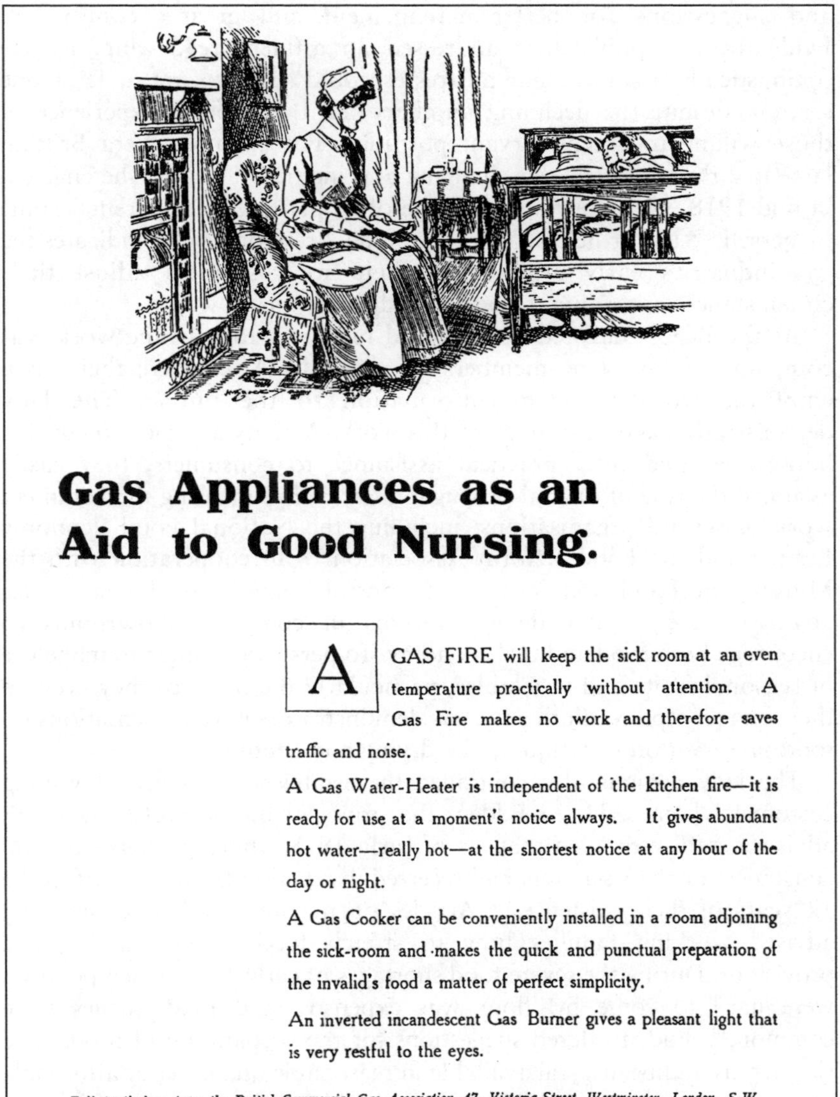

4.4 'Gas Appliances as an Aid to Good Nursing', British Commercial Gas Association

and suggestions for better management and greater comfort'.[133] Evidently, this publication addressed an affluent readership, as she optimistically assumed that mistresses could still procure at least one servant, despite the declining numbers and the dubious experience of those willing to accept servant positions. For example, Vera Brittain lamented the series of 'ever-changing and inefficient maids' she engaged in mid-1918, with the result that she often managed her parents' home by herself.[134] Nevertheless, Mrs Cloudesley Brereton's book indicates the gas industry's early attempts to help its customers adjust their circumstances according to the social dislocations of war.

As the BCGA devoted its staff and resources to wartime work, gas companies who were members of the BCGA enlisted their own employees to disseminate information to the public. The lady demonstrators assumed much of this work. Already assigned to educate housewives and offer practical assistance to consumers, they easily assumed the role of official propagandists, complementing the volunteer work of several organisations, including the National Food Economy League and the Food Reform Association.[135] In cooperation with the Ministry of Food and local food control committees, home service advisors gave regular demonstrations in company showrooms to encourage both food and fuel economy. To persuade women to make use of seasonal fruits and vegetables, particularly the produce they grew in their own company allotments, gas demonstrators gave presentations on food preservation techniques, like drying and canning.

The lady demons also continued the established practice of visiting customers' homes. Helen Edden, the principal of the GLCC women's advisory staff, estimated that, as of early 1918, approximately 130,000 consumers or their servants had received instruction from her staff in the 12 years of their existence.[136] Articles on nutrition and food rationing advised working families how to stretch food budgets and scarce provisions. During the severe food shortages of early 1917, when potatoes were hard to come by, flour was expensive and food queues were common,[137] Edden offered suggestions for the preparation of foods, like rice, lentils and beans, still available in reasonable quantities at affordable prices. One reader warned Edden that to publicise the economy of any one food was to ensure that it either disappeared from store shelves or dramatically rose in price. Nevertheless, she maintained it was worth the risk of price hikes to call attention to the nutritional and economic value of substituting grains and legumes for potatoes and wheat, reasoning that housewives were unfamiliar with these foods and had requested information on their preparation.[138] To economise on cooking time and fuel, of particular importance to working women, Edden recommended the use of the homemade 'hay-box', a makeshift slow-cooker, for finishing dishes like oatmeal porridge and stews.[139] As Lord Devonport's staff preached the importance of the home front, the lady demons' routine assumed an overtly patriotic dimension.[140] On behalf of the gas companies, they worked directly with British housewives and local food committees, passing on official propaganda as a way of trying to ease the mounting tensions over food and fuel shortages. Their community-based wartime work addressed the needs and concerns of ordinary housewives. In this respect, the home service advisors extended their public relations role from corporate representatives to community educators, working directly with women in their own districts.

Unfortunately, there is little evidence to suggest how successfully the demonstrators transmitted the official 'domestic front' propaganda. Previous studies have shown that, once in place, the Ministry of Food's official rationing programme curbed the worst problems of wartime food shortages by stabilising prices and ensuring fair distribution.[141] The home service advisors played an unacknowledged role in this process. Already a familiar liaison group who represented both producers and consumers, the lady demons took on the added role of government representatives. How useful their recipes were for one-pot meals of boiled butter beans, or warming-pan 'hay-boxes' remains anyone's guess, just as it is impossible to judge how many housewives actually attended their showroom demonstrations. According to Martin Pugh, housewives took little notice of the food propaganda while 'its chief function was to give status and employment to those who purveyed it'.[142] While this statement may hold some truth, it does not apply equally to all the groups concerned with food supplies. Edden criticised the recipes 'for artful concoctions of strange comestibles' that appeared in the women's pages of newspapers. As a professional home economist, Edden believed that part of her job, and that of her staff, was to counteract this barrage of bad advice with rational 'truly economical' plans for wartime domestic management.[143] With this in mind, the home service advisors effectively extended their official mandate, their duties and their professional reputations beyond that of the first lady demons who had drawn curious crowds at gas exhibitions some 30 years earlier.

The gas industry's publicity organisation assumed the role of domestic propagandists, complementing the work of the demonstrators. As noted above, the BCGA wartime advertising illustrated the gas industry's contributions to the war effort, fighting on the home front by producing essential chemicals for munitions and maintaining services for domestic consumers. BCGA executive chairman, Francis Goodenough, hoped that, by stressing the industry's war service work, consumers would overlook rising prices and lower fuel quality. Money spent on wartime propaganda was an investment in post-war sales.[144] But the advertising was more than crass self-promotion, and the BCGA used its well-organised publicity services to educate its employees and customers on domestic issues, including the national importance of food and fuel conservation.

In some instances, the gas industry's food campaigns pre-dated similar plans that were later adopted by the government. The 'National Food Preservation Campaign' provides one such example. In the summer of 1917, before the formal implementation of rationing or government-sponsored food preservation schemes, the BCGA published information on food preservation and home canning in an effort to help consumers preserve that year's abundant harvest.[145] Gas companies were particularly interested in food preservation; their employees were encouraged to supplement food scarcities by growing their own produce on company allotments. Unfortunately, the industry's attempts to initiate local food preservation projects failed to prevent much of that year's bountiful fruit crop going to waste or glutting urban markets. After conducting its own research on food preservation, in November 1917, the Ministry of Food and the Board of Agriculture's Department of Food Production officially launched their own plans for home canning aimed at the following year's growing season.[146] Possibly to promote the adoption of home canning, Labour MP J.R. Clynes, parliamentary secretary to the Ministry of Food, suggested that civilians would soon be unable to purchase jams and preserves since that year's stocks were requisitioned for the armed forces. Housewives had to produce their own preserves, or go without.[147] The government officials responsible for the project enlisted the BCGA to assist with public education through its extensive publicity network.

The BCGA readily joined the government campaign. To promote the patriotic need for food preservation, the staff designed a poster for the government, donating 5,000 copies of the 7,500 that were printed. In addition to the poster, the BCGA re-issued 30,000 copies of the previous year's BCGA pamphlets on home canning for distribution through local food committees; its staff also circulated over half a million government leaflets and booklets on food preservation. In addition to the distribution of official propaganda, the BCGA staff acted in an advisory capacity to the Ministry of Food by supplying instructions on food preservation for future government publications.[148] Stressing the nutritional importance of fruit and vegetables as a substitute for meat, the BCGA literature offered technical instructions for operating home canning equipment using the 'sugarless methods' of bottling and drying. To distribute the costs of the required equipment, the BCGA suggested that consumers form cooperative 'canning clubs' or that grocers purchase canners to lend to their customers.[149] The BCGA answered letters from consumers and cooperated with the education committees of county councils. Directly involved with consumers at the local level, it sought solutions to the wartime food crisis by initiating projects like food preservation. After the government launched its own food preservation campaign in late 1917, BCGA staff became unofficial information officers, distributing literature and answering customer queries on the government's behalf.

It was not an entirely satisfactory situation for the gas industry publicists. Although the BCGA endorsed the government food preservation programme, the executive objected to the 'red tape regulations' they encountered in coordinating a programme that answered to three separate agencies: the Board of Education, the Board of Agriculture' Department of Food Production, and the Ministry of Food.[150] Apparently, the Board of Education published information on food economy and home gardening, the Department of Food Production staff promoted 'canning and bottling', while the Ministry of Food handled the information on food drying. Frustrated consumers seeking assistance directly from government departments wrote letters to the BCGA complaining that their enquiries never seemed to reach the appropriate government office. Gas managers and lady demons offered solutions to the information problems by working with secretaries of agriculture sub-committees, arranging local exhibitions and lectures and setting up government information bureaus in company showrooms.[151]

While cooperating with three government agencies on a single project, the BCGA also complained that the Department of Food Production was entirely out of touch with the needs of average housewives. For its food preservation campaign, the Department of Food Production promoted a single canning apparatus, the 'Royal Home Canner', even though the device was far too large and too expensive for ordinary home use. To reach the greatest number of small fruit growers and encourage housewives to try home canning, the BCGA designed and patented smaller and less expensive equipment. The Department of Food Production approved the designs, authorising their immediate manufacture. The most ingenious of these inventions was the 'Wash-Copper Canning Set' which inserted into any ordinary household copper, whatever the fuel, converting the family washing machine into 'an efficient and economical canning and bottling appliance'.[152] In the absence of special equipment, the industry literature also included instructions for drying and preserving food using only the boiling rings and the oven of an ordinary gas cooker.[153] Experienced in both industrial design and dealing directly with consumers, the gas industry lent these practical skills to the government campaigns for food conservation. Unfortunately, the 1918 food production campaign coincided with a poor fruit crop, given that year's hot, dry growing conditions. Nonetheless, the Department of Food Production felt that the effort was justified, if only because it laid the foundations for future campaigns, to the credit of the gas industry's efforts.

Although the overall success of the food preservation campaign remains unclear, it does indicate several things about the gas industry and wartime food policy. First, it illustrates that the gas industry initiated corporate-sponsored programmes for food conservation in an effort to assist its employees and customers in overcoming the wartime hardships of food shortages. To carry this out, the gas industry relied on the contributions of its home service advisors, working with consumers in the home and through local food committees, and its publicity organisation, the BCGA, which produced and distributed information to consumers and employees. When called upon by the government to support public projects, the gas industry adapted its organisation to this purpose, despite the apparent frustration of dealing with government bureaucracies. As a business organisation that worked directly with consumers to address the wartime food shortages, it provides another example that extra-governmental programmes influenced government activity, as Margaret Barnett maintains.[154] But the gas industry's motives were not entirely altruistic. While the BCGA promoted domestic propaganda for itself and the government, as noted above, the executive committee chairman, Francis Goodenough, readily admitted that public service constituted the best form of wartime advertising, inevitably leading to increased domestic gas sales once the conflict ceased.[155] Gas managers eagerly anticipated the housing boom that post-war reconstruction would surely bring.

'A gas cooker in every home'

The plans, the promises and the priorities for reconstruction all developed throughout the war years. By implication, the term 'reconstruction' did not mean simply a return to the past, but compensation for the present and hope for the future. In many respects, the vision of a reconstructed society became a vague incentive to keep the British people behind the war effort, especially during the 'three bitter years' after 1916.[156] Preparations for the after-war years began in March 1916, when Asquith named a cabinet committee on reconstruction to investigate the problems that might occur. After 12 months, Lloyd George reorganised the same group as the Reconstruction Committee. In August 1917, he established a Ministry of Reconstruction, headed by Christopher Addison, to conduct research and shape public policy concerning the anticipated problems of dealing with demobilised soldiers, redundant munitions workers and replacement labour released from wartime jobs.[157]

Separate administrative branches dealt with finance, manufacturing, transport, labour and industrial organisation; the Ministry of Reconstruction's branch for social development included sub-committees, on health, education and housing. Advisory councils appointed committees and sub-committees to study specific problems, report their findings and make recommendations back to the Ministry of Reconstruction.[158] Throughout the remainder of 1917, a committee, headed by A.S. Comyns Carr, studied the proposals for extended unemployment and health insurance; committee, members William Beveridge, Mona Wilson and Arthur Greenwood agreed on the need for a Ministry of Health to administer such programmes. H.A.L. Fisher's educational committee addressed the problem of adult education, drafting proposals for a new education bill. Lord Haldane's Coal Conservation Committee cooperated with the new Department of Scientific and Industrial Research to consider plans for a national electrical supply system that was both efficient and inexpensive.[159] It was a complex bureaucracy, intended to draft a plan for the post-war society that would be able to compensate for the hardship of the war years and soothe existing class tensions.

Housing became a flash-point for debates over reconstruction and government intervention. Given the wartime building moratorium on new houses, and the resulting shortages and rent strikes that marked the war years, the need for more and better housing quickly became a central reconstruction issue. Moreover, by 1917, building costs had risen approximately 100 per cent from pre-war levels, with prices expected to remain high for several years after the war. Private builders hesitated to initiate new construction because of the anticipated capital loss, particularly given the rent controls imposed by the 1915 Rent and Mortgage Restriction Act.[160] In June 1916, the Local Government Board (LGB) first outlined this problem to the Reconstruction Committee, advocating government subsidies for local authorities and private builders to offset the costs of post-war housing construction. Later, in May 1917, the LGB continued its investigations into the housing problem, conducting interviews with numerous interested groups, including the National Housing and Town Planning Council, the Garden Cities and Town Planning Association, the Workmen's National Housing Council and representatives of private enterprise.[161] Labour groups, women's organisations, architects, businessmen and politicians all debated how to build enough homes for all of Britain's heroes.

The housing debate was highly politicised on several levels. In terms of labour relations, Lloyd George's government used promises about reconstruction housing policy to curb wartime industrial unrest on several occasions. The housing promise came up in mid-1917 just after the May engineers' strike and also in early 1918, as labour reacted against the enlistment of as many men as possible, including skilled workmen, for that spring's planned German offensive.[162] Moreover, to avert a post-war unemployment crisis, the Minister of Reconstruction anticipated that house construction would absorb demobilised soldiers and unemployed munitions workers. In the closing months of the war, the government's support for the Tudor-Walters Report (1918) was equally calculated to defuse social unrest. This was a housing study undertaken by the Ministry of Reconstruction, recommending that working-class housing reflect 'garden-city' model planning of the type previously built for occupants with higher incomes, usually members of the middle class.[163]

Disagreements over governmental jurisdiction, in terms of responsibility for housing policy, reflected another aspect of the housing debate. The Local Government Board supported the view that, as in the past, local authorities assumed control over their own housing programmes, changed only by the addition of government assistance. In contrast, the Ministry of Reconstruction and the Treasury envisioned a housing programme that was integrated into an overall plan for reconstruction, one that required direct government supervision and was funded by Treasury loans. Addison favoured the appointment of regional housing commissioners by the LGB to oversee the work of local authorities, empowering the government to intervene if local councils failed to meet their housing obligations.[164] In return, the local authorities demanded limited loan liability to protect themselves against financial losses over housing construction. In early 1919, due to renewed labour unrest in London and Glasgow, including strike threats by coal miners and railway workers, the housing campaign assumed national significance. Discontent stemmed from the pressures of demobilisation and the continued high cost of living. Moreover, the Wages (Temporary Regulation) Act, signed in November 1918, lifted the prohibition against lock-outs and strikes, officially reinstating the bargaining power of the unions, including the militant Amalgamated Society of Engineers and the Triple Alliance of miners, railway men and transport workers. The example of the Bolshevik's October Revolution and the Labour Party's adoption of a modest socialist constitution added to this sense of working-class unrest.[165]

By mid-March 1919, the Housing Bill was before the House of Commons. The bill's supporters, PM Lloyd George, Chancellor of the Exchequer Austen Chamberlain and Auckland Geddes, the former president of the Local Government Board, all emphasised the social and political importance of not simply more houses, but better housing standards. Anticipating the counter-revolutionary effect of the housing programme, politicians, such as N.P. Billing of Hereford and J.D. Gilbert, MP for Southwark, asserted that all new houses, even those constructed for working-class families, needed hot water systems, bathrooms and gardens. By building houses to higher standards and by limiting housing density, the government could effectively demonstrate that working-class aspirations could be met under the existing political system.[166]

The gas industry fully appreciated the social and economic significance of housing reform. Public housing schemes, whether subsidised by government grants or loans, represented a tremendous commercial opportunity for gas companies.[167] The advertising literature produced by the BCGA, however, considered the housing debate in terms of class and gender, stressing the social implications of gas technology in the reconstruction years. As early as 1915, reflecting the wartime housing crisis, Mrs Maud Adeline Cloudesley Brereton asserted that hot water was not a luxury reserved for the affluent, but an essential for all members of the community:

> It is manifestly unfair to expect a working man's home to be kept clean if no means are provided whereby an adequate supply of hot water can be readily obtained. The hot-water supply in working-class dwellings will at no distant date become a question for the united consideration and mutual suggestion of housing committees, sanitary officials, municipal engineers, and all interested in the moral welfare of the poor.[168]

Mrs Cloudesley Brereton accurately predicted that domestic technology would be a contentious social issue in the reconstruction years ahead. Several years before the reconstruction housing reports, she published a long list of recommendations for builders and architects in the BCGA press insisting that all new dwellings contain provisions for hot water boilers and 'penny-in-the-slot cookers'. In addition to increased facilities in individual homes, she recommended central laundry and bathing facilities. She particularly expressed her enthusiasm for that recent innovation, the 'spray bath', describing it as hygienic and cost efficient.[169] Mrs Cloudesley Brereton regretted that some landlords and builders still refused to install gas services for their tenants, under the wrongful assumption that the property owner was financially liable for the 'carcassing' and equipment rental, a situation which could be remedied through advertising and education. Given her background in public health and interest in child welfare, she clearly supported a general increase in building standards.

Her views coincided with those expressed by wartime housing advisory groups, including the Workmen's National Housing Council (WNHC) and the Women's Sub-committee on Housing. Since 1915, the WNHC had lobbied the LGB for improved standards to municipal housing made possible using government subsidies; they recommended a minimum of three bedrooms per house, a separate bathroom with hot and cold running water, and enforcement of minimum space standards for housing density.[170] In late 1917, under pressure from the Women's Labour League, Addison's Ministry of Reconstruction appointed a women's committee to consult with housewives on the question of post-war housing. In October 1918, the Women's Sub-committee on Housing issued its interim report, recommending provisions for gas cookers, hot and cold water and bathrooms in separate rooms from the scullery as standard necessities in both private and public housing.[171] Sub-committee chairman, Lady Gertrude Emmott, a long-time acquaintance of Mrs Cloudesley Brereton and the wife of a prominent Liberal MP from Oldham, shared Cloudesley Brereton's conviction that the government should respond to women's demands for higher housing standards and labour-saving domestic technology.[172] Echoing Cloudesley Brereton's earlier comments made before the BCGA in 1913, Lady Emmott's report reasserted the positive potential of technology on women's lives. Again, she also stressed the importance of hot water:

> The everyday work of the house as well as the laundry work is doubled by the lack of a proper supply of hot water. The extra strain on the woman's strength, coupled with the waste of time, leaves her without the opportunity or energy to attend to other household tasks or secure any form of recreation for herself.[173]

Although Mrs Cloudesley Brereton made no direct reference to the recommendations of the WNHC, she nevertheless endorsed the women's committee interim report in the BCGA press, quoting passages from committee members, Lady Emmott and Mrs C.S. Peel, a domestic economy specialist and wartime consultant with the Ministry of Food. For example, having read numerous letters from working women, Mrs Peel concluded that women wanted 'a gas cooker in every house' to save labour and fuel costs.[174] The women's report advocated labour-saving design features, like easy-to-clean floor surfaces, 'built-in fitments' for cooking utensils, and drain-boards for sinks. These same design elements had featured in Mrs Cloudesley Brereton's remodelled kitchen designs presented at the Institute of Sanitary Engineers in 1914. In many respects, the women's sub-committee reiterated the recommendations that Cloudesley Brereton had articulated since joining the BCGA seven years earlier.

The official housing document, the Tudor-Walters Report, reaffirmed the need to raise working-class housing standards above pre-war levels. The final report, which appeared in November 1918, represented the views of committee members, such as architect Raymond Unwin, a founder of the Garden City Movement. The report recommended several variations of the traditional terraced cottage, but with wider frontages, gardens, three bedrooms, separate sculleries, bathrooms and indoor toilets.[175] It called for the construction of 500,000 homes within three years. Although some of the recommendations of the Women's Sub-committee on Housing were missing from the official reconstruction document, including the emphasis on labour-saving designs and provisions for children's playgrounds and nurseries,[176] the Tudor-Walters Report still marked a radical departure from the housing schemes of the pre-war era, in terms of building standards, design and construction. Whereas previous building schemes reduced standards to cut building costs, the Tudor-Walters Report suggested that economy depended on efficient design, like the avoidance of dormer windows and the practice of grouping chimney flues. The government endorsed the report, modelling the 1919 housing legislation, the Addison Acts, on its ambitious recommendations.[177]

In 1919, the coalition government passed two housing acts, initiating the new policy of government-supported housing, under the supervision of Addison, president of the Local Government Board, and, later, Minister of Health. The first of these, the Housing and Town Planning Act (July 1919) required local authorities to survey their districts and submit plans for housing development to the Ministry of Health. The Treasury agreed to cover all building expenses for approved housing schemes, over a token penny rate, ensuring that local authorities and ratepayers were not fiscally responsible for the entire cost of working-class housing. Rents were fixed independently of costs, reflecting the rent controls of existing working-class houses, at least until the reconstruction housing shortage was resolved.[178] The second of the Addison Acts, the Housing (Additional Powers) Act (November 1919), was intended to stimulate private builders into activity after the slow start made by local authorities. It granted a lump-sum subsidy of £120–130 per house, not over a certain size, built for purchase or for rent. There was no restriction on the price of the house or who might occupy it; the measure was simply a cash incentive to stimulate the building trade.[179]

The gas industry enthusiastically supported the government's commitment to provide 500,000 new homes. They equally supported the Tudor-Walter Report's ideals that good design saves money, as this meant installing gas instead of electric lights and coal stoves. While this had always been a standard theme of gas advertising before and during the war, the gas managers were preparing to do battle with the electrical industry for the domestic market.[180] Gas men, like W. Dunn, argued that electricity was still too costly for housing schemes, even for basic lighting, and, therefore, it was a waste of government funds. Taking this point to the Royal Institute of British Architects, he contended that gas companies had the right to install mains in the new streets, thereby leaving homeowners and tenants the option of connecting to their services.[181] Gas men had to ensure that as many of the half a million new homes as possible were fitted for gas.

Design competitions and housing conventions followed the legislation, as local authorities, architects and builders considered their alternatives, now that the government directives were in place. The BCGA participated in these public debates, enlisting its own housing expert, H.H. Creasey, to lobby builders and architects on the gas industry's behalf. An employee of the GLCC in Francis Goodenough's gas sales department, Creasey was the company's special housing consultant, who dealt with the enquiries of businessmen and professionals, such as doctors and architects.[182] Creasey monitored the housing conferences, quoting the high-profile housing experts, to bolster arguments that gas appliances were both cost effective and labour-saving in working-class homes. The BCGA assessed the winning designs of the 1919 Ideal Home Competition, redrawing several of the architectural layouts and substituting gas fixtures for the coal stoves and electrical lights favoured by some architects. Creasey challenged the growing perception shared by members of the architectural profession that gas lighting was outmoded by electricity.[183] He quoted the opinions of several of the 'Ideal Homes' judges, including Labour Party member Mrs Sanderson Furniss, formerly of the Women's Sub-committee on Housing, social reformer Seebohm Rowntree and architect Sidney D. Adshead, who advocated the 'simplification' and 'standardization' of the mass-produced 'Standard Cottage'.[184] Their combined support for gas fires and stoves provided an unofficial endorsement for the industry.[185] Illustrated advertisements reflected the new emphasis on working-class consumers, particularly women, by marketing the gas stove as the 'Housewife's Friend'.[186] Architect Edwin J. Sandgrove, the president of the Royal Society of Architects from 1916 to 1919, recognised that women's wartime participation in the labour market had forever changed feminine expectations of work and family. In an article in the BCGA press, Sandgrove asked his readers:

> Does anyone imagine that now that the war is over, and our fighting forces have – more or less – returned, these women are going back to the old conditions of humdrum house-drudgery? Of course not. Why should they?[187]

Sandgrove continued that women rightfully demanded 'better treatment than the earlier generation of wives received in the matter of the planning and equipping of the home'. Because many women would remain in the workforce after the war, and, because of the unavailability of domestic help, Sandgrove asserted that labour-saving devices were critical to domestic and even national efficiency. By aligning itself with progressive housing reformers, who represented the interests of workers and women, the BCGA stressed its industry's capacity for beneficial public service.

In tune with the post-war problems of economic constraint and continued shortages, the BCGA emphasised the cost efficiency of gas. An advertisement in *Punch* ran the caption 'Intelligent Economy and the Government Housing Scheme', claiming that by installing gas fires, housing authorities could save £30 per house on bricks alone, with a total saving of £15,000,000 of public money.[188] The BCGA corroborated its findings by consulting with architects interested in government housing and garden city planning. W.J. Swain, the architect for Seebohm Rowntree's model housing estate (Rowntree Garden Suburb Estate, York), verified that the substitution of coal fires for gas would indeed provide savings in both materials and manpower.[189] Considering the issue of domestic labour, Swain contended that, while builders, politicians and architects debated the merits of different services, 'the women should be the final judges'.[190] Women recognised the importance of gas services, as Lady Emmott's committee report had revealed. Gas slogans adopted the optimistic language of reconstruction; the BCGA advertising phrase, 'Housing in the New Era', integrated the dual promises of efficiency and economy, while gas engineers reminded builders that the installation of gas equipment during, not after, housing construction was an 'essential preliminary' indicative of economical design.[191] While this was good news for local authorities, the gas companies directed their message at private builders, businessmen like themselves, who wanted to stretch their construction subsidies as far as possible. To illustrate their point, individual gas companies, like the SMGC, presented 'all gas' model rooms for the 'typical "Government" home' at reconstruction trade exhibitions, including the Building Trades Exhibition, at Olympia, and the Victory Exhibition, held at the Crystal Palace.[192]

Despite the optimism of architects and businessmen, including Sandgrove and Creasey, within six months Addison's housing programme was experiencing difficulties. The immediate post-war boom ended abruptly with a sharp economic downturn in 1920, blamed in part on the collapse of European trade and increasing unemployment due to demobilisation. That, in combination with the continued shortage and high cost of building materials, halted the government's ambitious housing plans, eventually undermining the optimistic expectations of housing reformers including Raymond Unwin and Ashton Webb.[193] Municipal politics and labour problems also affected housing construction. Local councils were slow to initiate housing schemes while private builders avoided participation in public housing projects because they could make higher profits on commercial projects like warehouses, department stores, cinemas and offices. In addition, the post-war building industry suffered labour shortages, particularly for skilled trades, like bricklayers, masons, joiners and slaters. The builders' unions resisted efforts to increase the numbers of workmen in the skilled trades and interpreted increased apprenticeships and the upgrading of unskilled men as a form of labour dilution. In 1920, Lloyd George blamed the failure of the government's housing programme on the obstruction of the building trades.[194]

Inflation and the high interest rates of the post-war slump also prevented municipalities and private builders from undertaking projects, despite the promised government subsidies for contractors. In February 1921, submitting to pressure from Austen Chamberlain, the chancellor of the exchequer, Addison agreed to reduce the proposed housing expenditures, amending the housing targets by half to 250,000. One month later, Lloyd George replaced Addison as minister of health with the conservative businessman, Alfred Mond; this decision sealed the fate of the government's housing initiatives. By July 1921, the Cabinet Finance Committee suspended assistance to local authorities for housing construction, while still honouring its commitment to complete 176,000 houses already underway. Addison resigned from the cabinet, his ambitious housing programme sacrificed in the interests of fiscal austerity as the coalition covernment, increasingly dominated by budget-conscious conservatives like Austen Chamberlain, retreated from its promised social programmes.[195]

After 1921, in the interests of providing cheaper homes, the municipal housing authorities abandoned the building recommendations, outlined in the LGB's *Housing Manual*, regarding the size, construction and density of housing estates. In 1919, the LCC drew up plans for 29,000 dwellings to be constructed over five years, to house 145,000 residents; but only 8,799 were completed before the suspension of the Addison Acts in mid-1921. However, those tenants lucky enough to receive houses built under the Addison Acts reported their satisfaction with the designs and amenities.[196] In the interests of saving money, municipal councils still opted for gas installations, even for gas lighting, at some urban estates, such as Roe Green in Hendon. While gas managers described this as a victory for the industry, housing experts were disappointed that model dwellings were still equipped with gas lighting and not electricity. Nevertheless, electric lighting was making rapid progress. In terms of domestic technology, local councils constructed homes with multiple services. Some flats came equipped with gas fires, coal stoves and electric lights, as if no one service could be trusted with the entire domestic workload. Gas managers, uncomfortable with this coexistence, feared the eventual takeover of electricity. They acted accordingly, as will be seen in the following chapter, by modernising the industry's image, reaching out to women in the community, emphasising the value of customer service and actively participating in the interwar housing movement.

Conclusion

The years of war and reconstruction transformed the gas industry in a number of ways. The industry suffered material losses, in terms of damaged and destroyed colliers and the neglect of structural maintenance at the works. These losses were offset, however, by the new chemical production facilities erected with government funds. Production expenses were compensated somewhat by price hikes for consumers, while the gas managers blamed the decline in product quality and service on the exigencies of war. Rightfully, consumers were disgruntled, but left with little recourse.

The industry promised, however, that all would be well in peacetime. To this end, the BCGA participated in the reconstruction housing debates. 'Housing in the New Era' of peace and prosperity would mean the availability of gas services for all classes, resulting in reduced domestic labour for women and improved health for British citizens in general. By identifying with progressive housing reformers, BCGA spokespeople, like H.H. Creasey, helped politicise the industry's commercial message. The attention which housing committees directed towards women's needs in the 'domestic workshop' placed Mrs Cloudesley Brereton and her brand of gas and water feminism at the forefront of labour-saving design. In the reconstruction years, housing represented more than a shortage of adequate homes. The promised 'homes for heroes' signified the dissolution of pre-war class barriers, the anticipation of material and democratic equality and new expectations for national efficiency after the individual and collective sacrifices of the war. Demobilised men expected jobs, dignity and decent wages. Demobilised women, who remained in the workforce, expected the same and resisted the return to menial low-paying jobs, like domestic service. In terms of housing standards, men and, particularly, women expressed the need for significant improvements, including utility services like hot water, gas stoves and indoor bathrooms. Post-war women valued their own labour more highly than before, as they had become wage-earners whose economic worth had been constantly reaffirmed in the national press throughout the war years. And, more than ever, they resented the wasted time and effort of domestic drudgery, especially the menial work of hauling water and coal.

Throughout the war years, the numbers of female gas employees increased in all areas of the business, as female labourers proved themselves capable of even the most arduous tasks. Although the male trade unionists needlessly feared the encroachment of women at the works after the war, some female employees remained, but usually as office staff and technicians in the company laboratories. No longer the only women in the gas industry, the lady demons – or home service advisors – retained a certain degree of prestige as the first women with permanent jobs in the industry. Despite the gas managers' undying praise for their female workers, the demonstrators were the only ones whose jobs were not called into question by post-war reconstruction. Because of their already public role, they easily assumed the responsibilities of distributing wartime domestic propaganda and, in this capacity, further enhanced both their professional image and public credibility. During the interwar years, the subject of the following chapters, the gas industry relied still further on the home service women to mediate between the corporation and the female consumer in the home. After all, if Britain's heroines were being asked to trade their overalls for aprons, the very least they could expect were efficient 'domestic workshops'.

Notes

1 'Keep the gas fires burning!', *Co-Partners' Magazine*, February 1916, 30.

2 Susan Pederson, *Family, Dependence and the Origins of the Welfare State: Britain and France, 1914–1945* (Cambridge, 1993), 82–83; see also R.J.Q. Adams and Philip P. Poirier, *The Conscription Controversy in Great Britain, 1900–1918* (Basingstoke, 1987), 55, 65.

3 James Wilkinson and H. Stuart Hughes, *Contemporary Europe*, 8th edition (Englewood Cliffs, NJ, 1995), 54–55.

4 Mrs J.S. Lee, 'The ladies' corner, Westminster', *Co-Partners' Magazine* (February 1916), 30; for additional articles and photographic essays about women gas workers see 'The women's part', *Co-Partners' Magazine* (March 1917), 35–38; 'Lamplighters – New Style', *Co-Partners Magazine*, (March 1918), 38; 'Wartime workers', *Co-partnership Journal* (February 1917), 25–28; 'War workers in the meter shop, Old Kent Road', *Co-partnership Journal* (March 1917), 60–61; 'Women's work in the gas industry', *ATAOUFG* (February 1918), 3–19.

5 Gail Braybon, *Women Workers in the First World War: The British Experience* (London, 1981); Angela Woollacott, *On Her Their Lives Depend: Munitions Workers in the Great War* (Berkeley, 1994); Arthur Marwick, *Women at War, 1914–18* (London, 1977); Martin Pugh, *Women and the Women's Movement in Britain, 1914–1959* (London, 1992), chap. 2; Deborah Thom, 'Women and Work in Wartime Britain', in Richard Wall and Jay Winter (eds), *The Upheaval of War: Family, Work and Welfare, 1914–1918* (Cambridge, 1988), 299–326.

6 Chris Wrigley, *David Lloyd George and the British Labour Movement* (Hassocks, England and New York, 1976), 98–99, 102–109, 113–114.

7 Existing studies of women and World War I tend to focus on female munitions workers and women's experiences in the military and nursing. See Gail Braybon, 'Women and the War', in Stephen Constantine, Maurice W. Kirby and Mary B. Rose (eds), *The First World War in British History* (London, 1995); Braybon, *Women Workers and the First World War*; Woollacott, *On Her Their Lives Depend*; Marwick, *Women at War*. Estimates of the number of women in the gas industry vary between 5,000 and 7,000. See the Women at Work Microfilm Collection, Imperial War Museum, Lambeth, Reel 40/Emp 27.6/*Substitution of Women for Men During the War, Reports by H.M. Inspectors of Factories showing the Position in Certain Industries at the end of 1918* (London, 1919), 132. According to this report, as of July 1918, 7,000 women were employed in the gas industry in Britain, excluding the numbers of female clerical workers who had also replaced enlisted men. A similar study reported the lower figure of 5,000 female gas workers; see *Report of the War Cabinet Committee on Women in Industry* (1919), Cmd. 135, xxxi: part 2, chap. III, 96. During the same time period, almost 1 million women worked in government arsenals, and the metal and chemical industries; see Pedersen, *Family, Dependence and the Origins of the Welfare State*, 90, fn. 20.

8 Arthur Marwick, *The Deluge: British Society and The First World War* (New York, 1965), 39; Samuel J. Hurwitz, *State Intervention in Great Britain: A Study of Economic Control and Social Response, 1914–1919* (London, 1968), 63.

9 E.M.H. Lloyd, *Experiments in State Control at the War Office and the Ministry of Food* (Oxford, 1924), 19–22.

10 Marwick, *The Deluge*, 56–62, 76–85.

11 Sidney Pollard, *Britain's Prime and Britain's Decline: The British Economy, 1870–1914* (London, 1989), 32–34; Keith Vernon, 'Science and Technology', in Stephen Constantine, Maurice W. Kirby and Mary B. Rose (eds), *The First World War in British History* (London, 1995), 81–82.

12 Ministry of Munitions, *History of the Ministry of Munitions* (1920–24), 7, ix, 22–23.

13 Ibid., 7, i, 5.

14 Ibid., 7, iv, 2, 12–13.

15 Ibid., 7, iv, 17.

16 Ibid., 7, i, 5; 7, ix, 2–4; appendix, 7, 89–92.

17 Ibid., 7, iv, 15–17.

18 Guy Hartcup, *The War of Invention: Scientific Developments, 1914–18* (London and Washington, 1988), 47.

19 *History of Munitions*, 7, i, 7.

20 Keith Middlemas, *Politics in Industrial Society: The Experience of the British System Since 1911* (London, 1979), 75; Maurice W. Kirby, 'Industry, Agriculture and Trade Unions', in Stephen Constantine, Maurice W. Kirby and Mary B. Rose (eds), *The First World War in British History* (London, 1995), 53–55.

21 Hartcup, *War of Invention*, 47–48.

22 *History of Munitions*, 7, iv, 21.

23 Ibid., 7, iv, 18. In January 1915, Lord Moulton's Committee of High Explosives became part of the Explosives Supply Branch of the War Department; five months later, this organisation was taken over by the Ministry of Munitions, and known thereafter as the Explosives Supply Department, with Lord Moulton appointed as director-general of explosives supply.
24 Ibid., 7, iv, 21–22.
25 Ibid., 7, iv, 22–23; 7, appendix, viii, 93.
26 Ibid., 7, appendix, x, 94.
27 Ibid., 7, iv, 15–17.
28 Matthews, 'Laissez-faire and the London gas industry', 258–253; Chatterton, 'State control of public utilities', 177–178.
29 'Keep the gas fires burning', *Co-Partners' Magazine* (February 1918), 30.
30 Everard, *History of the GLCC*, 304–305.
31 Wrigley, *David Lloyd George and the British Labour Movement*, 189.
32 'Sir Henry Roscoe's visitor', *Co-partnership Journal* (May 1916), 110; 'Our women workers', *Co-Partnership Journal* (August 1918), 129.
33 'Women and gas supply work', *JGL*, 20 June 1916, 656.
34 Everard, *History of the GLCC*, 305–307.
35 *Select Committee on Gas Undertakings (Statutory Prices) Report* (1918), 37–50, quoted in Williams, *History of the British Gas Industry*, 52.
36 LMA/B/GLCC 45/ *Minutes of the Board of Directors*, 16 October 1914, 395; GLCC, *Minutes of the Board of Directors*, 5 March 1915, 456. Only two months into the war, the GLCC raised its charges. As of mid-October 1914, penny-in-the-slot meters gave only 24 cubic feet of gas per penny, compared to the usual rate of 25 feet, reflecting the new price of three shillings and sixpence per 1,000 cubic feet of gas. By March 1915, 'ordinary' consumers also saw their rates increase from two shillings and eight pence to three shillings for 1,000 cubic feet of gas.
37 'South Metropolitan Gas Company', *GJ*, 6 August 1918, 250.
38 Cmd. 1381 xvi 717/ Board of Trade, *Reports by the Standing Committee (and Sub-Committees) on the Investigation of Prices, Gas Apparatus – Sale and Hire* (May 1921), 5–6.
39 Board of Trade, *Gas Apparatus – Sale and Hire*, 6.
40 'Prices of stoves and meters', *GJ*, 6 August 1918, 255.
41 'The general meeting of the company', *Co-Partners' Magazine* (March 1918), 1.
42 Mrs C.S. Peel, *How We Lived Then 1914–1918* (London, 1929), 101–102.
43 'Keep the gas fires burning', *Co-Partners' Magazine* (February 1916), 30.
44 *History of Munitions*, 7, iv, 23.
45 'Labour dilution in retort houses', *GJ*, 6 August 1918, 255.
46 Diana Condrell and Jean Liddiard, *Working For Victory? Images of Women in the First World War, 1914–18* (London, 1987), 7.
47 Woollacott, *On Her Their Lives Depend*, 17.
48 Wrigley, *Lloyd George and the British Labour Movement*, 98–99.
49 Braybon, *Women Workers in the First World War*, 53; Woollacott, *On Her Their Lives Depend*, 91–92.
50 'Index reading in war time', *Co-partnership Journal* (August, 1915), 181; 'The gas woman', *JGL*, 29 June 1915, 742; 'Women and gas supply work', *JGL*, 20 June 1916, 655–656; 'Rental department', *Co-Partners' Magazine* (November, 1915), frontispiece.

51 'Women's work in the gas industry', *ATAOUFG* (February 1918), 3–19.
52 'Women workers in the gas industry', *GJ*, 5 February 1918, 239.
53 NGA, Warrington/ SMGC, London, *Occupation and total number of women employees*, 25 November , 1917.
54 *War Cabinet Committee on Women in Industry*, 96.
55 Ibid., 96, 143.
56 'Manning retort-houses with women', *GJ*, 22 January 1918, 156.
57 LMA/B/GLCC 48, *Minutes of the Board of Directors*, 27 April 1917, 283; young male workers between the ages of 18 and 21 earned a weekly war bonus of nine shillings and sixpence; one shilling more than any adult female worker.
58 *War Cabinet Committee on Women in Industry*, 142–143.
59 'The ladies' corner, Westminster', *Co-Partners' Magazine* (February 1916), 30; 'Women's work in the company', *Co-Partners' Magazine* (March 1917), 36.
60 'Women's work in the company', *Co-Partners' Magazine* (March 1917), 36–38.
61 Letter from Miss J. Tewkesbury to Imperial War Museum, 21 October 1975, quoted in Marwick, *Women at War*, 79.
62 IWM, *Women at Work Collection*, Reel 57, Food, 4/39, 'Works canteens', *Bulletin*, 25 January 1918; Reel 57, Food 4/37, 'Canteens for women workers', *Oxford Times*, 19 May 1917.
63 LMA/SMGC/*Minutes of the Board of Directors*, 25 November 1915, 352; 17 May 1916, 416; *Co-partnership Journal* (March 1916), frontispiece.
64 LMA/SMGC/*Minutes of the Board of Directors*, 17 April 1918, 169.
65 *War Cabinet Committee on Women in Industry*, 129.
66 IWM/*Women at Work Collection*, Emp. 37/2, Interview with Miss Stubbington, South Metropolitan Gas, 28 October 1918.
67 Health of Munitions Workers Committee, *Final Report*, Cd. 9065, vol. 5, part 3, 34, 161. My thanks to Myra Baillie for bringing this reference to my attention.
68 'Women employees' outing', *Co-Partners' Magazine* (November 1917), 176.
69 'Beckton Station', *Co-Partners' Magazine* (September 1918), 121.
70 IWM/*Women and Work Collection*, 39/18, Women war workers' procession of homage to King George V and Queen Mary at Buckingham Palace, 29 June 1918.
71 'Royal Silver Wedding: presentation of the address of the war women workers at Buckingham Palace', *Co-Partners' Magazine* (August 1918), 98.
72 'Protective legislation for women', *New Statesman*, 1 March 1930, 662–663. For the reasons in favour of protective legislation for women workers see: Standing Joint Committee of Industrial Women's Organisations, *Protective Legislation and Women Workers* (London, 1927).
73 *War Cabinet Committee on Women in Industry*, 129.
74 IWM/*Women at Work Collection*, *Subcommittee on Women's War Work*, Emp 37/2, Interview with Miss Stubbington.
75 Home Office, *Substitution of Women for Men During the War 1919: Reports by H.M. Inspectors of Factories Showing the Positions in Certain Industries at the End of 1918* (London, 1919), 132.
76 'Women's work in the gas industry', *ATAOUFG* (February 1918), 3, 5.
77 'Editorial notes: retort-house women', *GJ*, 29 January 1918, 174.
78 'Manning retort houses with women', *GJ*, 22 January 1918, 153–155.

79 'Women's latest field: in the retort-house', *GJ*, 2 April 1918, 17.
80 'Manning retort houses with women', *GJ*, 22 January 1918, 153.
81 IWM, *Women and Work Collection, Subcommittee on Women's War Work*, Emp 37/2, Interview with Miss Stubbington.
82 'Women's latest field – in the retort-house', *GJ*, 2 April 1918, 17.
83 'Vauxhall sends forth women stokers', *GJ*, 1 October 1918, 19.
84 Woollacott, *On Her Their Lives Depend*, 63–65.
85 'Manning retort houses with women', *GJ*, 22 January 1918, 154.
86 LMA/GLCC 32/*Minutes of the Board of Directors*, 21 June 1872, 51; 27 September 1872, 173. Beckton gasworks, which opened in 1871, included a canteen where workers could obtain hot food, drinks and a daily ration of one pint of beer per worker.
87 'Vauxhall sends forth women stokers', *GJ*, 1 October 1918, 19; 'The women gas-workers of Woking', *GJ*, 3 December 1918, 514; 'Women workers in the retort-house', *GJ*, 17 December 1918, 611–12.
88 'Women workers in the retort-house', *GJ*, 17 December 1918, 611.
89 'Women's work in the gas industry', *ATAOUFG* (February 1918), 3.
90 *Report of the War Cabinet Committee on Women in Industry (Physical Sub-Committee)*, Cmd. 135, 21 November 1918. Thanks to Myra Baillie for this reference.
91 Home Office, *Substitution of Women in Non-Munition Factories During the War* (London, 1919), 22.
92 'Manning retort-houses with women', *GJ*, 22 January 1918, 153; 'Editorial notes: retort house women', *GJ*, 29 January 1918, 174.
93 'Labour dilution in retort-houses', *GJ*, 6 August 1918, 247.
94 Kapp, *Air of Freedom*, chaps 1 and 3; J.T. Ward and W. Hamish Fraser (eds), *Workers and Employers: Documents on Trade Unions and Industrial Relations in Britain Since the Eighteenth Century* (London, 1980), 127.
95 'Female labour in gas-works', *GJ*, 6 August 1918, 255.
96 'Labour in gas works next winter', *GJ*, 30 July 1918, 203.
97 *War Cabinet Committee on Women in Industry*, 129.
98 'Vauxhall sends forth women stokers', *GJ*, 1 October 1918, 19.
99 'Women workers in the retort-house', *GJ*, 17 December 1918, 611–612.
100 Ibid., 612.
101 'Women gas workers of Woking', *GJ*, 3 December 1918, 514.
102 'Women workers in the retort-house', *GJ*, 17 December 1918, 612.
103 'Women and workers and unemployment', *Co-partnership Journal* (January 1919), 5.
104 Braybon, *Women Workers in the First World War*, 67–82.
105 Kapp, *Air of Freedom*, 84–85.
106 *Report of the War Cabinet Committee on Women in Industry, Appendices: Summaries of Evidence* (1919), Cmd. 167, xxxi: Evidence of Mr John Jones, 41.
107 Ibid., 41.
108 Braybon, *Women Workers in the First War*, 76–80.
109 *War Cabinet Committee on Women in Industry, Appendices*, Evidence of Mr John Jones, 41.
110 Braybon, *Women Workers in the First World War*, 100–101; Harold L. Smith, 'The issue of "equal pay for equal work" in Great Britain, 1914–19', *Societas*, 8 (Winter 1978), 39–51.
111 Woollacott, *On Her Their Lives Depend*, 105–109.

112 *War Cabinet Committee on Women and Industry*, *Appendices*, Evidence of Mr Albert Stokes and Miss R. Stubbington, 129.
113 Woollacott, *On Her Their Lives Depend*, 109.
114 *War Cabinet Committee on Women in Industry*, 130.
115 Braybon, 'Women and the war' in Constantine et al. (eds), *First World War in British History*, 157–160.
116 L. Margaret Barnett, *British Food Policy During the First World War* (Boston, 1985), 118.
117 Ibid., 120.
118 James Bishop, *The Illustrated London News Social History of the First World War* (London, 1982), 107.
119 Wrigley, *Lloyd George and the British Labour Movement*, 191–193.
120 Barnett, *British Food Policy*, 130–131.
121 Ibid., 131; Royden Harrison, 'The War Emergency Workers' National Committee, 1914–1918' in Asa Briggs and John Saville (eds), *Essays in Labour History 1886–1923*, (London, 1971), 227, 232, 237–238.
122 Barnett, *British Food Policy*, 136.
123 Ibid., 140–142.
124 Ibid., 148–189; also see Burnett, *Plenty and Want*, 245–247.
125 Lloyd, *Experiments in State Control*, part III; Burnett, *Plenty and Want*, 246–248; Marwick, *The Deluge*, 192–194.
126 Burnett, *Plenty and Want*, 246–247; Sir William H. Beveridge, *Economic and Social History of the World War: British Food Control* (London, 1928).
127 Barnett, *British Food Policy*, 143–145, 153–157.
128 *Punch*, 16 June 1915, iv.
129 *Punch*, 17 February 1915.
130 Hurwitz, *State Intervention in Great Britain*, 175–176.
131 'Wartime cookery', quoted from Capt. C. Bathurst, MP, *Co-Partners' Magazine* (April 1917), 63.
132 Horn, *Rise and Fall of the Victorian Servant*, 166–167; Pugh, *Women and the Women's Movement in Britain*, 21.
133 M.A. Cloudesley Brereton, *The One Maid House* (London, 1916), frontispiece.
134 Vera Brittain, *Testament of Youth*, reprint of 1933 edition (London, 1980), 429.
135 Pugh, *Women and the Women's Movement*, 13.
136 'Women's advisory staff', *Co-Partners' Magazine* (March 1918), 30.
137 Marwick, *The Deluge*, 191–192.
138 Helen Edden, 'War time cookery and rations for working people', *Co-Partners' Magazine* (April 1917), 63–64.
139 Ibid., 63–64. See also Mrs C.S. Peel, *The Daily Mail Cookery Book* (London, 1920), 16–24. In her reconstruction cookbook, Mrs C.S. Peel, of the Ministry of Food, included a section on 'haybox' cookery indicating its continued use.
140 'Women's advisory staff', *Co-Partners' Magazine* (March 1918), 30.
141. Burnett, *Plenty and Want*, 246–252; Barnett, *British Food Policy*, chap. 5.
142 Pugh, *Women and the Women's Movement in Britain*, 14.
143 Edden, 'War time cookery', 63.
144 'Seventh annual report of the general committee', *The Bulletin* (October 1918), 160.
145 'Fruit and vegetable conservation', *ATAOUFG* (August 1917).

146 'The Association and the food preservation (canning, bottling, drying) campaign', *Bulletin* (October 1918), 172.
147 'Special food conservation number', *ATAOUFG* (July 1918), 3–4.
148 'Report of the executive committee of the BCGA', *Bulletin* (October 1918), 173.
149 'Communal canning', *ATAOUFG* (July 1918), 4–6.
150 Hurwitz, *State Intervention in Great Britain*, 218–219; 'The Association and the food preservation (canning, bottling, drying) campaign', *Bulletin* (October 1918), 172.
151 'The Association and the Food Preservation Campaign', *Bulletin* (October 1918), 172.
152 'How the gas industry is helping the government to save the country's fruit and vegetables', *ATAOUFG* (July 1918), 8.
153 Ibid., 11–12, 15.
154 Barnett, *British Food Policy*, 147–157.
155 'Seventh annual report of the general committee', *Bulletin* (October 1918), 160.
156 Paul Barton Johnson, *Land Fit for Heroes: The Planning of British Reconstruction, 1916–1919* (Chicago, 1968), 4.
157 Ibid., 71–74. Addison was formerly the Minister of Munitions in the Lloyd George coalition government. His credibility was damaged after the engineers' strike in May 1917 and he was moved to the new Ministry of Reconstruction.
158 Hurwitz, *State Intervention in Great Britain*, 287–290.
159 Johnson, *Land Fit for Heroes*, 79–80.
160 John Burnett, *A Social History of Housing 1815–1985*, 2nd edition (London, 1986), 221.
161 Swenarton, *Homes Fit For Heroes*, 67–70.
162 Ibid., 70–77; Wrigley, *Lloyd George and the British Labour Movement*, 230.
163 Swenarton, *Homes Fit For Heroes*, 95–96; Burnett, *Social History of Housing*, 225.
164 Swenarton, *Homes Fit for Heroes*, 75–76.
165 Eric Wigham, *Strikes and the Government 1893–1974* (London, 1976), 47–51.
166 Swenarton, *Homes Fit For Heroes*, 80–87.
167 'Housing schemes', *GJ*, 13 August 1918, 295.
168 'Domestic hot water from the cottage to the mansion', *ATAOUFG* (September 1915), 3–4.
169 Ibid., 4.
170 Swenarton, *Homes Fit for Heroes*, 91.
171 Ravetz, 'View from the Interior', in Attfield and Kirkham (eds), *A View From the Interior*, 194–197.
172 Ministry of Reconstruction, Women's Housing Sub-committee, *First Interim Report, Subcommittee of the Advisory Council of the Ministry of Reconstruction* (1918), Cmd. 9166, x: 629, 4–6.
173 Ibid., 5.
174 'Government housing schemes and the gas industry', *Bulletin* (October 1918), 180.
175 Burnett, *Social History of Housing*, 224–226.
176 Ravetz, 'A View From the Interior', in Attfield and Kirkham (eds), *A View from the Interior*, 196–97.
177 Swenarton, *Homes Fit for Heroes*, 109–111.

178 Kenneth O. Morgan, *Consensus and Disunity: The Lloyd George Coalition Government 1918–22* (Oxford, 1979), 89–90.
179 Burnett, *Social History of Housing*, 226–227.
180 'Electricity supply memoranda', *GJ*, 17 December 1918, 613.
181 'Electricity too costly for housing schemes', *GJ*, 26 November 1918, 984.
182 Everard, *History of the GLCC*, 300, 322.
183 Creasey, 'Government housing schemes and the gas industry', part 1, 185–186.
184 S. Martin Gaskell, *Model Housing from the Great Exhibition to the Festival of Britain* (London, 1987), 76–80.
185 NAL/ *Daily Mail Ideal (Workers) Homes Competition Catalogue* (1919); Creasey, 'Government housing schemes and the gas industry', part 1, 178, 181; Creasey, 'Government housing schemes and the gas industry', part 2, *ATAOUFG* (December 1919), 193, 195–196.
186 'Suggestions for Local Advertisements', *ATAOUFG* (September 1919), xxiii; *ATAOUFG* (October 1919), xxxi.
187 Edwin J. Sandgrove, 'How shall we build the new England?' *ATAOUFG* (September 1919), 157.
188 *Punch*, 5 November 1919, xix.
189 Creasey, 'Government housing schemes and the gas industry', *Part 1*, 196–197.
190 W.J. Swain, 'The house we want', *ATAOUFG* (March 1920), 26.
191 'Housing in the New Era: An Essential Preliminary', *The Bulletin* (October 1919), xxxvi.
192 'Gas installation for the small home', *Co-partnership Journal* (May 1920), 176–177; 'Gas exhibits at the Crystal Palace', *Co-partnership Journal* (September 1920), 294.
193 Swenarton, *Homes Fit for Heroes*, 91–99, 155–161.
194 Morgan, *Consensus and Disunity*, 90–92.
195 Ibid., 97–104.
196 Burnett, *Social History of Housing*, 230–231.

CHAPTER 5

Marketing Modernity: Gas versus Electricity between the Wars

> There is no question that in everyday life our surroundings have a large psychological effect, and beauty – like cleanliness – is a material producer of such effect, and of the very best order. Therefore, it is to our interest to cultivate beauty in our gas-using appliances, so long as it is not subversive of utility and therefore of economy and efficiency ...[1]

In an address to the BCGA, Sir Lawrence Weaver, a chief organiser of the 1924 British Empire Exhibition, considered the relationship between art, industrial design and commerce. An architect, journalist, director of the advertising agency the London Press Exchange and member of the Rainbow Circle, Weaver stressed the value of artistic beauty in everyday life.[2] Moreover, he emphasised the importance of educating the producers and consumers of manufactured objects to appreciate the aesthetic importance of design, even in mass produced objects used for the most ordinary purposes.[3] This aesthetic ideal combined the design legacy of the British Arts and Crafts Movement, popularised by William Morris and Walter Crane, with the more recent modernist school of industrial design which originated on the continent with architects such as Walter Gropius at the Bauhaus school.[4] Although the machine aesthetic of industrial modernism had few strict adherents in England, forward-thinking gas men, like Francis Goodenough and David Milne-Watson, agreed that their industrial engineers had to achieve a necessary balance between 'desirable efficiency and attractive appearance' when designing consumer durables, like cooking stoves and gas fires.

That gas managers turned their attentions to 'the Influence of Beauty in Commerce and Industry', the title of Weaver's address, indicated that some realised that their industry had entered a new commercial era characterised by intense competition with electricity. To meet this aggressive competition and maintain their levels of domestic consumption, the BCGA executive and editorials in the *Gas Journal* preached the necessity of devoting greater attention not only to customer service, but also to consumer tastes. With this strategy in mind, despite the threat posed by electricity, Milne-Watson and Goodenough remained confident that their industry was 'very much alive' with a 'great future before it'.[5] But they realised that this 'great future' depended, to a large extent, on the persistent cultivation of the domestic market. It depended on the support of female customers who depended on the gas industry to make their homes cleaner and more comfortable, and their communities safer, healthier places in which to live.

This consumer-oriented message was evident throughout the interwar period. As both the Conservative and Labour governments turned their respective attentions to developing a cheaper national electrical service,[6] the gas industry responded by directing its resources towards the retention and expansion of its domestic market through customer service, intense advertising and design improvements to gas appliances. The campaign to electrify British homes began during the interwar years, bolstered by the construction of the national grid between 1926 and 1933.[7] Economic histories of the interwar period have documented this process, as have studies of domestic technology and design.[8] The use of electricity increased between the wars; however, levels of domestic gas consumption also grew over the same period, although this last fact has received much less attention. Sue Bowden and Avner Offer note the slow adoption of electrical 'consumer durables' in interwar Britain, despite the popular image of the newly electrified nation. Their findings suggest that although homeowners and tenants adopted electric lighting, women still preferred gas for heating, cooking and washing because it was 'clean, easy, quick and economical'.[9] Contemporary sources reveal that, even in electric homes, some tenants continued to use candles and oil stoves, because of the high cost of electric power.[10] Based on their 'cursory reading' of interwar women's magazines, Bowden and Offer write that, in comparison to their electrical competitors, the gas industry's advertising appealed to women of every class, in a concerted effort to develop and retain their domestic market.[11] Although studies of the interwar period generally overlook the social and economic impact of the gas industry, it was never eclipsed by electricity between the wars.

This chapter takes up Bowden and Offer's general observation regarding the promotional strategies of the gas industry. The large metropolitan companies, like the GLCC and the SMGC, and the industry's publicity group, the BCGA, worked in several mediums to present their message to female consumers. Some methods were traditional, like the trade exhibitions, but others were costly innovations, such as documentary films and a corporate-sponsored housing development. While the electrical industry's propaganda addressed affluent consumers, gas advertising stressed the affordability of gas and was directed at a broad consumer base that included lower-middle- and working-class women. Despite the perceived view of the gas industry as conservative and hopelessly outmoded, for their advertising campaigns Milne-Watson and Goodenough enlisted the assistance of well-known social progressives, including the filmmaker John Grierson and architect Maxwell Fry. In an effort to appeal to female shoppers and maintain the loyal custom of English housewives, they consulted feminist designers and housing specialists, including Elizabeth Denby and Dorothy Braddell. To appreciate how the gas industry attempted to modernise and feminise its corporate image, let us begin by looking at the gas industry's largest trade extravaganza of the interwar period: its contribution to the British Empire Exhibition in 1924.

Wembley and the world of women

Despite the royal pageantry and cultural show-and-tell of the national pavilions, the 1924 British Empire Exhibition was about buying and selling. It was the largest international exhibition on British soil since 1851. King George V opened the event on 23 April, St George's Day. His entourage motored from Windsor to North London, completing the journey by carriage, arriving at Wembley in traditional processional style and waving to the crowds who lined the new town's streets. Edward, the popular Prince of Wales and president of the exhibition, attended the opening ceremonies to the delight of the crowd. Despite the dull grey skies, over 50,000 people crowded into Wembley stadium to witness the official commencement of the six-month-long event. Plans for an imperial trade exhibition had been proposed in 1910, but the war and reconstruction had delayed the British Empire Exhibition until the mid-1920s.[12]

Wembley, in north-east London, was an obvious location. A rapidly expanding suburb, Wembley was easily accessible by rail and road, an important consideration since organisers expected an average of 75,000 visitors daily.[13] Set in 220 acres of grounds, spectators could have experienced the wonders of the entire empire in a single day. One might have seen examples of exotic architecture from the east represented by a replica of a Burmese pagoda, a rebuilt Malaysian mosque and an ersatz Taj Mahal. There were also the more everyday sights of distant dominions and colonies, including a mock-up of an Australian sheep ranch and an artificial street market representing commercial Hong Kong.[14] For adventurous pleasure-seekers, the exhibition contained a 40-acre amusement park of roller coasters and water chutes that reporters claimed made New York's Coney Island look like a 'back street'.[15] Visitors of 'a more serious turn of mind' might have visited the amusement park which included a replica of Tutankhamen's tomb and a 'full-sized model colliery' complete with a shaft and underground miniature railway for the coal cars.[16] At a cost of about £12 million, the exhibition brought the 'empire in little' to this north-east London suburb.[17] However, within the stirring speeches and the royal rhetoric of fraternal cooperation for the 'better development of the family estate',[18] the fundamental message of Wembley, like that of the Great Exhibition in 1851, was commercial development.[19] Displays of Australian butter, Canadian lumber and Indian carpets were cogent reminders to businessmen and consumers of the variety of goods available within Britain's domains, while strengthened inter-empire trade suggested prosperity for this metaphorical family of nations, with England leading the way.

It was the sort of event in which the publicists for the gas industry revelled. Well experienced in the practice of coordinating collective trade displays, the gas managers and manufacturers had staged their own industrial exhibitions in London since the Great Exhibition in 1851 with successful collaborative events in 1882, 1904 and 1913. In addition, the metropolitan gas companies were regular participants in a variety of publicity events, including the Universal Cookery and Food Association shows and the *Daily Mail* Ideal Home Exhibitions held at Olympia.[20] The chief organisers of the British Empire Exhibit included Goodenough and Milne-Watson of the GLCC and Charles Carpenter of the SMGC. These three alone had participated in numerous trade shows. Add to that the advertising acumen of the BCGA staff – still directed by William Mason and Mrs Maud Adeline Cloudesley Brereton, with Francis Goodenough retaining his chairmanship of the executive committee – and the gas industry could draw on years of accumulated ability in terms of raising funds, coordinating participants and constructing displays. The gas trade journal noted with satisfaction the large number of cooperative displays at Wembley that had been 'organized by industries and not separate firms' and were erected alongside their own display in the Palace of Industry.[21] In accordance with Sir Lawrence Weaver's organising concept for the commercial exhibition spaces, the Palace of Industry resembled a huge shopping arcade. Participating industries were allotted space to be used as self-enclosed 'shops'. Patrons gained access to the exhibition displays through decorative porticos that opened off the building's spacious corridors, the 'streets' and 'avenues' of the Palace of Industry (see Fig. 5.1). Weaver urged the trade associations to commission professional architects to design beautiful and imaginative stands that captured the essence of the commerce and modernity. Thus, Weaver hoped to avoid the usual 'garish muddle' seen at English exhibitions. Weaver also intended that manufacturers should pool their resources and produce unified displays since he reasoned that one impressive and cohesive exhibit, showing the range and power of an industry as a whole, was far more impressive than 50 individual kiosks of as many manufacturers. He noted that the gas industry exhibit truly demonstrated the positive outcome of a coordinated effort, 'more faithfully than anything in the palace'.[22]

5.1 The steel roof in the Palace of Industry, British Empire Exhibition, 1924. Sir Lawrence Weaver intended that the interior of the building should resemble an indoor street, with entrances to exhibits leading off the main 'avenues'

Unlike the other participating exhibitors, the gas industry had long used this type of collective promotional style. Thanks to the BCGA, the member gas companies were accustomed to pooling resources and talent to create larger, more impressive exhibition displays. The total cost of the gas exhibit was approximately £80,000, but Goodenough assured gas managers that 'the impression of progressive life and enterprise that is being created in the public mind' by their ambitious exhibit would 'more than repay every penny that is being spent on the enterprise'.[23] True to form, Goodenough upheld the persuasive power of advertising.

The British Empire Exhibition gas exhibit was designed with the consumer in mind, and, in particular, the female consumer. Occupying a large corner site on the main avenue of the Palace of Industry, the gas exhibit consisted of a large reception lounge surrounded by a series of model rooms decorated in various period styles, from early Regency to Queen Anne. The domestic section outweighed the industrial displays, with the focal point being a set of rooms which thematically addressed

'The Seven Ages of Woman'. Each chamber was decorated to represent a feminine stage in life, from the girl-child in the cosy nursery, to the schoolgirl cleaning up after a hockey-match, to a young working woman efficiently managing her own business, ending with the mature matron relaxing at home. Every room came fitted with the appropriate gas appliances and fixtures and furnished by Messrs Osbourne and Company of Grafton Street and Messrs Heal and Son of Tottenham Court Road. Female gas demonstrators occupied the model rooms to create the effect of a living tableaux.[24] Notably absent from the 'Seven Ages' were any scenes of married women caring for husbands and children. While it remains likely that women needed no reminder of that frequent domestic scenario, the exhibition designer may have avoided that setting for fear of alienating the many customers that were single women. By depicting the advantages of gas throughout the entire female life cycle, the industry stressed that gas technology offered numerous advantages to women of all ages, in all walks of life. Additional display rooms included models for working-class flats, of special importance considering the Labour government's intentions to boost housing development under John Wheatley's 1924 Housing Act. Finally, the exhibit featured a demonstration hall, with a platform for cookery demonstrations, several fully equipped model kitchens and rows of seating for about 150 patrons.

To capture the attention of female patrons, the gas exhibit organisers considered carefully the location of their display. They requested space, not in the Palace of Engineering where the electricity display appeared, but in the Palace of Industry, near the exhibits of silks, jewellery, pottery, food and music, those items 'of special interest to every woman'. Goodenough and Milne-Watson also relied on the expertise of professional architects and designers, including H. Austen Hall, for the layout of the exhibit. Hall, a recognised expert in the design of commercial buildings, had toured America studying American retail architecture. Upon his return, Hall was commissioned to design a new store for Peter Robinson's, an upscale clothier in Oxford Street.[25] Stressing the importance of colour and line in creating a harmonious 'restful atmosphere' for the gas exhibition, Hall selected a classical Georgian design complemented by a colour scheme of cool grey with gold and black accents.[26] Mrs Ethel M. Wood, the director of the Samson Clark advertising, agency, served as the design consultant for 'The Seven Ages of Woman' series.[27] Like Mrs Maud Adeline Cloudesley Brereton, Mrs Wood was an early participant in the field of advertising, possessing a keen interest in the female consumer. After World War I, she chaired a government inquiry into the problems of domestic service for women, as workers and employers.[28] Given her continued association with the BCGA, Mrs Cloudesley Brereton handled much of the exhibit's publicity that was directed at the feminine market. She was among a list of dignitaries and notable professionals invited to attend the British Empire Exhibition's opening ceremonies.[29]

By organising the entire gas exhibit around a large central lounge, Hall addressed one of the most troublesome aspects of Wembley: the apparent shortage of free rest areas. Critics of the Wembley exhibition complained that the expansive grounds were actually too large and therefore tiring for patrons on foot. In addition, the exhibition did not provide sufficient rest areas for fatigued visitors, unless one wanted to order food and sit in a café.[30] By including a comfortably furnished lounge in the gas exhibit, Hall succeeded in creating a space where people could gather (see Figs 5.2 and 5.3). Entering through the Georgian portico, visitors immediately found themselves in a spacious lounge. With its comfortable armchairs, end tables, telephones and gas fireplaces, the committee hoped to attract visitors via this appealing rest station, confident that it would be 'appreciated, especially by women, on a tiring day, which all exhibitions tend to provide'.[31] Milne-Watson knew that their lounge of upholstered chairs provided a haven for foot-sore spectators much like the rest areas

5.2 Entrance to the gas exhibit hall, designed by H. Austen Hall, at the Palace of Industry, British Empire Exhibition, 1924

5.3 The gas exhibit hall, designed by H. Austen Hall, at the Palace of Industry, British Empire Exhibition, 1924. The lounge was a meeting place for exhibition visitors, thereby drawing customers and also demonstrating that the gas companies cared about their customers' personal comfort

for women found in the major department stores. As spaces that were specifically designed to make shopping a pleasant, restful activity, the department store lounge also imitated the middle-class sitting room, with its chintz chairs and cosy fireplace. By creating a public space that offered the benefits of the private home, the exhibit communicated to women that gas meant domestic comfort and ease.

Although some gas men, notably Sir Arthur Duckham, objected that the exhibit under-represented the industrial uses of gas, Milne-Watson defended its focus on domestic gas applications. It was a critical promotional strategy, designed to remind consumers, particularly women, that gas was a modern source of heat, light and power. Not only was gas reliable and economical, it was compatible with 'any scheme of decoration and furnishing, 'modern' or period, costly or inexpensive'.[32] No doubt this was why the gas exhibit organisers went to the fashionable West End shops, particularly Heal and Son's, to outfit their display with attractive curtains, furniture and floor coverings, spending over £2,000 in the process. Possibly, Goodenough and Milne-Watson determined that Heal's reputation was worth the expense, since Heal's was also outfitting several of the prize-winning model rooms selected by the *Country Life* design competition.[33] In some cases however, Heal's furnishings may well have overshadowed the gas appliances. For example, an article in the *Journal of Decorative Art* praised 'Messrs. Heal and other firms' for their 'very charming' series of model rooms entitled 'The Seven Ages of Woman'. The author apparently did not realise that Heal's had merely provided the furniture and was not responsible for executing the entire display.[34] *Homes and Gardens* was more appreciative of the gas industry's efforts, praising the model nursery for its 'warmth, hot-water supply and pleasant lighting by gas equipment of the latest pattern', a surprising recommendation considering the enthusiasm of most designers for electrical lighting. Despite the growing popularity of electricity, the promoters were anxious to assure female consumers that the golden age of gas had not faded with the pre-war Victorian era. Gas still offered a viable alternative to costly electricity, even for lighting purposes.

The politics of publicity

The exhibition at Wembley enabled gas managers and engineers to assess their recent progress and plan for the future. Before his assembled colleagues, Milne-Watson reported on the present state of the industry. In 1924, the gas industry throughout the United Kingdom employed approximately 150,000 workers, servicing 8 million homes and businesses, consuming a ton of coal every two seconds. Milne-Watson stated that the industry was growing, providing more jobs each year despite the introduction of hydraulic equipment and the modification of work processes to reduce heavy work. He was proud that his industry was not 'aggravating in any way the serious problem of unemployment' which remained around a million in June 1924 despite the gradual economic recovery which followed the 1920–1921 slump.[35]

Those had been troubled years for the gas industry. Immediately after the war, production and maintenance costs continued to increase dramatically because of the escalating costs of labour, coal and gas equipment, such as stoves and meters, forcing gas companies to raise their prices for gas and equipment rentals.[36] At the same time, new government legislation affected the way consumers were charged for their gas, and, while gas engineers welcomed the changes, the new policies came into effect during the depths of the reconstruction slump. Consumers interpreted the new billing system as a clever means of concealing price increases and marketing sub-standard gas.

The Gas Regulation Act (1920) began a new method of measuring and charging for gas based on calorific heat units, or British Thermal Units (BTU),[37] as opposed to the previous system of measuring gas by its candle power and charging by the cubic foot. The therm, which equalled 100,000 BTUs, became the new standard unit for gas production and sales.[38] Under the new system, consumers only paid for combustible gas and not inert constituent gases, such as oxygen, carbon dioxide and nitrogen. In addition, the 1920 Act deregulated the gas production process, permitting companies to take advantage of new methods and additives, but stipulating the minimum permissible levels for pressure and calorific value, although these varied slightly throughout the country. The gas companies were required to inform consumers about the new thermal method of calculating gas consumption, notifying them of any changes in the pressure or composition of the gas. If the composition of the gas or its pressure varied, the companies had to adjust consumers' meters and fixtures, at the company's expense, to ensure that the gas burned safely and efficiently.[39]

Generally, gas engineers applauded the terms of the 1920 Act, although the accountants dreaded the extra book-keeping of switching from one method of price calculations to another. They also warned that consumers would find the change confusing unless they were properly informed.[40] Unfortunately, the 1920 Act, which reinstated the pre-war sliding scale for pricing and dividends, coincided with the high production costs of the reconstruction slump. To avoid the penalties of raising gas prices, the companies hiked up the equipment rental fees for stoves and meters. Consumers demanded a public inquiry into the price increases, complaining that the companies used the new thermal system to unfairly increase rates, while maintaining dividends to their shareholders.

The gas companies defended their price increases before the Board of Trade. Responding to a 1921 public inquiry on increased rates for equipment rentals, the GLCC representative declared that their production costs were between two and three times higher than pre-war levels: in 1919, production costs exceeded pre-war levels by 80 per cent, rising an additional 64 per cent by 1920. They blamed the post-war increase on higher wages after the absorption of demobilised men and the reduction of the work week from 50 to 47 hours. Capital costs for stoves and meters had doubled between 1913 and 1918, rising another 50 per cent by 1919. The companies responded by raising the rental charges since an increase in gas prices would have necessitated a dividend reduction. After 1920, consumers paid 50 per cent more for their existing stoves and even higher rates for new stoves, as much as 200 per cent above pre-war equipment rental charges.[41] Coal shortages incurred by the miners' strike, which lasted from April to July 1921, pushed production costs even higher. During the work stoppage, Stanley Baldwin, then head of the Board of Trade advised the gas companies to conserve coal supplies by reducing gas pressure and 'stretching' the coal gas by adding 'water gas', a lower-quality gas made from coke and steam.[42]

Following the deregulation of gas processing under the terms of the 1920 Gas Regulation Act, the companies extended coal gas with 'water gas' even though the later product contained high levels of carbon monoxide. While this solution conserved coal stocks, the end product, a mixture of coal and water gas, was potentially harmful because of its dangerous levels of odourless toxic carbon monoxide. After 1921, the sudden increase in accidental carbon monoxide poisoning deaths aroused a public outcry against the gas companies.[43] Members of the scientific community, including Professor W.A. Bone of the Fuel Research Board, joined the editor of *The Times* in accusing the gas companies of endangering consumers. The SMGC's chairman, Charles Carpenter, publicly defended the industry by responding on behalf of the Gas Companies Protection Association.[44] The Board of Trade and the gas industry still contended that it was unnecessary to restrict the permissible levels for carbon monoxide in domestic coal gas because the 1920 Act already enforced a system of regular gas testing by a team of impartial experts. However, to help assuage public fears, Baldwin appended a special order to the 1920 Act stipulating that no gas containing carbon monoxide could be sold to domestic consumers unless it had been treated to smell like pungent coal gas.[45] Unsatisfied, the press continued to demand a full investigation.[46]

The public's agitation over gas prices and carbon monoxide levels eventually resulted in the 1921 Board of Trade inquiry. Following two years of public meetings, interviews with gas company representatives and municipal officials, the investigating committee determined that the method of calculating gas usage by the therm was not responsible for increased gas consumption and subsequent increased charges. Nevertheless, the Board of Trade committee was not satisfied that consumers received the best-quality gas appliances, reporting that many homes used older model gas cookers that had been in service for 20 years, a situation that failed to justify the 50 per cent increase in cooker rental fees. The committee recommended a policy of regular appliance renewal, such as that initiated after the war by Francis Goodenough at the GLCC.[47] It also recommended better communication with consumers, noting that the companies which invested in educational publicity, like the SMGC, received fewer customer complaints. Regarding carbon monoxide levels, the committee maintained that the gas companies were 'free to supply gas of whatever quality and composition' they found suitable, trusting that the gas testing referees and the scenting of water gas offered protection enough for consumers.[48] A few criticisms notwithstanding, the Board of Trade inquiry favoured the gas industry and helped to restore its damaged public image.

Although the report absolved the companies of any direct blame for the recent increases in carbon monoxide deaths, the LCC undertook its own study into the matter. In January 1923, the council passed a motion that the Public Control and General Purposes Committee investigate whether legislative action was necessary to prevent the gas companies from spreading 'illness and death by poisonous fumes as consequence of imperfect gas or faulty mains or pipes'.[49] Following a two-year study, the LCC report concluded that, although electric lighting remained the 'best means at present available for artificial illumination', electrical wiring was too expensive to install in all of the new council houses and schools. Electrical installations cost approximately £20 per house, compared to the gas companies' fee of from £1 to £3 for carcassing.[50] Several prominent public officials, including Herbert Morrison and Susan Lawrence, both members of the LCC and the Labour Party strongly favoured electricity. Morrison condemned the council's 'retrograde' decision not to install electric lighting in the new East End public housing developments, like Beacontree and Collingwood.[51] Curiously, the issue of gas toxicity and customer safety, the original purpose of the inquiry, was missing from the LCC final report, being overshadowed, it seems, by cost comparisons of gas and electricity. Like the earlier Board of Trade report, the LCC investigation offered reassuring news for the gas industry, given its conclusions that gas was safe and inexpensive. Nonetheless, the report suggested that electricity – presented as modern, clean and convenient – was the new fuel of choice, and only fiscal restraints prevented the council from adopting it more extensively.[52] On the other hand, gas services – outmoded, but cheap – could still be tolerated as a sorry second-best, at least until a national electrical system reduced the prices for local authorities and consumers.

Plans for reorganising the electrical industry dated back to 1917, when PM Asquith appointed a wartime Coal Conservation Committee, under Lord Haldane, to investigate the problems of fuel shortages, including the question of electricity supply. Lord Haldane's sub-committee reported that the existing system of small, local power companies should be restructured to incorporate large generating plants with regional distribution areas controlled by a national board of electricity commissioners. Addison's Ministry of Reconstruction inherited the earlier proposals of Lord Haldane's committee. Its Electricity Supply Committee asserted the critical importance of a national electrification plan to the post-war reconstruction of industry, especially if the government wished to check the spread of small private generating stations that only hindered industrial efficiency.[53] Reorganisation proposals ranged from suggestions for voluntary cooperation between the existing electricity companies to Sir Archibald Williamson's bold plans for centralised control and complete public ownership under the management of the Board of Trade.[54]

The draft bill of the Electricity Supply Act (1919) reflected Williamson's vision of a unified system controlled by district boards and electricity commissioners. The bill's proposals for compulsory purchase and national management met with powerful opposition from municipal and private undertakings, led by electrical industrialist George Balfour and the Federation of British Industries. Drastically amended in both the House of Commons and the House of Lords, the final Electricity Supply Act fell short of its original proposals. For example, it sanctioned the appointment of a central group of commissioners mandated to encourage greater cooperation between local interests but left the overall organisation of the electrical industry virtually unchanged. Several trade organisations resulted from the 1919 Act, including the establishment of the British Electrical Development Association (EDA).[55] The EDA developed common sales and advertising material for the industry in much the same manner as the gas industry's BCGA. Although the Joint Electricity Authorities, formed under the 1919 Act, made some strides towards standardised power voltages and interconnection schemes between individual undertakings, plans for a national electricity grid did not materialise until the 1926 Electricity Supply Act. This latter piece of legislation finally initiated the widespread electrification of Britain.

As members of the Federation of British Industries, it is possible that the gas companies approved of George Balfour's bid to undermine the 1919 Electricity Act. Balfour used the fear of nationalisation to garner support from other industrial sectors nervous of the possibility of any parallel schemes for their industries. In addition to the concern about the trend to government management of industry, the gas engineers were decidedly nervous at the prospect of competing with a reorganised government-supported electrical industry. Despite the watered-down proposals of the 1919 Electricity Act, the gas industry still faced an aggressive and expanding competitor, one that was working to provide better service to consumers in an effort to ward off plans for public control. To counter this, gas men and women had to rebuild their industry, secure as much of the new housing market as possible and restore consumer confidence in their services, particularly after the public relations fiascos of the carbon monoxide fears and consumer complaints of profiteering. The costly display at Wembley's British Empire Exhibition in 1924, advertising the modernity, reliability and economy of gas in the home, illustrated this image-making process at work. By the mid-1920s, as the electrical industry's EDA imitated the promotional strategies of their rivals in the gas industry, gas men fully appreciated that sales, service and showrooms were integral to the survival of their industry. This point became more apparent following the construction of the 'national grid' between 1926 and 1932. Competition with the electricity industry intensified over which utility would supply the municipal housing schemes.

'Service first'

Economic historians of the interwar period generally agree that the electrical industry ranks as one of the most significant 'growth sectors' in Britain between the wars.[56] Describing the electrical industry as the symbol of the new industrial Britain, Sydney Pollard maintains that electricity freed industry from its traditional dependence on the coalfields in the north and west. This precipitated the expansion of light industry in the Midlands and the south-east, another 'growth sector' of the interwar years. Manufacturers favoured these areas because of the easy access to the consumer markets and consequent reduction of distribution costs.[57] This relocation of jobs and capital from the north to the south exacerbated the increasing problem of regional imbalance. While the southern counties experienced an economic recovery after the mid-1920s, the northern and western regions continued to decline because of their dependence on depressed 'export-oriented' staple industries, notably coal, iron, steel and ship-building.[58]

Despite the 1919 Electricity Supply Act, which encouraged company amalgamations, the industry remained inefficient and expensive. Numerous small undertakings serviced handfuls of customers throughout the country. With no system of standardised currents for either electrical producers or appliance manufacturers, the electrical companies offered consumers a variety of voltages, depressing the demand for electrical services and consumer goods. In addition, electricity was still far more expensive than either coal or gas. The 1925 Weir Committee Report and the resulting Electricity Supply Act of 1926 introduced a series of reforms designed to make the industry more efficient, accessible and cheaper. Under the 1926 Act, a Central Electricity Board controlled all the wholesaling of electricity with powers to concentrate production in larger 'super-stations'; also included were plans for the construction of the 'national grid'. Completed in 1934, this network of high-tension transmission cables and pylons facilitated intra-regional connection schemes. As a result of the grid system, electrical production and consumption increased exponentially throughout the 1930s. Electrical output in 1925 was 6,600 million kilowatts, rising to

11,413 million kilowatts in 1931, to 22,877 million in 1937. By 1937, the industry serviced 9 million consumers, compared to only 2 million in 1923.[59] Electrification stimulated the demand for a range of consumer goods, including cookers, vacuum cleaners, irons, radios, heaters and even refrigerators, available at retailers and electric showrooms and offered on the popular hire-purchase payment system. Assisted wiring schemes helped homeowners and landlords pay the initial installation costs of electrical wiring, usually for lighting and a couple of low-power sockets. Deferred payment arrangements and mass advertising stimulated consumer demand, while mass production reduced prices, placing electrical consumer durables within the reach of a growing segment of the population.[60]

Undoubtedly, the electrical industry made great strides during the interwar period. Yet, Avner Offer and Sue Bowden have concluded that, despite the growth in the number of British households wired for electricity (73 per cent by 1937), only a small, affluent urban minority enjoyed the full benefits of domestic electrification.[61] Offer and Bowden contend that, although electric lighting grew in popularity, the vast majority of households did not possess the full range of electrical labour-saving appliances, such as cookers, 'Hoovers' and washers, because these articles were expensive to purchase and costly to operate. Assisted wiring schemes provided for a minimum of electrical fixtures. The wiring schemes did not provide sockets of sufficient voltage to accommodate high-current appliances, like heaters and cookers. Although these consumer durables were available through hire-purchase schemes, middle- and lower-income families tended to use the hire-purchase system to acquire 'non-gendered' leisure appliances, especially wireless sets. Electric lighting was another 'non-gendered' convenience as it could be enjoyed equally by all members of the household. The purchase of 'gendered' domestic equipment, labour-saving devices for women's use in the home, had less claim on the family budget because of the low value attached to female domestic labour. Besides that, there were cheaper substitutes, like gas.[62]

Leslie Hannah makes a similar point in his study of the electrical industry before nationalisation. By 1939, two-thirds of British homes had electrical service, but it was often restricted to a lighting circuit and a single plug socket. Appliance ownership levels remained low until after 1945, with the exception of the electric iron and the radio.[63] During the interwar years, electricity did not radically transform the British home or significantly reduce women's domestic labour. According to Hannah, ready-made clothing, packaged food, reduced family-size through birth control and the amenities of running water and lavatories included in the new housing estates, were more significant labour-savers than electrical services.[64]

An interwar report on the electricity industry in Britain substantiates the above argument. In its 1936 study, the Political and Economic Planning (PEP) group concluded that the number of domestic electricity consumers had risen by three times in just ten years while the cost per unit of electricity had declined by about 60 per cent. Despite these factors, electrical usage in the home was confined to electric lighting.[65] This resulted in a 'low "average" consumption per consumer', even with a socket for a wireless set and an electric iron. Consumers shied away from the heavy-current heat-generating appliances like water heaters and cookers. According to the report, sales of electric cookers, boilers and heaters remained low because of the high cost of electrical appliances, the lack of standardised voltages and the expense of switching from one fuel to another. For example, the study found that of the 6.6 million domestic electricity consumers, only 650,000 used electric cookers and a mere 120,000 had electric wash boilers. They blamed the slow adoption of electric cookers on the problem of standardisation and the high cost of the special 'machine-bottomed' cooking utensils required for use with the solid hot plates.[66] Nowhere does the report acknowledge the overall efficiency of gas appliances over electrical ones or consumer preferences. As will be seen, this was also an important factor that depressed sales of electrical goods and services.

Though homeowners and tenants might have preferred a single fuel source, be it gas or electricity, throughout the interwar years the average English home relied on several fuels: electricity for lighting, gas for cooking, a coke boiler for hot water and the open coal range for heat and some cooking in winter. Although gas lighting was losing popularity, the market held steady because of the increased demand for cookers and, to a lesser extent, for boilers, gas fires and gas refrigerators.[67] By far the most common domestic appliance was the gas cooker.[68] A 1939 study of the gas industry, undertaken by the PEP group, revealed that 75 per cent of families in Britain used gas cookers, in approximately 9 million homes, compared to the 1 million electric cookers then in service. The report indicated that gas cookers consumed 75 per cent of gas used in the average household. Gas fires and heaters met only 5 per cent of home-heating requirements while gas boilers accounted for only 20 per cent of water heating needs. The combustion of solid fuels, such as coal and coke, made up 80 per cent of water heating; however, this figure included the popular coke boiler, a product heavily promoted by the gas industry. Domestic consumers were the heaviest gas users; between 60 and 70 per cent of all gas produced went to home consumption.[69] The vast majority of gas consumers used the penny-in-the-slot meter payment system. Of the 11 million homes that used gas, 7 million had pre-payment meters.[70] Gas cookers, and the women who used them, sustained the industry throughout the interwar years. It was a market that the gas industry recognised, cultivated, appreciated and rarely took for granted.

Since the advent of the BCGA in 1911, influential individuals with the industry – notably Francis Goodenough, David Milne-Watson, Corbett Woodall and Charles Carpenter – had stressed the commercial value of advertising and dedicated customer service. Due to the industrial priorities of the war years, domestic appliance design failed to keep pace with this spirit of enthusiastic retailing. In 1920, the gas companies promoted virtually the same models of gas cookers, in terms of external appearance and the features that they had in 1895. After the war, however, manufacturers devoted their attentions to improving appliance designs, taking into consideration mechanical efficiency and consumer tastes. This movement towards better design was motivated by a post-war preoccupation with improving national efficiency and health standards. Design groups, such as the Design and Industries Association, adopted the views of American efficiency experts, such as Frederick Taylor, that rational design eliminated wasted effort and ensured greater overall productivity in the workplace and at home.[71]

Technical innovations developed during the war, such as metal enamelling, found new commercial applications. The widely used of vitreous enamel metal surfaces for stove construction, usually in white or cream with blue or green trim, were a design innovation of the early 1920s that replaced the old cast-iron models. Increased concern for home sanitation resulted in higher domestic standards for work undertaken by both housewives and hired domestics. Not only were enamelled cooking stoves simpler to keep clean, their smooth surfaces appeared more sanitary and efficient. Thermostatic heat control, first introduced in 1923 with the Radiation Company's New World H16 cooker, was another significant mechanical innovation because it allowed cooks to control oven temperatures, ensuring better results with less attention during cooking.[72] Enamelled surfaces and the provision of stove legs were important labour-saving features as increasing numbers of women did their own housework, because of the continued shortage of domestic servants.[73] Mrs Cloudesley Brereton praised the easy-clean surfaces of the new enamel stoves in the BCGA press. Her ideal kitchen of 1924 included an enamelled 'Regulo' Radiation New World cooker, set in a tiled recess. She assured readers that the cooker's 'simple lines' were designed 'to reduce cleaning operations to a minimum and to prevent the accumulation of dust and dirt'. While the raised legs simplified cleaning beneath the appliance, the New World cooker was more convenient than the older legless models, as it allowed 'the cook to inspect the food without much stooping' (see Fig. 5.4).[74] In this New

5.4 'I can do my Shopping whilst the dinner is cooking', advertisement for the New World cooker, Radiation Company

World advertisement, with its contrasting illustrations of the quiet home and the bustling high street, the cooker saves both labour and time by enabling the housewife to do her shopping while the dinner cooks itself, an important feature for the servantless housewives of the middle classes. Technology was finally eradicating the distinction between women's public and private worlds, as they were now able to occupy both spaces simultaneously.

During the 1920s, given the expense of the enamelled cookers, stove manufacturers continued to produce the cast-iron models for working-class penny-in-the-slot meter customers. In 1920, R. & A. Main Ltd marketed their cast-iron 'Surrey' model as a 'housing' cooker because of its small dimensions and low price.[75] It was designed for borough public housing. There was also a discrepancy between the cooking stoves available for hire-purchase on the payment plans and those cookers that were simply hired appliances offered to working-class consumers. Housing authorities overseeing public housing projects continued to install cast-iron cookers, such as the 'Surrey' or the standard Richmond 'Bungalow' cooker, presumably to keep installation and rental costs low but also because the cast-iron models were practically indestructible.[76] By the early 1930s, however, manufacturers developed compact enamelled models for the working-class market, including the Radiation Company's 'Kingsway' New World. In 1934, R & A Main Ltd replaced its black-iron cookers with the cream-coloured enamelled 'GLC O', designed for the GLCC's penny-in-the-slot meter customers. This model, equipped with oven thermostats and constant pressure controls, was the first gas cooker produced on an assembly line. In 1934, the company turned out 210,000 of these cooking stoves.[77]

Given the high cost of electricity, gas was still popular with housing authorities until the mid-1920s. In 1926, Mrs Cloudesley Brereton estimated that of the 28,500 houses constructed under the LCC housing schemes, 22,000 were lit by gas and the borough had installed gas cookers in 'almost every one'. There were 6,000 houses served by a combination of gas and electricity, and only 500 had been wired exclusively for electric lighting.[78] While some municipal councillors, notably Morrison and Susan Lawrence, regretted that electricity was not more widely used in working-class dwellings; in general, the tenants favoured gas because it was cheaper to use and was a familiar domestic fuel source.[79]

The 1926 Electricity Supply Act had an immediate effect on the gas industry, and, particularly, on the use of gas lighting. After the 1926 Act – with its promise of less expensive and more available electrical services – private and municipal housing developments abandoned gas lighting in favour of 'all-electric' flats or a combined service of electricity for lighting, and gas for cooking and heating.[80] Tenants openly debated the merits of each utility service. In one instance, tenants mounted a public protest against electrical services. There were 500 women and men, occupants of the Woolwich Borough Council's estate at Eltham Hill, who complained that their all-electric flats were too expensive, signing petitions to this effect. In particular, the tenants requested the removal of the electric cookers which they claimed had been installed, at a cost of £10,000 to the corporation, to increase the consumption of electricity generated at the Woolwich municipality's own electrical station. Tenants complained that an electric cooker cost about seven–eight shillings per week to operate, and although the SMGC offered to install penny-in-the-slot meter gas cookers free of charge, the occupants protested that, under the terms of the house rental agreement, they were prohibited from removing the corporation cookers. A public meeting with the Woolwich Borough's Housing Committee, where a group of tenants requested that the gas prohibition be rescinded, resulted in a rowdy debate between those who favoured the all-electric houses versus those who desired a combined service. The ensuing shouting match between tenants prevented a motion from being put to the floor allowing for consumer choice, and the Eltham Hill residents were left with their costly electric cookers.[81] The Eltham Hill tenants' protest indicates the politicised nature of the debates over municipal services. Adopting a form of political protest, the public petition, the women and men demanded freedom of choice as consumers and as citizens of Woolwich Borough. That the meeting degenerated into a show of temper indicates the level of public tension that existed, particularly when consumers believed that their municipal authorities were motivated by a desire to increase electricity sales and not to provide the cheapest service for working-class families.

Perhaps learning from the Eltham Hill incident, other councils were more amenable to combined services. Islington Borough Council agreed that the GLCC should install a gas supply, cooker and one lighting point in each of the 188 tenements at the Hornsey Rise Housing Estate, at a cost of £1 per dwelling. At the same estate, the borough installed electric lighting at a total cost of £3,320.[82] Combined services of gas and electricity were increasingly common after 1926. By 1927, the LCC provided both at some of its housing estates. Although the post-war developments at Becontree, Roehampton, Belling, Downham, Norbury, White Hart Lane and Old Oak had no electricity for any purposes, the portions of Downham and Becontree completed after 1927 had wiring for electrical lighting and included one socket per flat. At the request of tenants, electrical suppliers installed electrical wiring at several existing housing estates, including Becontree and Old Oak, while the suppliers wired the new estates throughout south London at Kennington, Kennings, Bermondsey and Watling. At the LCC's Watergate Street estate, tenants received a combined gas cooking/electric lighting service; however, the borough denied the SMGC its permission to 'carcass' the new houses for gas lights.[83] Similarly, Lewisham Borough Council prevented the SMGC from installing gas lighting services at its new housing scheme in Grove Park, since the housing committee desired a combined service of gas cooking and heating with electric lighting. Although the gas company protested that consumers were being denied their 'choice of illuminant' and simply requested permission to carcass fully the new homes, at no expense to the municipality, Lewisham Borough Council maintained its decision not to allow carcassing for gas lights.[84]

Significantly, the reaction against gas lighting by the municipal housing authorities occurred after the 1926 electricity legislation. Until then, municipal housing schemes favoured all-gas installations to cut building costs. After the new legislation and its provisions for assisted wiring schemes, housing authorities opted for combined services of gas for cooking and heating, and electricity for lighting. It was a trend that the gas engineers observed somewhat nervously, perhaps suspecting that the decline of domestic gas lighting would be followed by similar reductions to the heating and cooking loads. Moreover, to encourage the use of electricity, the LCC Housing Committee advised the aggressive promotion of hire-purchase electrical appliances, available in showrooms from electrical undertakings. The LCC also supported a programme of customer education using female demonstrators and an efficient maintenance service to promote the domestic consumption of electricity.[85]

The above plans for customer education using women demonstrators resembled the promotional programme that the gas industry had consistently developed since the Edwardian years. The early addition of lady demons to the gas sales departments and the formation of the BCGA to oversee the industry's public relations activities suggest that the gas industry had led the way in this type of promotional activity. In anticipation of the commercial impact of the 1926 Electricity Supply Act, Goodenough warned gas men that the legislation which promised to standardise and develop a national electrical supply would either stir them to fresh activity or precipitate the industry's demise. According to Goodenough, a 'satisfactory supply of good gas' in itself, was insufficient 'to win and hold the custom of the community unless accompanied by complete and satisfactory service' made known to the customer 'by persistent publicity of all kinds, local and national'. He predicted that the future prosperity of the gas industry depended on how well they succeeded over the next five-year period.[86] Slightly over a decade later, in 1939, the report of the PEP group revealed that the gas industry had successfully enlarged its domestic sales despite vigorous competition with electricity.[87]

Whether the increased gas sales throughout the 1930s were a direct result of the industry's advertising schemes remains difficult to conclude, since it is impossible to measure the real economic return of elaborate showrooms, customer service and advertising. It is certain, however, that the gas industry invested heavily in these forms of publicity between 1926 and 1939 in a conscious effort not simply to increase sales, but to modernise its public image by addressing social issues, like child welfare, housing and pollution. All were topics of direct concern to women, as housewives, citizens and consumers. Thus, women, as buyers and sellers of domestic technology, were essential to this image-building process because of the industry's concern to present itself as a progressive corporate citizen, maintaining and protecting the interests of female consumers both inside and outside the home.

Popular modernism and the gas industry

In a 1924 editorial reflecting on the success of the Wembley exhibit and its lessons for gas promoters, an anonymous gas engineer concluded: 'There is no better means of advertising than practical demonstration'.[88] Although, for most products, advertising meant a catchy phrase, a cute jingle or a comic poster, marketing an 'essential commodity' like gas was a much more 'complicated matter'. The engineer added that just as gas was the most comprehensive of all the different branches of engineering, so the industry had to 'consider, adopt, reject, or adapt all the known methods [of advertising] and if necessary, evolve new ones of its own' to ensure that the capabilities of the industry reached as many consumers as possible. While the editorialising engineer supported the ongoing publicity work of the BCGA, he concluded that a well-staffed and properly appointed 'permanent exhibition' – the company showroom – was a superior, albeit underrated, promotional method.[89] It was a familiar message, particularly for those who had attended the BCGA conference in 1912 where the topic of improved showroom facilities had been raised before the assembled gas men.[90] In that early address, George Clarry, of the Cardiff Gas Company, urged managers to consider the economic advantages of spacious sales facilities in high-traffic locations staffed by knowledgeable salesmen and lady demons. He expressed the importance of window displays, showroom design and cookery classes as strategies for attracting business.

As previously noted, the large metropolitan undertakings, such as the GLCC and the SMGC, opened their first company showrooms in the 1890s.[91] Gas managers usually set up their showrooms, such as they were, near the accounts desk in the company offices. The displays usually consisted of dangling clumps of lighting fixtures and rows of assorted meters, cookers and gas fires, arranged so that regular customers who came to the company office to settle their quarterly payments were forced to walk by, through and under the new appliances. Perceptive as this scheme was, it also enabled the accounts clerks simultaneously to watch the showroom from their desks, thereby pushing these men to assume two separate jobs. In the smaller gas undertakings, the general manager had more than one role: running the works as the senior gas engineer and also managing the showroom and the sales business, albeit in an often 'half-hearted way'.[92] With the advent of the lady demons and the cookery shows, gas managers increasingly viewed the showroom as not simply an adjunct to the office, but a specialised retail space. By enlarging their central showrooms, constructing permanent lecture theatres and installing model kitchens, gas managers were able to attract female consumers on a regular basis.[93] The SMGC adopted the additional promotional tactic of cooperating with local furniture merchants to create joint displays. Together, the SMGC and furniture salesmen leased and outfitted model homes in residential neighbourhoods so that consumers saw the appliances correctly installed in fully-furnished, decorated houses, not within the constructed artifice of the office-showroom or display window.[94]

The suspension of normal trade during the war years affected the role of gas company showrooms. As we have already seen, throughout the duration of the war, these commercial spaces doubled as public meeting places, used by women's groups and local food councils distributing information on food and fuel rationing. After the war, the showrooms resumed their former commercial purpose. However, managers wished to retain that image of the gas showroom as a local meeting place for women by developing what Victoria de Grazia termed a 'feminine public sphere' that catered to both the social and material needs of the interwar consumer-housewife.[95] Reflecting interwar trends in commercial architecture, the feminine culture of shopping and the increasing importance of advertising, the gas company managers reconstructed their showrooms during the interwar years to modernise the sales atmosphere while simultaneously offering female consumers entertainment, information and leisure. The gas showroom was not simply a house of business, but a public gathering place.[96]

Under Milne-Watson, the GLCC introduced a policy of updating its company showrooms while raising its standards of architectural design and display. Architects H. Austen Hall and Walter Tapper devised 'a look' for the GLCC offices and new showrooms. In the interests of what was then considered 'good taste in commercial architecture', Milne-Watson favoured plain brick exteriors, with the company name in clean bronze lettering.[97] This company 'look' applied to the large urban premises, such as the remodelled central offices in Horseferry Road, or the expansive new offices and showrooms constructed in districts throughout London, including Islington, Kensington, Golders Green, Hackney, Barking and Stratford.[98] For all its apparent conservatism, the historicism of the 1920s was in fact a modern interpretation of Georgian classicism, considered by 1920s interior designers, antique dealers and architects to be the essence of 'good taste' and a clear rejection of the 'fussiness' of Victorian pastiche.[99] Although the GLCC still favoured the classical forms and traditional materials of the 1920s for its central offices, by the 1930s the company had adopted the machine aesthetic of streamlined art deco for its smaller retail showrooms. This new commercial style was particularly evident in exterior lettering and lighting designs, however, while Milne-Watson commissioned architectural plans that incorporated bright metal showroom frontages of 'Stay Brite' steel.

According to Penny Sparke, art deco, with its stylised imagery and rounded geometric forms, was 'the first truly popular modern style'.[100] Named after the new style presented at the 1925 Paris Exposition des Arts Decoratifs, art deco included numerous stylistic variations, from the brilliantly coloured geometric patterns of the 1920s to the streamlined metallic modernism of the 1930s. In America and Europe, mass-produced items, including radios, dishes, women's toiletries and costume jewellery, adopted the art deco style, making this design and its associations with elegant modernism available to consumers of even modest means.[101]

Art deco also inspired new designs for commercial architecture and retail displays. American department store owners appreciated the connections between visual display and consumer desire, turning to modern store design as a means of attracting business in a highly competitive retail market. According to cultural historian Neil Harris, even in the 'boom years of the 1920s', department stores experienced dwindling net profits. In the drive to increase sales, and having exhausted the possibilities of 'increased efficiency', retailers turned to the 'architecture of merchandising'. Because of its associations with modernity, efficiency and personal comfort, designers of modern retail interiors and display windows adopted the art deco style for the major commercial establishments in large cities. Architects redesigned store interiors to create a modern look by adding new lighting, streamlining exteriors and removing excess counters. Retailers hoped the new look would generally stimulate customer imagination. By surrounding the shopper with new objects, tastefully and dramatically displayed, department stores outpaced museums and art galleries as the arbiters of popular taste.[102]

While American architects and designers embraced the art deco style, throughout the 1920s, their British counterparts favoured the conservative modernism of design historicism inspired by the seventeenth and eighteenth centuries.[103] This trend was evident in retail store design. Austen Hall's 1924 British Empire Gas Exhibit at Wembley, with its mock-pillared 'early Georgian' central lounge in cream and gold, and model rooms of 'period reproduction' furniture, aptly characterised this design trend. Hall repeated the same design layout for the GLCC's Islington premises, which was completed the following year.[104] The new office-showrooms at Kensington reveal a similar vernacular historicism. Hall executed the square three-storey red brick building in the 'English Renaissance' style using Portland stone dressings, arched portals and cornices as a tribute to nearby Kensington Palace. The building's spacious columned entrance hall and display rooms adopted the favoured Georgian classicism, 'redolent of good taste and quiet charm'.[105] Using its offices and showrooms, the gas industry conveyed the impression that it was stylish and tastefully modern, the embodiment of 'good taste and quiet charm'. This image of tranquil efficiency was a far cry from the one that had appeared only a few years earlier as a result of the public inquiries into gas poisonings and company profiteering.

By the end of the 1920s, the historicist trends in British design gradually gave way to the modernist lines of art deco as an increasing number of young architects adopted the new style. While Britain lacked the dramatic art deco skyscrapers that transformed the large American cites, new hotels, theatres and office buildings incorporated art deco motifs in facades and foyers.[106] Preferences for antique and reproduction furniture shifted towards modernist designs in chrome, steel and leather in settings enhanced with mirrors, concealed lighting and rich carpeting. Cinemas featured dramatic art deco interiors, familiarising the general public with this architectural design style.[107] Architects favoured streamlined art deco for anything associated with technology and speed, from airport design to the interior of deluxe railway trains and ocean liners. Frank Pick, vice chairman of the London Passenger Transport Board and an enthusiastic supporter of the modern style, selected simple geometrical shapes for the new tube stations north and west of the city, creating what some contemporaries considered 'the most satisfactory series of modern buildings in England'.[108] Under Pick's direction, the London Passenger Transport Board developed a public image based on a comprehensive design strategy which included transport stations, signs, posters and a corporate logo. Pick and his design team translated the avant-garde forms of cubism and de Stijl into commercial design for the masses. Retailers of quality merchandise, such as haberdashers Peter Jones in Sloan Square and Simpson's department store in Piccadilly Circus, selected modernist designs for their respective new stores, under construction between 1935 and 1939.[109] By the mid-1930s, a small group of progressive metropolitan retailers were using artistic design to attract consumers, paying increased attention to lighting, mirrors, colour and line.[110]

The lesson was not lost on the gas industry. Mrs Cloudesley Brereton, still editor-in-chief at the BCGA, continued to oversee the production of the BCGA's advertising material, and it cannot be said that she failed to keep up with design trends. She appreciated the value of imaginative, creative work, and commissioned layout designs from contemporary artists.[111] Some of this advertising reflected a modernist style, exemplified by Clive Gardiner's cubist urban landscape of fractured asymmetrical forms and flattened perspectives.[112] In 1931, the public relations department at the GLCC furnished the industry with its most popular mascot, Mr Therm. The company commissioned Eric Fraser, a freelance commercial artist who supplemented his income by teaching at Camberwell School of Art, to design a figurative image that anthropomorphised the physical properties of the live gas flame. Fraser had a reputation for producing unconventional designs, and his bizarre cartoon-like creation, Mr Therm, certainly sustained that image.[113] Within months of his first appearance, Mr Therm's flat, schematic face, haloed head and stylised limbs appeared on hoardings, motorbuses, trade exhibition posters and in showroom windows (see Fig. 5.5). Mr Therm provided the industry with a popular publicity motif, a means of product branding practised by other successful manufacturers. An earlier attempt at finding a product motif to suit the industry had been an utter failure. Artist Septimus Scott had designed the 'Spirit of Coal', a sinewy giant emerging from a flame, for the 1924 British Empire Exhibition in Wembley. This tormented-looking figure appeared sporadically in advertisements and at trade shows, but was never adopted as a universal product symbol.[114] In contrast, Mr Therm had personality. In advertisements, the illustrators often cast Mr Therm as the resident expert, answering questions from several women at once, while, in another advertisement, he teamed up with kitchen expert Dorothy Braddell; Mr Therm was easily recognised by consumers, reflected a contemporary artistic style and added a playful element to the gas industry's usually serious image. As a public relations motif, Fraser's cartoon had many strange incarnations, but perhaps the most bizarre tribute to Mr Therm was the GLCC's Therm Band, formed in 1934. The musicians wore ridiculous bulky cardboard costumes with cut-out holes for their faces, impersonating the ubiquitous mascot.[115] As a product symbol, Mr Therm represented the gas industry until the 1960s.

Showroom design in the 1930s also reflected the industry's increasing attention to commercial modernism. The Oxford Street showrooms,

5.5 The cheerful Mr Therm teams up with Dorothy Braddell to advise customers on kitchen design. 'Plan Your Kitchen to save your legs!', British Commercial Gas Association

opened in January 1937, aptly illustrate the gas industry's adoption of commercial art deco. Using clean, bold lines for the store's frontage and company lettering, in conjunction with another art deco inspiration, concealed external lighting, architect Michael Tapper created an attractive retail facade, well suited to central London's busiest shopping district. Once inside, the consumer entered a bi-level showroom in monochromatic blue. On the main level, six pillars of blue glass supported cluster gas burners with reflectors floodlighting the showroom displays of the latest gas appliances, including refrigerators, enamelled cookers, gas fires and water heaters. The lower level, reached by blue-carpeted stairs, contained model kitchens and a fully-equipped café bar, complete with chrome stools and red upholstery.[116] A report of the opening ceremonies noted the showroom's proximity to another modernist landmark, the BBC's new Broadcasting House, completed in 1932. The GLCC had also used modernist elements, such as 'Stay Brite Steel' and 'strip lighting' at its new Queen Street showroom which opened in 1936.[117] Modern art deco schemes appeared in the new suburban showrooms in Twickenham, Kingsbury and Edgware Road.[118]

At Leytonstone, in the north-east of London, the company commissioned architect George Grey Wornum to design its 62nd showroom.[119] A well-known advocate of the modernist school, Wornum designed the Royal Institute of British Architects' new Portland Place building which was completed in 1932. He was later known for his innovative designs of ocean liner interiors, notably the 1938 Cunard luxury ship, the *Queen Elizabeth*.[120] Wornum's modernist storefront design for the GLCC's showrooms relied on symmetrical lines and rectangular shapes to create a machine-like austerity. The undersized company lettering above large showcase windows suggested that, by themselves, the tasteful commodity displays sufficiently identified the premises to passing consumers. This reflected a modernist retail aesthetic which viewed the store window as a 'silent loudspeaker', in recognition of the immediate impact of visual information[121] (see Figs 5.6 and 5.7). By adopting a modernist architectural style for its new showrooms, the GLCC consciously presented itself as a corporation that was efficient and instep with the times.

London's second largest gas concern, the SMGC, also expanded its showroom facilities between the wars, opening new premises in the suburbs and rebuilding existing showrooms.[122] Following current retail designs, the company selected streamlined art deco typography for its new showroom frontages located, respectively, at Charlton and Walworth Road in south-east London. Both opened in 1937. Despite this design innovation, the SMGC appeared unsure of the selling value of minimalist modernism as their interior display areas at the Charlton and

5.6 Leytonstone showrooms (exterior view), London, designed by G. Grey Wornum

5.7 Leytonstone showrooms (interior view), London, designed by G. Grey Wornum

Walworth premises appeared packed and congested, reflecting the old 'more is better' retail aesthetic.[123] While the SMGC adopted a cautious acceptance of art deco retail design, they experimented with different types of retail space, including the patronage of Brixton's new Granville Arcade. Opened in June 1937, this one-acre shopping centre, described as the 'largest and most modern of its kind', provided spaces for 112 lock-up shops. The SMGC leased two spaces: one for displaying merchandise and the other for regular daily cookery demonstrations with seating for 50 people.[124] Although this marked a departure from the usual high street showroom, the SMGC showroom managers anticipated brisk business because the new marketplace was the shopping hub for housewives from a three-mile radius.

Striking the right balance between modern and traditional elements in gas showroom design was a difficult problem. As A.V. Bishop of Ipswich observed, few other businesses catered to consumers from such a broad social and economic spectrum. Bishop observed that 'high-class' shop owners arranged their windows with uncrowded simplicity because affluent consumers required less immediate product information, relying instead on the retailer's reputation. The 'poor-class' retailers displayed as much merchandise as possible, with prices written clearly, to allow the windows themselves to do the selling. Because the gas industry appealed to consumers of different classes and economic backgrounds, they needed to gear their showrooms accordingly. Bishop believed that 'a "tony" showroom' deterred the poorer class of customer, while a display lacking polish and distinction failed to attract the better-class consumer. Yet, gas showrooms had to cater to both types.[125] According to Bishop, 'tony' stores favoured artistic, modernist designs while 'cheap' shops relied on a more old-fashioned retail aesthetic. But modern attractive retail design need not be an indicator of high prices. Bishop preferred the strategy of frequently-rotated 'neat displays' of variously priced goods, with a permanent display of 'cheap lighting fittings' in a separate, but visible, section of the showroom. While the SMGC experimented cautiously with modernist store layouts, the GLCC, under David Milne-Watson, utilised contemporary design more extensively, confident that art deco showrooms would attract, not intimidate, its less affluent consumers. As streamlined art deco appeared in an increasing number of public spaces and left its imprint on a wide variety of mass-produced consumer goods, art deco design epitomised a fashionable, feminine modern style, made accessible to consumers of even limited means through the popular hire-purchase plans.

As noted earlier, gas appliances eventually succumbed to the influence of streamlined art deco designs. Dorothy Braddell, an interior design specialist and the wife of architect Darcy Braddell, designed one of the most unusual gas cookers of the interwar years. In the 1920s, Dorothy Braddell established her reputation as a commercial artist, producing advertising material for clients that included Shell Mex Petrol, Crosse & Blackwell and Viyella Textiles. Examples of her work appeared at the *Daily Mail* Ideal Homes Exhibitions. She also contributed to the British Academy's Art in Industry exhibition in 1935, the Dorland Hall exhibitions of British industrial art (1933, 1934) and the British Pavilion at the Paris Exhibition of 1938. In terms of her interior designs, Dorothy Braddell specialised in kitchen and bathroom planning, the greatest design challenge facing architects between the wars. Her 'common-sense' kitchen designs relied on labour-saving materials with a functionalist aesthetic that she claimed was equally important in the working-class flat as it was in the luxury home.[126] Dorothy Braddell concerned herself with efficient designs, and her layouts did not reflect a partiality to electricity, gas or solid fuel. For example, she produced several large design layouts to promote coal-fired Aga cookers. However, the gas industry also made extensive use of her talents, and she wrote the foreword for their publication on kitchen planning.[127] In cooperation with the Radiation Company, she also designed the innovative Kabineat gas cooker, a compact stove that featured a folding cabinet top and front door so that, when it was not in use, this kitchen appliance resembled an upright radio.

The new consumer status symbol the gas refrigerator, which appeared on the market in 1926, exemplified the best of domestic art deco with its gleaming enamelled surfaces, rounded curves and sleek Bakelite and chrome handles. That gas appliances could be stylish, even beautiful, was a remarkable alteration. In the early years of gas stoves, the lady demons listed the benefits of gas technology as fuel economy, reduced domestic labour and culinary efficiency, but made no mention of interior design. The cast-iron cookers of the pre-war era lacked aesthetic appeal, but it mattered little, since appliances were hidden away in tiny sculleries or basement kitchens. The enamelled cookers developed in the mid-1920s changed all that, as did the new priorities in domestic interior design. As a result of the housing investigations of World War I, including the Tudor-Walters Report, architects designed working-class cottages and flats with larger kitchens by combining the scullery and living/dining room areas. Regarding homes for the more affluent, interwar architects transformed the basement kitchen/scullery – that hidden netherworld of servants – into a main-level workspace for the servantless housewife.[128]

Gas appliances, in white or cream enamel with coloured trim, enabled housewives to choose installations to match the colour schemes of their homes while the showroom interiors of cream, blue and green complemented the displayed enamelled appliances in the same tones. Showroom sales managers encouraged women to view their gas cookers with the same 'critical eye' as they would their carpets and furniture, replacing domestic equipment when it became worn or out-of-date. Because women were responsible for outfitting the family home, they were the natural target audience for showroom services. Women were also more able to appreciate a stylishly appointed showroom than male consumers. According to GLCC general manager, R.W. Foot, women's awareness of the whims of fashion and popular design made them better domestic consumers because they were 'more susceptible to change in their household arrangements than their men-folk who were wont to put up with what they had'.[129] The advantage of updating gas appliances, in comparison to other redecorating projects, was that the company fitters did the replacement work while payments were easily managed through hire-purchase plans. The gas undertakings encouraged customers, particularly homeowners, to select the hire-purchase option, as opposed to simply hiring the cooker, by offering a wider selection of models and colours under those terms. The issue of ownership was of key importance, as the gas managers believed that if a customer owned her cooker she was more committed to fuel type and therefore much less likely to switch to electric cooking.

By the 1930s, however, there were still thousands of straight-hire cast-iron cookers in use. An editorial in *Gas World* expressed concern that the continued use of the 'old black cookers' left a poor impression of the industry on the public, as consumers wanted the latest and most efficient appliances. Although, the cast-iron cookers were unfashionable and out-of-date, the majority of the smaller gas undertakings complained that they lacked the resources to replace the old, black hire cookers with those of a more modern design. The writer complained that the smaller companies with the outdated cookers were being unfairly stigmatised as they still offered their customers serviceable appliances. Responding to these arguments, Milne-Watson urged his fellow gas managers to consider stove replacement as an investment in future sales, despite the initial expense of 'ruthless scrapping'. Under Milne-Watson, the GLCC assumed the lead in this area, and, in 1935, the company replaced 92,000 old and obsolete cookers and also supplied an additional 364,000 new cookers.[130] Despite this campaign, according to Frances Goodall, the 'old black cookers' survived in some districts until nationalisation in 1949. Still, the debate over cooker replacement indicates that gas managers were acutely aware of the importance of female consumer taste and appliance design. While the old black cast-iron cookers might have sufficed, these appliances gave the impression that gas fuel belonged to the Victorian era. To counteract this negative impression, the gas companies had to replace as many of the old cookers as possible. The industry's public image and potential sales depended upon it.

Housing Problems and corporate solutions

In the mid-1930s, the gas industry undertook its most progressive advertising projects to date, a direction which earned gas men the reputations of being humanitarian businessmen and supporters of the arts.[131] To win such public acclaim, the BCGA commissioned a series of documentary films dealing with social issues, the best known being *Housing Problems*, *Enough to Eat* and *Children at School.* At the same time, the governors of the GLCC sponsored the construction of a modern housing estate in North Kensington. Kensal House provided affordable housing for 64 working-class families relocated from slum houses destined for demolition. Despite the evident philanthropic merits of these projects, the films and the housing project were essentially expensive advertisements for the gas industry – projects that were motivated by competition with the electrical industry.

As a new advertising medium, documentary film fitted the needs of the commercial market perfectly, offering a combination of entertainment, education and publicity, which were David Milne-Watson's three fundamentals of 'good propaganda'.[132] The gas industry was not the first to explore the commercial possibilities of film. Shell Mex Petrol, Imperial Airways and Anglo-Persian Oil were also sponsoring their own newsreel 'infomercials'.[133] Stanley Baldwin's Conservative government also appreciated the potential of motion pictures by funding a film department at the Empire Marketing Board. A government department that existed between 1926 and 1933, under later Labour and national governments, the EMB promoted the consumption of empire-made goods and foodstuffs. Guided by the imaginative leadership of civil servant, Sir Stephen Tallents, and the creative direction of John Grierson, a philosophy graduate interested in the relationship between the film industry and public opinion, the EMB film group produced several documentaries intended to educate British consumers about the economic and social benefits of Commonwealth trade. The EMB film unit and its successor group at the General Post Office attracted young filmmakers who shared Grierson's ideas that films should educate citizens, raise public awareness about the need for social reform and dignify the labour of ordinary workers.[134] The gas industry's social documentaries, produced between 1935 and 1939, adhered to that agenda.

In 1935, the BCGA commissioned its first series of publicity films. These black and white productions, of 10–30 minutes in length, addressed a range of subject matter from domestic problems and cookery tips to visual explanations of the gas manufacturing process and social concerns related to housing and public health.[135] It was this latter group, the films that explored social problems, that attracted critical acclaim which earned the BCGA and the GLCC a respected place in the British documentary film movement. For example, *Housing Problems*, the most outstanding piece in that first film series, considered the problems of urban slum clearance and re-housing. The film consisted of a series of interviews with a group of Stepney tenants, the majority of whom were women, describing their deplorable housing conditions and the personal frustrations of slum life: the rats and insects, the falling plaster, the constant damp, the overcrowding, the absence of sanitary conveniences.[136] A second group, recently relocated in a new housing estate, expressed their satisfaction with their new homes compared to the dirt and vermin of their previous quarters. None of the women mentioned the usual problems associated with re-housing in the suburbs: the loss of community and the distance from work and shops. The film also featured architect's maquettes of proposed new housing estates to replace existing slums. The models showed a miniature Quarry Hill, in Leeds and the GLCC's own project, Kensal House, in North Kensington.

From its first screening in Stepney town hall in late 1935, *Housing Problems* attracted immediate public attention.[137] Although other major corporations were already making promotional films, according to film supporters including Paul Rotha and Grierson, *Housing Problems* was the first corporate-funded film that was not directly about the sponsor or its product.[138] In addition, it was also the first filmed interview with working-class men and women in their own homes, allowing them to relate their experiences directly to the camera.[139] Apparently, John Grierson's sister, Ruby Grierson, an associate of Anstey and Elton at the film unit, deserves much of the credit for the film's realistic quality. She interviewed and pre-selected the participants prior to filming, and it was she who questioned the subjects on camera. Film historians have suggested that Miss Grierson's presence probably helped put the interviewees at ease, especially the working-class housewives.[140] Moreover, the housing film focused attention on a systemic social problem – namely, the need for affordable homes for working-class families – alluding to the national government's inability to effectively address this issue.

Throughout the 1930s, housing remained a highly politicised issue. The housing programmes of the 1920s had little effect on the living conditions of the lowest sector of the population, the working-poor, who still resided in overcrowded decaying houses. Private enterprise built few houses for rent, and the main beneficiaries of Austen Chamberlain's housing subsidies were middle-class homeowners. The new council dwellings built under Wheatley's 1924 Housing Act served the housing requirements of a lower income group, including small clerks, tradesmen and skilled workmen in stable employment. Rents for council houses averaged around 15 shillings per week, still exceeding the means of the mass of poorer workers.[141] While families with stable incomes were able to migrate to the new suburbs, the urban poor had few options but to remain in the older neighbourhoods, occupying the vacated homes of their social betters which landlords subdivided for maximum occupancy and profit. As long as there was an actual shortage of homes, landlords and councillors did not demolish existing dwellings, however decrepit or unsafe for habitation they were. By 1928, Baldwin's Conservative government pledged itself to slum clearance, a topic of concern to all parties in the 1929 election.[142]

Ramsay MacDonald's second minority Labour Government of 1929–31 tackled the problem in 1930. Arthur Greenwood, the minister of health, presented a Housing Act that awarded subsidies for slum clearance based on the number of persons re-housed. The Greenwood Act also specified that local governments submit five-year plans for slum clearance; however, the deepening economic crisis of 1931 and the subsequent collapse of the Labour government by the end of the year diverted the attention of MacDonald's new national government away from housing and towards unemployment and stabilising the currency.[143] The national government returned to housing legislation in 1933, maintaining grants for slum clearance programmes and introducing a building society scheme to facilitate cheap finance for investment in rental property by private enterprise. The final major piece of interwar housing legislation, the 1935 Housing Act, continued municipal subsidies for slum clearance and introduced an additional grant to eliminate overcrowding by the construction of blocks of flats.[144] Slum clearance legislation served as an important political gesture, but the 1935 Housing Act contained a fatal flaw: the omission of clear guidelines for local authorities as to what constituted a slum in need of demolition and replacement. In effect, slum clearance depended entirely on local political initiative.[145] Still, on a more positive note, progressive borough councils undertook ambitious five-year plans to eradicate their worst houses. The LCC proposed to re-house 5 per cent of its population, involving the demolition of 33,000 dwellings and their replacement with 60,000 new ones.[146]

By the mid-1930s, when the BCGA commissioned the film *Housing Problems*, there existed in Britain an overwhelming disparity in housing conditions. The middle-class housewife in the new suburbs, living in a three-bedroom semi-detached cottage, enjoyed the comforts of a modern home: electric lights, hot water, private bathroom. In contrast inner-city slum dwellers, renting several rooms with no water, no WC and no kitchen, lived much like their Victorian ancestors.[147] This widening gap in material standards of living became a major preoccupation of the gas industry that ultimately resulted in the production of a documentary on this subject. The construction of a model housing complex, Kensal House, was another response to the growing housing crisis. Although these projects constituted an elaborate advertising programme designed to improve the gas industry's image and cultivate customer loyalty among the urban working class, these projects nevertheless tackled an acute social problem and proposed a possible solution.

Film historians concur that the BCGA and the GLCC were motivated by commercial interests, agreeing that the filmmakers persuaded the businessmen to produce socially relevant films.[148] According to Edgar Anstey and Arthur Elton of Associated Realist Films, they 'succeeded in persuading the gas industry' to identify their organisation with a social cause, specifically slum clearance, 'to the advantage of both'. In Anstey's words, the executive at the BCGA appreciated the subject's importance because the gas industry had a 'liberal tradition, a non-conformist tradition'.[149] As young documentary filmmakers, who had worked with Grierson at the EMB and the GPO, both Anstey and Elton accepted the 'Grierson notion' that no corporation, either public or private, could dissociate itself from national social issues, particularly if it provided a public service. But considering the personalities involved in the gas industry's decision to make films like *Housing Problems*, and the BCGA's existing commitment to advertising, it seems doubtful that the gas men required all that much persuasion by filmmakers like Grierson, Anstey and Elton.

The idea to produce documentary films may have originated with A.P. Ryan, an Australian who worked alongside Grierson at the EMB. Milne-Watson hired Ryan in 1931 to manage the GLCC's new publicity department, taking over where Francis Goodenough left off after his retirement that year. Ryan anticipated that documentary film was a medium perfectly suited to the promotional needs of the gas industry because of its ability to reach a wide audience. Not only did the industry have to persuade individual consumers of various economic circumstances, but, more importantly, it had to convince municipal politicians, architects and builders that gas was still a viable alternative to electricity. Ryan's successor at the GLCC, S.C. Leslie, another Grierson disciple formerly with the advertising agency, the London Press Exchange, was equally enthusiastic about film.[150] Milne-Watson hired both these men on the basis of their creative talents and because they ranked among the most progressive individuals in the world of advertising.

But even if the medium was new, the message was not. The strategic use of socially progressive advertising had always been the mandate of the BCGA under the guidance of William Mason, Mrs Cloudesley Brereton, Milne-Watson and Goodenough. Goodenough, the 'born publicist', served as BCGA executive chairman from 1912 to 1936. The production of *Housing Problems* was one of the last major projects completed while the BCGA was still under his direction.[151] As we have already seen, Milne-Watson shared this preoccupation with advertising and corporate image-making. He served as the GLCC director from 1919 until his death in 1945 and, throughout that time, he was the general secretary for the BCGA executive. As previously noted, Milne-Watson undertook an extensive modernisation programme after World War I that included updating company showrooms, expanding customer service facilities and creating a public relations department under A.P. Ryan.[152] Both Goodenough and Milne-Watson recognised the importance of creating a positive public image for both the GLCC and the industry in general, effectively using the BCGA for this purpose. And, in addition to the objectives of Goodenough and Milne-Watson, Mrs Cloudesley Brereton had consistently identified the gas industry with social welfare causes related to housing and nutrition throughout her 20-year career with the BCGA.

In accordance with this mandate, the BCGA promotional literature upheld a progressive position on a variety of social issues. Between the wars, housing remained a predominant theme.[153] Anticipating the reconstruction building boom under the quotas of the Addison Acts, the gas industry had enthusiastically supported government promises for state-funded housing schemes, stressing that the working classes had as much right to comfortable and fully-serviced accommodation as the middle and upper classes. Naturally, their trade literature stressed the fuel and cost efficiency of gas stoves, lights, water heaters and fires, all of which were made affordable to the working classes through hire-purchase payment agreements and penny-in-the-slot meters.[154] Throughout the 1920s, the BCGA aggressively promoted its services to builders, architects and local politicians.[155] Therefore, in light of this sustained interest in housing policy and construction, it was not surprising that the BCGA sponsored films about housing and nutrition. It is likely, therefore, that Anstey and Elton had no trouble convincing the BCGA executive of the importance of these topics, from both a social and a promotional perspective.

But, despite the BCGA executive's enthusiasm for film, some gas men expressed their scepticism. According to one critic, *Housing Problems* was 'overdone' in its graphic depictions of slum life, contained 'too little of the gas message' and was far less effective than an actual demonstration of modern gas appliances in use. Moreover, critics of the film worried that the images of crumbling walls and dingy rooms gave audiences the impression that gas technology was a symptom of bad housing, that it was part of the problem instead of being part of the solution.[156] Responding to this criticism, the BCGA reported that working-class audiences in London received the film enthusiastically by applauding after each screening, leading the association to plan a series of public showings throughout the country, from Newcastle to Liverpool, and also in the south-west.[157] In addition, Milne-Watson claimed that the film clearly indicated the modern use of gas in the working-class flat, and made no direct references to old-fashioned gas lighting or cookers.

Milne-Watson, S.C. Leslie and the BCGA executive appeared entirely satisfied with the film's reception to the extent that they planned a documentary series on various social themes. In 1936, Anstey produced *Enough to Eat*, a film about malnutrition. Dietary scientist Sir John Boyd Orr participated in the production which was narrated by the renowned biologist Julian Huxley. Like *Housing Problems*, the nutrition film contained interviews with working-class women, who related their continual struggles to stretch their food budgets. Government representatives, such as Herbert Morrison of the LCC, acknowledged that malnutrition constituted a national heath crisis. He described public initiatives intended to address the problem, including school meal programmes and milk rations for nursing mothers at welfare centres.[158] In 1937, the Associated Realist Film unit and Grierson's Film Centre produced several additional social documentaries for the BCGA. *Children at School*, a 'review of the public education system', was an indictment of the continued use of unsafe and badly-equipped school buildings. Contrasting the British system with images of Hitler's Germany, the film urged the national government to act quickly and invest in child health and welfare, for the sake of Britain's next generation, but also to counteract the threat of fascism.[159] *The Smoke Menace* also addressed national health issues by portraying the problems and ill-effects of unchecked air pollution.[160] Throughout this series, the gas industry and the filmmakers exploited the theme of British racial decline as a result of shoddy housing, urban pollution, inadequate nutrition and poor education. It was a means of attracting attention to social issues and indirectly suggesting that the gas industry offered a solution. Although the BCGA's more commercial productions outnumbered its social documentaries by about four to one, it was the latter group that earned them the reputation of being progressive businessmen and committed supporters of the fledgling British documentary film movement, a reputation of which Milne-Watson was exceedingly proud.[161]

Curiously, the gas industry's largest and most expensive advertisement has gone unnoticed by film scholars and business historians. Kensal House, shown in *Housing Problems* as an architectural model, was the real advertisement for gas, the cure for slum clearance, for child malnutrition and for smoke pollution. This all-gas model block of flats, designed by a committee of modernist architects and constructed near Ladbroke Grove, North Kensington, addressed the urgent need for slum clearance and offered a clear alternative to the 'problem moments' portrayed in the documentaries. Kensal House itself represented an enormous promotional investment aimed at municipal authorities, at the authors of public housing policy and at the consuming public, particularly women. Moreover, it was an expression of social and economic planning, a demonstration that the public and the private sectors could effectively cooperate to solve social problems as complex as slum clearance and re-housing.

Kensal House is rarely viewed in this manner, that of the GLCC's greatest advertising experiment and an expression of the interwar planning debate (see Fig. 5.8). Architectural historians of the interwar years usually consider the building from a design perspective. In Kensal House, Peter Dean sees the expression of the modernist ideals which motivated

5.8 Kensal House, Ladbroke Grove, designed by Maxwell Fry. View of the low, semi-circular nursery school and play area with apartments behind

Maxwell Fry and his peers in the Modern Architectural Research Society (MARS): the belief that society itself might be rebuilt through efficient, progressive design.[162] According to Alistair Service, Kensal House exemplified the 'international modernism' inspired by Le Corbusier, characterised by the flat roof, the clean lines and use of industrial materials, including reinforced concrete and prefabricated steel.[163] S. Martin Gaskell described Kensal House as a transitional building, representative of the shift from the nineteenth-century working-class tenements to the modern flats constructed after World War II.[164] But Kensal House was significant for other reasons beyond its design innovations. It was transitional in the sense that it presented a desirable alternative to past high-density housing schemes for working-class tenants, such as barracks of model flats constructed by the Peabody Housing Trust. In retrospect, Kensal House had been a more desirable alternative to the rows of council flats built during the 1960s. To the architects and businessmen responsible for its construction, Kensal House demonstrated that the modernist style was not reserved for elitist luxury flats and Surrey estates. It demonstrated that working-class housing could be modern, attractive, comfortable and cost efficient.

At the opening ceremonies for Kensal House, in March 1937, Milne-Watson outlined the reasons why his company built a block of flats. According to him, it was pointless to re-house working-class people in new houses or flats with 'old-fashioned laborious kitchens and equipment'. Kensal House demonstrated that working-class tenants could be provided with every gas amenity that middle-class families demanded, and without abnormal capital outlay or prohibitive operating costs. It presented a solution to 'the problem of providing the right living conditions for re-housed slum dwellers'.[165] Considering the GLCC's successful 1933 legislative challenge against 40 local housing committees who had restricted the use of gas lighting in their districts,[166] the construction of Kensal House was a clear, but costly, celebration of this political and commercial victory.

To undertake the project, in early 1935, the governors of the GLCC invited submissions from a group of established architects, including Robert Atkinson, C.H. James, Wornum, Michael Tapper and Maxwell Fry. Although both Tapper and Wornum later produced showroom designs for the GLCC,[167] Fry won the housing design competition. According to him, it was not that Milne-Watson was a committed follower of the modernist MARS aesthetic, but rather that Fry's winning design incorporated the circular curve of an old gasholder, although it was not specified as available for the building scheme (see Fig. 5.8).[168] This curved wall was a design feature of the crèche, while the remaining circle provided the outline for the play yard. The GLCC also enlisted the assistance of Elizabeth Denby, as housing manager. A journalist and recognised authority on working-class housing and interior planning, Denby had collaborated with Fry on an earlier project, another modernist housing complex, Sassoon House, built in south London.[169]

In Kensal House, both Denby and Fry expressed their ideal of what working-class housing could and should be. Looking back on the project, Fry noted that it was 'no ordinary block of flats, but a community in action'.[170] Denby possessed definite views about high-density urban use, expressed in her 1938 book on public housing and city planning in Europe. For example, she advocated the development of small-scale, carefully planned multi-storey urban communities. She strongly condemned the wasteful fashion of 'garden city' speculative building, the type of suburban sprawl that absorbed valuable agricultural land and encouraged road construction, and left the urban core empty and 'rotting'. Although Denby made no direct reference to her employer's housing complex, Kensal House stood in obvious contrast to the failings of interwar British housing developments that she cited in her study.[171]

What was so special about a reinforced concrete block of 68 flats, built on land donated by private industry, jointly funded by public subsidies and corporate investors? There are several instances of businessmen funding model housing for the working-classes. Notable examples date from the late nineteenth century: Port Sunlight, built by the soap magnate W.H. Lever; or the town of Bourneville, constructed by the Quaker cocoa manufacturer George Cadbury. At the turn of the century, Joseph Rowntree enlisted Raymond Unwin to design his model estate, New Earswick, outside York. Port Sunlight was a company town, a model village for Lever's employees and an expression of his version of profit-sharing; Bourneville and New Earswick provided employee housing, but also accepted suitable tenants who did not work for the village patron. In common, they shared the 'garden city' aesthetic of the planned cottage estate with its semi-detached houses, winding lanes and private gardens. The model villages were intended for the middle classes, tradesmen and the skilled sectors of the working classes in regular employment.[172]

In contrast to the garden lanes and low-density cottage dwellings of earlier corporate-sponsored housing, Kensal House was a space-efficient 'urban village', built and priced for slum dwellers. Like Raymond Unwin and Sir Ernest Simon,[173] a Liberal housing reformer then undertaking the Wythenshawe development south of Manchester, Maxwell Fry aspired to design homes that provided privacy, comfort, fresh air and sunlight for their occupants. He was convinced that these specifications were possible within a multi-storey complex of flats. Internal stairways and private balconies offered tenants a sense of privacy, while the complex's cafe, recreation rooms and communal gardens helped to create a community atmosphere. The on-site nursery school, with its classrooms and fenced playground, was one of the first day nurseries in Britain, and, according to Fry, the first to be constructed in conjunction with a housing scheme. Staffed by a head teacher, several assistants and a cook, the crèche provided full- and part-time daycare for children between the ages of two and five, to assist working women or simply provide some relief for busy mothers. When it opened in January 1937, there were 66 children enrolled.[174] Although the school was on the estate, the Board of Education and the LCC jointly supervised its operations. Children of the tenants were given priority; however, school vacancies were filled from the 'overcrowded and badly housed families in the immediate neighbourhood'.[175]

Although it was not a garden estate, Fry and Denby attempted to establish a cohesive community at Kensal House using physical design and social facilities. Likewise, Elizabeth Denby expected the tenants to 'take responsibility for the order and good atmosphere of the estate'. For example, there was a tenants' executive committee, made up of elected representatives from each floor of the building and chaired by Denby. The committee dealt with such issues as noise control, club activities in the common rooms and the preservation of the estate's trees, flowers and grass.[176] In this manner, Kensal House was 'no ordinary block of interwar rental flats', but anticipated all the elements of the modern housing cooperative.

The housing boom of the 1920s depended on private speculation by builders. The majority of the homes constructed were too expensive for families in the lowest income brackets, with the result that the middle and skilled working classes were the chief beneficiaries. Although there existed a crucial need for high-density affordable housing, few modernist architects responded to this problem. A number of architects, such as the members of the Tecton group, experimented with designs for what Le Corbusier termed the 'vertical garden city'. These designs, notably Berthold Lubetkin's Highpoint One, at Highgate, were for luxury flats with rents from £130 to £234 per annum. The complexes included service flats and maids' bedrooms.[177] Even modest fully-equipped flats, such as the block constructed along Broad Street in Lambeth, or Evelyn Court in Hackney, commanded rents of between 15 and 18 shillings per week, a price still beyond the means of most slum tenants.[178]

In contrast, Kensal House was deliberately constructed to assist families with very low incomes. Modelling its administrative organisation on the Peabody Housing Trust, the directors of the GLCC formed a subsidiary company, the Capital Housing Association Limited, to oversee the construction and management of Kensal House.

The Capital Housing Association's board of directors included Milne-Watson (who also sat on the government's National Housing Committee), Stephen Lacey (the company's chief controller of sales) and S.C. Leslie, the GLCC's publicity manager. Housing activist and Kensal House manager, Elizabeth Denby, also sat on the board. Although the Capital Housing Association was a limited company of public shareholders, the GLCC controlled the scheme, having funded the housing venture with a capital loan of £58,892.[179] The company also made up the nursery school's annual £200 shortfall and loaned money to the Capital Housing Association for its social clubs.[180] In addition to the corporate funds, the association received an undisclosed amount from the Treasury, under the terms of the 1935 Housing Act. In consultation with Kensington Borough Council, the board of the Capital Housing Association selected 380 slum dwellers, including 244 children, to occupy the new flats. The rent for a three-bedroom flat with a separate kitchen, living room, bathroom and two balconies was 11 shillings and sixpence per week; the rent for a two-bedroom version was nine shillings and sixpence per week – roughly four shillings less than the average weekly rent of a London council house. The gas company stressed that while Kensal House rents were within the means of the unskilled labouring class, tenants also saved on fuel costs, ideally leaving more of the weekly budget for food.[181] In this respect, the company addressed the interrelated problems of poor housing and inadequate nutrition, which were social ills that feminist investigators increasingly defined as women's health issues.[182]

Women's pressure groups had long noted the effects of poor housing on women's health. In her study of pre-war Lambeth, Maud Pember Reeves described the difficulties of working-class families from the perspective of the women who managed their homes on slim budgets and lived in overcrowded, unserviced houses.[183] The wartime Women's Sub-committee on Housing, chaired by Lady Emmott, consulted working-class women regarding their housing priorities for the reconstruction years, while the Women's Co-operative Guild campaigned for higher standards in the construction of low-rent working-class housing. In 1933, the Women's Health Inquiry Committee, chaired by Gertrude Tuckwell, questioned 1,250 married working-class women about their health, housing, children and family budgets. In her report on the committee's findings, secretary Margery Spring Rice of the North Kensington Women's Welfare Centre, stressed the relationship between women's health and housing. Because the 'house is a mother's workshop', women felt the effects of poor living conditions more than their husbands and children, who spent more time outside of the home. In her introduction, Dr Janet Campbell identified the connection between women's poor health, bad housing and excessive domestic labour.[184] The committee recommended affordable fully-serviced houses that were private and easy to maintain. They also noted the health benefits of even a few hours of weekly leisure time for working women, identifying the need for community facilities where women might gather outside of the home, without spending much money.

Kensal House addressed these feminist concerns. Using new materials and rational design, Maxwell Fry and Elizabeth Denby constructed flats that would minimise the 'soul-killing drudgery' that robbed women of their time, strength and overall health. In addition, the provision of a nursery school for toddlers addressed the childcare needs of women who worked outside of the home, while it also offered some relief to mothers with several small children. The community centre met women's need for leisure time by providing a meeting place that was convenient and inexpensive. In her 1939 publication, Spring Rice includes an illustration provided by the GLCC of the Kensal House community centre, where a group of women are seated at tables, visiting and sewing over cups of tea.[185] Evidently, the Women's Health Inquiry Committee was familiar with the gas company's housing experiment. As feminist housing reformers lobbying on behalf on working-class women, they recognised that Kensal House offered solutions to many of the health issues that they attributed to poor housing.

In addition to the social welfare component of this housing project, there can still be little doubt that Kensal House was a corporate-funded advertisement, and a very expensive one. Perhaps the most obvious point in this regard was the building's use of switch-controlled gas lighting. This artificial means of illumination was suffering a rapid decline in favour of electricity despite technological improvements, such as the development of the wall-switch system. Even the GLCC's 1933 parliamentary actions, resulting in an act that prohibited municipal governments from restricting the use of gas services in their council dwellings, did not significantly increase the numbers of new gas lighting installations. Although gas remained popular for cooking and heating, few architects and builders seriously considered gas lighting for new buildings after 1930.[186] Ignoring the apparently retrogressive lighting arrangements, Kensal House was favourably received by the architectural community, even by the editors of the pro-electric journal, the *Architectural Review.* In particular, they noted the building's many labour-saving design features, including the painted walls, prefabricated kitchen cabinets, steel counter tops and linoleum floors. The complete elimination of coal, for either cooking or heating, further reduced domestic labour and the need for fuel storage in the flats.[187] In particular, the architectural press appreciated that the 'urban village' concept of

Kensal House presented a possible solution to the problems of slum clearance and urban renewal.

Contemporary housing reformers, progressive architects and well-intentioned businessmen hoped that Kensal House provided the 'right living conditions' for working-class families. But did the tenants agree? A survey of the building's occupants, conducted in 1942 by the publicity journalist Marjorie Milne, helps to shed some light on this question. A.E. Sylvester, the managing director of the GLCC, hired Mrs Milne to interview the Kensal House residents and compile the results of those interviews in a confidential report intended for the board of directors of the gas company. Milne was the former editor of *Modern Living* and *Hearth and Home*, two publications produced by the GLCC and circulated to 1.5 million female customers as part of its local publicity. *Modern Living* was for 'housewives of moderate means' while *Hearth and Home* was for women 'who had to consider every penny'.[188] Given her experience with the GLCC, Mrs Milne knew her clients well and was accustomed to overstating the positive attributes of gas. That was not the intention with this assignment, however, and, in the report's opening statement, Mrs Milne wrote that she had adhered to her employer's wish that she write an entirely frank report. According to Milne, 'facts favourable to gas have not been magnified, nor have unfavourable opinions been suppressed or even toned down'. To better convey the sentiments of the Kensal House occupants, Mrs Milne made, in her view, every effort 'to present the arguments objectively through the mouths and in the words of the tenants themselves'. She titled the document simply: 'What the tenants think of Kensal House'[189].

In preparation for the report, Milne interviewed 68 tenants, one from every flat in the building. Of the 68, only ten were not original tenants; the remainder had lived in the building since it had opened six years earlier. Although Milne did not identify the interviewees by name or gender, the responses suggest that she spoke with women, given the frequent references to children, the size of kitchens, concerns about dampness and dust and the problems of drying laundry and husbands' work clothes. Of the 68 persons interviewed, 61 felt that generally they were 'better off' at Kensal House. Tenants responded positively when questioned about some of the design features. For example, 62 respondents liked and used the sun balconies that nearly compensated for the lack of a garden.[190] With the exception of the lighting arrangements, the tenants were happy with the gas services offered, which included the hot water heaters, the coke fires in the sitting rooms, the gas coppers for laundry and the gas cookers, although 'one or two' tenants thought their cookers were too small to prepare meals for a large family.[191] Regarding the lighting arrangements, 36 of the tenants preferred electric lighting, because it was cleaner, cheaper and easier to maintain, as the gas switches were frequently out of order. There were 28 who were satisfied with gas lighting fixtures, but only three tenants had used both and preferred gas.[192]

Even though the majority of the tenants interviewed admitted that Kensal House was an improvement over their former residences, there were still a number of recurring complaints. From a design perspective, the size of the kitchens appeared to be a common irritation. According to Marjorie Milne, 'Kensal House flats were planned to enable the housewife and her family to eat their meals "away from the pots and pans"'; however, she found that nearly half of the tenants preferred to dine in their kitchens despite the limited space. There were 32 tenants that wanted a kitchen large enough for a table to seat the entire family for meals, thereby allowing housewives to keep the sitting room tidy. Of the 21 tenants that always took their meals in the kitchen, Mrs Milne noted that some used the ironing board or the shelf by the dining hatch for seating.[193] Mrs Milne sympathetically reported the views of the working women she interviewed, stating that their desire for an eat-in kitchen should not be dismissed as an indication of their inability to 'rise' to their new, more modern surroundings. She put the problem thus:

> If the kitchen is so small that everything except the actual cooking and washing has to be done in the big room, if the family eats there (and in some families meals are almost continuous), has to work and play there, if Father mends his boots, does his bit of carpentry in the sitting room, changes his working clothes in it and then hangs them before the coke fire to dry, it is not surprising that the imitation Persian strip looks like a rag and that the furniture is 'all broke up'. Nor is it surprising that a woman with any feeling for order longs either for a kitchen which can take some of the strain or a sitting room that can be kept decent for special occasions.[194]

Dampness and condensation in the building appeared to be the greatest structural problems, and the housing manager thought that building contractors should be called in to make recommendations.[195] Noise, dust and the lack of privacy were also sources of discomfort and even conflict between tenants. Some residents complained about the excessive noise of the children in the complex, or of overhearing the smashing crockery of their neighbour's domestic disputes. The following quote is representative of several in the report: 'The one frightful drawback is the terrible noise. You can hear every sound through the walls. If people next door quarrel or raise their voices you hear every word.' While some attributed the noise problem to poor design and low ceilings, others believed that the constant racket was due to the 'uncontrollable behaviour of other tenants, chiefly their children'.[196] Of the 68 persons interviewed, 25 tenants hoped one day to have a small house of their own, with a garden, perhaps a basement and bit more room for drying clothes.

The Kensal House survey is instructive on several levels. According to Marjorie Milne's tabulations, it indicated that housewives were quite satisfied with gas technology in every area except lighting. However, in terms of practical design, Maxwell Fry and his team had misunderstood the fundamental priorities of working-class women, particularly with regard to the allocation of space for the kitchen and living areas. Unfortunately, regarding the most progressive features of Kensal House, the children's nursery school and the women's recreation centre, Mrs Milne's report was silent.

Conclusion

Although the gas and electrical industries had coexisted warily for decades, it was not until the interwar years, and the implementation of parliamentary measures to encourage electrical development, that the two power industries became serious competitors for the domestic market. Gas men and women, working in sales, advertising and public relations, realised the urgency of the situation, and responded by undertaking sophisticated promotional strategies for cultivating and maintaining customer loyalty. Design historians have portrayed the gas industry as conservative and retrogressive compared to its more modern competitor. This assumption rests on the obvious advance of electrical lighting compared to the demise of gas as an illuminant.

Despite the decline of gas lighting, the electrical industry did not supersede gas during the interwar period. Although there was a rapid increase in the numbers of electrical consumers between 1926 and 1939, the gas industry retained its competitive edge in the domestic market for home heating and cooking, even expanding into the new field of artificial refrigeration. In a concerted effort to modernise its corporate image, while retaining and developing a loyal customer base, the gas industry, particularly the BCGA and the GLCC, adopted a series of sophisticated costly promotional strategies. Their use of a modernist retail design aesthetic in showrooms and at trade exhibitions, their experiments with documentary film and model housing construction indicate that the gas industry was far from old-fashioned in its approach to public relations and advertising. Amid the planning debates of the mid-1930s,[197] the housing and film projects offered a subtle political commentary on the possibilities of greater cooperation between businessmen and politicians to remedy Britain's chronic social problems related to housing and health.

Within all of these projects, those individuals who were responsible for the industry's public relations policies, including Milne-Watson, Goodenough and Mrs Cloudesley Brereton, considered the impact of the advertising on female consumers in the home. More than ever before, the industry paid close attention to the 'women's point of view', directing advertising and sales campaigns at women and remodelling showrooms in fashionable colour schemes. This dedication to the feminine market is clearly illustrated by the actions of the companies' home service departments between the wars, the subject of the final chapter. Tensions did arise, however, as salesmen received greater professional training and recognition within the industry than women. Perhaps in an effort to assert the special nature of their roles and responsibilities, the women demonstrators cooperated with interwar feminist groups to increase the public profile of their industry, but also to better assist the female consumers they allegedly spoke for.

Notes

1 'Art in industry', *GJ*, 22 October 1924, 219.
2 Freeden, *Rainbow Circle*, 367.
3 'Art in industry', *GJ*, 22 October 1924, 219.
4 Penny Sparke, Felice Hodges, Emma Dent-Coad and Anne Stone, *Design Source Book* (London, 1986), 86–87; Penny Sparke, *An Introduction to Design and Culture in the Twentieth Century* (London, 1986), 45–49.
5 'The gas exhibit: first official visit', *GJ*, 21 May 1924, 526.
6 Hannah, *Electricity Before Nationalisation*, 88–104.
7 Ibid., 186–192.
8 Charles Loch Mowat, *Britain Between the Wars 1918–1940* (Cambridge, 1983), 342–343; Sidney Pollard, *The Development of the British Economy, 1914–1990*, 4th edition (London, 1992), 43–45; John Stevenson, *British Society 1914–45* (Harmondsworth, 1990), 109–110; John Stevenson and Chris Cook, *Britain in the Depression: Society and Politics, 1929–39* (London, 1994),17–18; Forty, *Objects of Desire*, 184–200; Worden, 'Powerful Women' in Attfield and Kirkam (eds), *A View From the Interior: Feminism, Women and Design*, 128–147.
9 Bowden and Offer, 'The Technological Revolution That Never Was' in de Grazia with Furlough (eds), *The Sex of Things*, 255–256.
10 'Dissatisfaction with electricity at Eltham', *GJ*, 15 September 1926, 613.
11 Bowden and Offer, 'The Technological Revolution That Never Was' in de Grazia with Furlough (eds), *The Sex of Things*, 267.
12 'Opening of the empire exhibition: pioneers of the scheme', *Daily Telegraph*, 24 April 1924, 13.
13 Ibid., 13; Mowat, *Britain Between the Wars*, 177–178, 227; Stevenson, *British Society*, 5, 387.
14 'Festival of empire', *The Times*, 23 April 1924, special supplement.
15 'Exhibition opened amid wondrous scenes', *Daily Express*, 24 April 1924, 3.
16 'General notes from the exhibition', *GJ*, 30 April 1924, 316.

17 'Exhibition opened amid wondrous scenes', *Daily Express*, 24 April 1924, 1.
18 Ibid., 1.
19 Richards, *Commodity Culture*, chap 1.
20 For reports of the gas industry's involvement in public exhibitions see Universal Cookery and Food Association, *Cookery Annual and Exhibition Souvenir* (London, 1894); 'The gas exhibition', *The Cookery Teacher* (November 1904), 136–138; 'The exhibition', *The Cookery Teacher* (March 1905), 70; NAL/*Daily Mail Ideal Home Catalogue* (April 1910), 34–35; NAL/*Daily Mail Ideal Home Catalogue* (October 1913), 32; 'The Ideal Home exhibition: gas well represented', *GJ*, 2 March 1923, 602; 'The Ideal Home exhibition', *GJ*, 5 March 1924, 546.
21 'A great exhibition', *Co-Partners' Magazine* (March 1924), 51; 'The British Empire exhibition', *GJ*, 30 April 1924, 314.
22 Sir Lawrence Weaver, 'Exhibitor's architecture', *Architectural Review*, 55 (1924), 224, 228.
23 'General notes from the exhibition', *GJ*, 30 April 1924, 316; Francis Goodenough, 'What Wembley means to the industry', *GJ*, 30 April 1924, 315.
24 'General notes from the exhibition', *GJ*, 30 April 1924, 317–318.
25 H. Austen Hall, 'The planning of some American department stores', *RIBA*, 20 March 1920, 250.
26 'General notes from the exhibition', *GJ*, 30 April 1924, 317.
27 Nevett, *Advertising in Britain*, 149.
28 'General notes from the exhibition', *GJ*, 30 April 1924, 319.
29 Ronald P. Jones, 'The exhibits', *RIBA*, 7 June 1924, 506.
30 Fred Boissonnas, 'A general view', *Architectural Review*, 58, (1925), 10.
31 'The rest lounge', *GJ*, 30 April 1924, 317.
32 'The gas exhibit: first official visit', *GJ*, 21 May 1924, 525–526.
33 AAD/1978/2/2/Heal and Son's Corporate Records, Minutes of the Board of directors, 9 January, 1924; Ambrose Heal reported that he had received an invitation from Sir Lawrence Weaver to execute the model rooms with the guarantee that Heal would receive equal publicity with the designers. Minutes of the Board of Directors, 30 January 1924; Heal reported to the board that the Associated Gas Industries placed an order valued at £1,580 for furniture, curtains and floor coverings and leased an additional £1,000 worth of furniture for its exhibition stand.
34 AAD/1978/2/1979/Heal Archives, Press cuttings of British Empire Exhibition, *Journal of Decorative Art*, July 1924; *Homes and Gardens*, November 1924.
35 'The gas exhibit: first official visit and luncheon', *GJ*, 21 May 1924, 526. For a discussion of the 1920–1921 slump and slow recovery after mid-1923, see Derek Aldcroft, *The British Economy. Vol. 1, The Years of Turmoil 1920*–1951 (Atlantic Highlands, NJ, 1987), 5–11, 12–14.
36 Board of Trade, *Gas Apparatus (Sale and Hire)*, 717, 4–5.
37 One BTU is the amount of heat required to raise the temperature of one pound of water one Fahrenheit degree. A therm equals 100,000 BTUs.
38 'Gas regulation act', *GJ*, 11 August 1920, 280, 292–293.
39 Fuel Research Board, *Report to the Board of Trade* (January 1919), Cdm. 108 xxii, 569, 4–9.
40 'Price and dividend movements on the therm basis', *GJ*, 29 September 1920, 385.
41 Board of Trade, *Gas Apparatus (Sale and Hire)*, 3–5.

42 CAB/24/127/C.P. 3133/*Coal Mines Stoppage, Report on Emergency Arrangement*, April–July 1921, 608–615, 641.
43 LMA/B/GLCC 49/ *Minutes of the Board of Directors*, 20 May 1921, 428, reported the deaths of three persons by coal gas poisoning; *Minutes of the Board of Directors*, GLCC, *Minutes*, 3 June, 1921, 434, reported another two deaths; 'Gas burner danger', *The Times*, 21 January 1922, 7, described the death of a Manchester family of three; 'Carbon monoxide danger', *The Times*, 4 February 1922, 4, reported proceedings of an inquest into the deaths by carbon monoxide poisoning of four members of a Liverpool family.
44 'Letter: Dr. Charles Carpenter', *The Times*, 2 January 1922, 6; 'Domestic gas difficulties: profiteering denied', *The Times*, 6 January 1922, 6; 'Letter: Charles Carpenter', *The Times*, 16 January 1922, 6; 'Letter: Dr. Charles Carpenter', *The Times*, 9 February 1922, 19; 'Letter: Dr. Charles Carpenter, *The Times*, 17 February 1922, 11.
45 'Board of Trade and Gas Regulation Act', *The Times*, 11 February 1922, 10; 'Carbon Monoxide in Domestic Gas Supply', *The Times*, 14 February 1922, 15.
46 In late 1922, the *Daily Mail* and the *Pall Mall Gazette* imputed charges against the gas companies regarding gas quality and pricing. *The Times* resumed its attack in early 1923 by printing an editorial by Professor Bone, a critic of the industry. Milne-Watson, Goodenough and Carpenter, who represented England's two largest metropolitan gas companies, defended current gas standards and pricing. See letters to *The Times*, 4 January 1923, 11; *The Times*, 8 January 1923, 10; *The Times*, 10 January 1923, 7; *The Times*, 12 January 1923, 9; *The Times*, 16 January 1923, 7; *The Times*, 17 January 1923, 5; *The Times*, 1 February 1923, 7; *The Times*, 15 February 1923, 18; *The Times*, 17 February 1923, 11; *The Times*, 26 February 1923, 15; *The Times*, 28 February 1923, 8; *The Times*, 5 March 1923, 13; *The Times*, 10 March 1923, 7, 11; *The Times*, 15 March 1923, 21.
47 Board of Trade, *Report of Departmental Committee, Methods of Charging for Gas on a Thermal Basis* (1923), Cmd. 1825 xi 569, 7, 13–16.
48 Ibid., 6–7.
49 LMA/LCC/Administrative County of London, *Meeting of Council, Report: gas versus electricity in the home*, 24 February 1925, 3.
50 Ibid., 5; 'Housing schemes', *GJ*, 17 September 1926, 417.
51 'Gas or electricity for housing schemes', *GJ*, 14 February 1923, 402; 'The King at Becontree', *GJ*, 13 June 1923, 666; 'London County Council housing schemes', *GJ*, 10 September 1924, 733; 'Gas and electricity in L.C.C. dwellings', *GJ*, 27 October 1926, 239.
52 LMA/LCC, *Report: gas versus electricity*, 3
53 Johnson, *Land Fit for Heroes*, 41, 191.
54 Ibid., 426–430.
55 Hannah, *Electricity Before Nationalisation*, 77.
56 Stevenson, *British Society*, 108–109; Stevenson and Cook, *Britain in the Depression*, 17; Mowat, *Britain Between the Wars*, 455; Pollard, *Development of the British Economy*, 44–45.
57 Pollard, *Development of the British Economy*, 43.
58 Aldcroft, *The British Economy*, 15–22.
59 Pollard, *Development of the British Economy*, 44; Hannah, *Electricity Before Nationalisation*, chap. 4; Mowat, *Britain Between the Wars*, 455.

60 Hannah, *Electricity Before Nationalisation*, 197–198; Stevenson and Cook, *Britain in the Depression*, 17.
61 Offer and Bowden, 'The Technological Revolution That Never Was', in de Grazia with Furlough (eds) *The Sex of Things*, 245.
62 Ibid., 250–259.
63 Hannah, *Electricity Before Nationalisation*, 205–206; 77 per cent of British homes owned electric irons; Offer and Bowden, 'Revolution That Never Was', in de Grazia with Furlough (eds) *The Sex of Things*, 260, Table II, indicate that 71.5 per cent had radios.
64 Hannah, *Electricity Before Nationalisation*, 208.
65 Political and Economic Planning Group, *Report on the Electricity Industry in Great Britain* (London, 1936), 76.
66 Ibid., 77–84.
67 Political and Economic Planning Group (PEP), *Report on the Gas Industry in Great Britain* (London, 1939), 5–10.
68 Elizabeth Roberts, *Women and Families, An Oral History, 1940–70* (Oxford, 1995), 29; Davidson, *Woman's Work*, 68; PEP, *Report on the Gas Industry*, 79.
69 PEP, *Report on the Gas Industry*, 5. Breaking down the uses of gas, the PEP report compares the 60–70 per cent domestic load, with 10–20 per cent commercial, 10–20 per cent industrial, 5 per cent public lighting. Unfortunately, the domestic load figure does not break down into specific percentages for domestic lighting, heating and cooking.
70 PEP, *Report on the Gas Industry*, 77–79.
71 Forty, *Objects of Desire*, 115–118.
72 Doreen Yarwood, *British Kitchen,* 100; Davidson, *Woman's Work*, 67; Hardyment, *From Mangle to Microwave*, 128–130.
73 Yarwood, *British Kitchen*, 97, 100.
74 'Up-to-date labour-saving kitchens', *ATAOUFG* (February 1924), 13, 15.
75 'Main's "housing" cookers', *GJ*, 1 June 1920, 495.
76 'Up-to-date labour-saving kitchens', *ATAOUFG*, 15; 'The King at Becontree', *GJ*, 13 June 1923, 666; 'Smokeless heating, cooking, water-heating and lighting equipment in housing schemes', *ATAOUFG* (October 1926), 89–94.
77 Barty-King, *New Flame*, 204.
78 'Smokeless heating, cooking and water-heating in housing schemes', *ATAOUFG* (October 1926), 96.
79 'Gas and electricity in LCC dwellings', *GJ*, 27 October 1926, 239.
80 Political and Economic Planning Group (PEP), *Report on the Electrical Industry in Great Britain* (London, 1936), 78–79.
81 'Dissatisfaction with electricity at Eltham', *GJ*, 15 September 1926, 613; 'Dissatisfaction with electricity at Eltham: noisy public meeting', *GJ*, 22 September 1926, 66–67.
82 'Housing schemes', *GJ*, 17 November 1926, 417.
83 LMA/ B/SMGC, *Minutes of the Board of Directors*, 29 June 1927, 159.
84 LMA/B/ SMGC, *Minutes of the Board of Directors*, 20 July 1927, 251; SMGC, *Minutes of the Board of Directors*, 17 August 1927, 261.
85 'Electricity supply memoranda: housing estates', *GJ*, 19 October 1927, 176.
86 Francis Goodenough, 'How to meet the electricity development scheme', *GJ*, 3 February 1926, 270.
87 PEP, *Report on the Gas Industry*, 3–5.

88 'The present position of the gas industry by a gas engineer', *GJ*, 30 April 1924, 309.
89 Ibid., 308–309.
90 George Clarry, 'Showrooms in relation to gas supply', *First Annual Report and Conference Proceedings of the BCGA*, 171–206.
91 Everard, *History of the GLCC*, 278, 283.
92 Clarry, 'Showrooms and Gas Supply', *First Annual Report and Conference of the BCGA*, 183.
93 Everard, *History of the GLCC*, 300.
94 Walter T. Layton, 'Gas in the home', *Co-Partnership Journal* (December 1909), 269–275.
95 Victoria de Grazia, 'Empowering Women as Citizen-Consumers', in de Grazia with Furlough (eds), *The Sex of Things*, 281.
96 'Gas Light and Coke Company's new central showrooms: service enterprise at Horseferry Road', *Co-Partners' Magazine* (May 1926), 128.
97 Everard, *History of the GLCC*, 319–321.
98 'Gas Light and Coke Company's new central showrooms: service enterprise at Horseferry Road', *Co-Partners' Magazine* (May 1926), 128–132; 'Opening of the company's new showrooms in Islington: the union of art and utility', *Co-Partners' Magazine* (December 1925), 310–313; 'Opening of the new Kensington showrooms', *Co-Partners' Magazine* (November 1926), 303–311; 'Gas service in east London', *GJ*, 30 July 1930, 252–254; 'Our new premises at Hackney', *Co-Partners' Magazine* (August 1931), 311–312; 'Our new showrooms and offices at Stratford', *Co-Partners' Magazine* (April 1924), 159–160.
99 Alison Light, *Forever England: Femininity, Literature and Conservatism Between the Wars* (London and New York, 1991), 34–35.
100 Sparke, *As Long As It's Pink*, 126.
101 Sparke et al., *Design Source Book*, 100–104, 119–125.
102 Neil Harris, 'Museums, Merchandising and Popular Taste: The Struggle for Influence' in *Cultural Excursions: Marketing Appetites and Cultural Tastes in Modern America* (Chicago, 1990), 70–74.
103 Sparke et al., *Design Source Book*, 120.
104 'Opening of the company's new showrooms in Islington', *Co-Partners' Magazine* (December 1925), 311.
105 'Opening of the new Kensington showrooms', *Co-Partners' Magazine*, 307–308.
106 Dan Klein, Nancy McClelland and Malcolm Haslam, *In the Deco Style* (London, 1987), 227–231.
107 Klein, et al. *Deco Style*, 211–212; Sparke et al., *Design Source Book*, 120–121; David Dean, *The Thirties: Recalling the English Architectural Scene* (London, 1983), 100–102, plates 18 and 19.
108 J.M. Richards, *An Introduction to Modern Architecture* reprint of revised 1956 edition (Harmondsworth, 1959), 99–100; Dean, *English Architectural Scene*, 89–92.
109 Dean, *English Architectural Scene*, 92–96; Richards, *Modern Architecture*, 29, 144.
110 Harris, *Cultural Excursions*, 361.
111 'Mrs. M.A. Cloudesley Brereton relinquishes the editor's chair', *GJ*, 29 June 1932, 864.

112 Alan Windsor, *Handbook of Modern British Painting 1900–1980* (Aldershot, UK, and Brookfield, VT, 1992), 109. A painter known for his cubist-inspired murals, Clive Gardiner was also principal of Goldsmith's College between 1929 and 1957.

113. Pat Hodgson and Bruce Cowen, *Eric Fraser: An Illustrator of Our Time, Exhibition Catalogue* (London, 1991), 9, 19; Harris, *Cultural Excursions*, 361. Harris names Fraser among a list of British artists, including Vanessa Bell, Henry Moore, Paul Nash, Tom Purvis and Barbara Hepworth who did advertising work between the wars.

114 'Gas: the spirit of coal', *Punch*, 6 May 1925, xi.

115 Cowen and Hodgson, *Eric Fraser*, 21.

116 'Opening of the new Oxford street showrooms', *Co-Partners' Magazine* (January 1937), 10–11.

117 'New Queen street showrooms', *Co-Partners' Magazine* (December 1936), 566–568.

118 'Mr. Therm at Twickenham', *Co-Partners' Magazine* (July 1934), 312; 'Up-to-date gas showrooms for growing N.W. London district', *GJ*, 10 June 1936, 776.

119 'New showrooms at Leytonstone', *Co-Partners' Magazine* (September 1937), 505–506.

120 Dean, *English Architectural Scene*, 82–83, 92.

121 Sparke, *As Long As It's Pink*, 128.

122 'Local offices and showrooms', *Co-partnership Journal* (May 1929), 127; 'Company's new showrooms at Court Road, Mottingham', *Co-partnership Journal* (June 1937), 168.

123 'Charlton showrooms', *Co-partnership Journal* (July 1937), 213; 'Walworth Road showrooms', *Co-partnership Journal* (August 1937), 231.

124 'Gas in the marketplace', *Co-partnership Journal* (June 1937), 181.

125 A.V. Bishop, 'The "tone" of showrooms', *GJ*, 15 February 1928, 37.

126 AAD/1980/2/11, Dorothy Braddell Papers, Sketches for Crosse & Blackwell, 1925; AAD/2/81, Plan for kitchen of weekend house, British Pavilion, Paris International Exhibition, 1937; AAD/2/81–AAD/2/110, Kitchen planning pamphlets, AAD/2/107/ 'Kitchens up to date', *Homes and Gardens*, May 1933, 569. Also see Patrick Abercrombie (ed.), *The Book of the Modern House* (London, 1939), 338–337, for Mrs Braddell's essay on kitchen and bathroom planning.

127 Gas Journal, *New Kitchens for Old*, foreword by Dorothy Braddell (London, 1937).

128 Burnett, *Social History of Housing*, 232–234, 274–276.

129 'New showroom at Ilford', *Co-partners' Magazine* (March 1936), 100.

130 Goodall, *Burning to Serve*, 210–212.

131 Rachael Low, *The History of the British Film, 1929–1939: Documentary and Educational Films of the 1930s* (London, 1979), 92–94; Forsyth Hardy (ed.) *Grierson on Documentary* (New York, 1947), 83–84, 171; Elizabeth Sussex, *The Rise and Fall of British Documentary* (Berkeley, 1975), 62.

132 'Our new films', *Co-Partners' Magazine* (November 1936), 515.

133 'The gas industry presents', *Co-Partners' Magazine* (November 1937), 611; Paul Swann, *The British Documentary Film Movement* (Cambridge, 1989), 106–107.

134 Erik Barnouw, *Documentary: A History of the Non-Fiction Film*, 2nd revised edition (Oxford, 1993), 89–95; John Baxendale and

Chris Pawling, *Narrating the Thirties: A Decade in the Making* (Basingstoke, 1996), 25–28.

135 British Commercial Gas Association, *Modern Films on Matters of Moment* (London, 1939), 3.

136 British Commercial Gas Association and Associated Realist Films, *Housing Problems* (London, 1935), directed by Arthur Elton and Edgar Anstey. See interviews with tenants Mrs Hill and Mrs Graves.

137 Swann, *British Documentary Film Movement*, 106–107; Brian Winston, *Claiming the Real: The Griersonian Documentary and its Legitimations* (London, 1995), 43–45; Sussex, *The Rise and Fall of British Documentary*, 61–64.

138 Hardy (ed.), *Grierson on Documentary*, 178–179; Paul Rotha, *Documentary Diary: An Informal History of the British Documentary Film, 1928–1939* (New York, 1973), 156–157; Swann, *British Documentary Film Movement*, 106–107.

139 Barnouw, *Documentary*, 95.

140 Sussex, *The Rise and Fall of British Documentary*, 64; Winston, *Claiming the Real*, 44.

141 Burnett, *Social History of Housing*, 238; Sean Glynn and John Oxborrow, *Interwar Britain: A Social and Economic History* (London, 1976), 223–224.

142 Stephen Merrett, *State Housing in Britain* (London, 1979), 49–50.

143 For a discussion of the currency crisis, see Robert Skidelsky, *Politicians and the Slump: The Labour Government of 1929–1931* (Harmondsworth, 1970); on the neglect of housing policy, see Merrett, *State Housing*, 50.

144 Keith Laybourn, *The Evolution of British Social Policy and the Welfare State* (Keele, 1995), 202–203; Merrett, *State Housing*, 50–51, 311–312.

145 Marian Bowley, *Housing and the State 1919–1944*, reprint of 1945 edition (New York and London, 1985), 244.

146 Ibid., 156–157.

147 Margery Spring-Rice, *Working-Class Wives: Their Health and Conditions* (Harmondsworth, 1939), 128–154.

148 Barnouw, *Documentary*, 94–95.

149 Sussex, *The Rise and Fall of British Documentary*, Anstey, interview, 62.

150 Swann, *British Documentary Film Movement*, 13, 23, 50, 99; Everard, *History of the GLCC*, 346–348.

151 Everard, *History of the GLCC*, 321; 'Twenty-five years service as executive chairman: presentation to Sir Francis Goodenough on his retirement', *GJ*, 14 October 1936, 120–121.

152 Everard, *History of the GLCC*, 324–326.

153 M.A. Cloudesley Brereton, 'Problems of domestic economy', *Papers Delivered to the Second Annual General Meeting and Conference of the BCGA*, 108–110; 'Laundries and their equipment', *ATAOUFG* (December 1913), 3–7; 'Domestic hot water from the cottage to the mansion', *ATAOUFG* (September 1915), 3–11.

154 Sandgrove, 'How shall we build the new England?', 155–158; Creasey, 'Government housing schemes and the gas industry' part I; Creasey, 'Government housing schemes' part II.

155 'Two domestic problems and their solution: cooking and refuse disposal', *ATAOUFG* (February 1924), 14–17; 'Labour-saving bungalows', *ATAOUFG* (April 1924), 34–44; 'Smokeless heating, cooking, water-heating and lighting equipment in housing schemes', *ATAOUFG* (October 1926), 90–97.

156 'Candid criticisms of the BCGA publicity films', *GJ*, 9 October 1935, 139.
157 'The gas films', *GJ*, 4 December 1935, 688.
158 'A review of the Gas, Light and Coke Company's films', *GJ*, 14 October 1936, 157–158.
159 British Commercial Gas Association and Associated Realist Films, *Children at School* (London, 1937), directed by Basil Wright. Also see BCGA, *Modern Films on Matters of Moment*, 6.
160 'The gas industry presents—', *Co-Partners' Magazine* (November 1937), 608–611; BCGA, *Modern Films on Matters of Moment*, 5–10.
161 'Letter: David Milne-Watson', *The Times*, 16 June 1938, 12; Hardy (ed.), *Grierson on Documentary*, 216; Low, *History of the British Film*, 92–94.
162 Dean, *The Thirties*, 22, 60–63.
163 Alistair Service, *The Architects of London and Their Buildings from 1066 to the Present Day* (London, 1979), 191–192.
164 Gaskell, *Model Housing*, 102–103; Dean, *The Thirties*, 60–63.
165 'Opening of Kensal House', *Co-Partners' Magazine* (April 1937), 181.
166 'The Gas, Light and Coke company's bill', *Gas World*, 8 April 1933, 348–349; 'Gas in council houses', *Co-Partners' Magazine* (August 1936), 375–376.
167 'Oxford Street showrooms', *Co-Partners' Magazine* (January 1937), 13; 'New showrooms at Leytonstone', *Co-Partners' Magazine* (September 1937), 505–506.
168 Maxwell Fry, *Autobiographical Sketches* (London, 1975), 143.
169 Service, *Architects of London*, 190–191; 'Opening of Kensal House', *Co-Partners' Magazine*, 181; Fry, *Autobiographical Sketches*, 142.
170 Fry, *Autobiographical Sketches*, 143.
171 Elizabeth Denby, *Europe Re-Housed* (London, 1938), 260–278; see also Elizabeth Denby, 'Plan the home', *Picture Post*, 4 January 1941, 21–23.
172 Burnett, *Social History of Housing*, 181–183; Gaskell, *Model Housing*, 65–69.
173 Stevenson, *British Society*, 222–223, 228, 233.
174 Fry, *Autobiographical Sketches*, 143–144; S.C. Leslie, *Kensal House: The Case for Gas is Proved* (London, 1937.); Madge Waller, 'Toys for wartime nurseries', *Fanfare*, 6, 4 (Autumn 1943), 4; NGA/*Minute Book, Capital Housing Association Ltd.*, 30 September 1937; NGA/*Minute Book, Capital Housing Association, Ltd.*, 30 September 1938.
175 'An all gas contribution to the new London', *GJ*, 24 March 1937, 780–781.
176 Ibid., 782.
177 Dean, *The Thirties*, 53–54; Service, *Architects of London*, 186.
178 'Gas in housing and slum clearance', *ATAOUFG* (June 1934), 2–8; 'Gas provides hot water requirements in 320 flats', *GJ*, 12 December 1934, 819.
179 NGA/*Minute Book, Capital Housing Association*, 30 September 1937.
180 NGA/*Minute Book, Capital Housing Association*, 30 September 1938.
181 Leslie, *Kensal House*, 6–13, 22.
182 Deirdre Beddoe, *Back to Home and Duty: Women Between the Wars 1918–1939* (London, 1989), 90.
183 Ibid., 91; Maud Pember Reeves, 'Family Life on a Pound a Week', in Fabian Women's Group, *Women's Fabian Tracts*, 200–223.
184 Spring-Rice, *Working-Class Wives*, 131; see also the book's introduction by Dr Janet Campbell, ix.
185 Ibid., illustration no.14, titled 'The community centre of a new block of flats in W. London, Gas, Light and Coke Company'. Although the author

does not name the building, it is Kensal House, given the building's location and the provenance of the photograph.

186 Davidson, *Woman's Work*, chap. 5.

187 'Nursery school and flats, Ladbroke Grove', *Architectural Review*, 81 (1937), 207–210.

188 NGA/GAL/2 CH/E/B/1, Resumé of Marjorie Bruce Milne, 1. *Modern Living* and *Hearth and Home* were both launched in the autumn of 1938; however, the GLCC suspended publication a year later due to the outbreak of World War II.

189 NGA/GAL/2 CH/E/B/1, Marjorie Bruce Milne, 'What the tenants think of Kensal House', opening statement.

190 Ibid., 4–5, summary.

191 Ibid., 10.

192 Ibid., 7.

193 Ibid., 9–10.

194 Ibid., 14.

195 Ibid., 5.

196 Ibid., 5–6.

197 Daniel Ritschel, *The Politics of Planning: The Debate on Economic Planning in Britain in the 1930s* (Oxford, 1997), 10–14.

CHAPTER 6

'A Women's Industry': Demonstrations of Difference and Equality

> The gas industry is a women's industry. It is the biggest industry which caters for the daily needs of women. Seventy per cent of the gas output in this country is sold for home use.... You have ten million women who are regular daily customers. Think what Gordon Selfridge would do if he could have all these millions of women on his books, making daily purchases from him! I can tell you the first thing he would do to make sure of keeping their custom: he would engage the best women he could find, so as to keep the woman's point of view in the forefront.[1]

In the early 1930s, the BCGA experienced a few changes, not the least being the retirement of Mrs Maud Adeline Cloudesley Brereton in 1932. As chief editor of the association's in-house publications, consultant to the mainstream press and advisor to professional groups, including architects and physicians, Mrs Cloudesley Brereton was said to have combined the 'qualifications of the expert publicist with the crusading zeal of the reformer'.[2] An important aspect of this 'crusading zeal' had consisted of drawing the industry's attention to the fact that women were their best customers and should be treated as such. In addition, Mrs Cloudesley Brereton had used her position to demonstrate that women were capable of success in the male world of business and commerce. As the new head of the BCGA's home service section, Mrs Eileen Murphy was no less forceful in this area. Murphy's inaugural address to the BCGA in September 1934 focused exclusively on women and the industry, as consumers and employees. According to Eileen Murphy, gas managers needed the assistance of both groups of women if their industry was to survive national electrification. The gas industry was, if not in production and distribution, at least in the crucial areas of buying and selling, a women's industry.

After decades of warily watching their competitors, gas managers, by the 1920s, recognised that the electrical industry posed a genuine threat to the continued existence of their own industry. In an effort to be better prepared to meet this challenge, they promoted the sales and service aspect of their industry by initiating training programmes for salesmen and maintenance inspectors. In addition, gas managers relied on their home service departments, staffed by the acknowledged female experts in personal customer service, who continued to represent the gas industry in women's homes and in the showrooms. Although the sales managers appreciated the women's efforts in this capacity, the home service advisors were unable to expand their responsibilities in a significant manner to include inspecting gas equipment in the home or holding senior sales positions in the showrooms. To some extent, this containment of female employment was due to the growing numbers of white-collar professional salesmen in the gas industry. It may also have been a response to interwar concerns about increasing rates of unemployment and the fears that women were competing with men in the job market.

Considered within the context of the interwar debates over gender and work, home service women successfully defended their existing professional status. Despite concerns that women were encroaching on male jobs, only a few women actually undertook the work that was traditionally done by men, such as working as gas inspectors. These women received high praise from their employers and other home service women, and were held up as role models within the industry. With the inauguration of the Women's Gas Council (WCG) in 1935, an organisation which brought together women gas employees and female consumers, the lady demons established professional contacts with a network of post-suffrage women's organisations, including the National Council of Women, the Federation of Business and Professional Women and the Women Citizens Association. Through these affiliations, women in the gas industry participated in a number of social welfare causes, including the campaigns for better health and housing, cleaner air and family allowances.

Historians have noted that the women's movement experienced a severe decline in popular support during the interwar years. Feminist debates over post-suffrage strategies and priorities created divisions within the old suffragist pressure groups, while feminism itself had little appeal for a younger generation of women, more concerned with fashion and entertainment than with political and social activism.[3] For example, the former suffragette Sylvia Pankhurst lamented that, for young women in the 1930s, feminism meant wearing shorts and smoking cigarettes.[4] In contrast to the dwindling support for the old suffragist groups, organisations that appealed to women at home thrived throughout this period. Among these were the Women's Institutes and the Townswomen's Guilds. As an organisation with a mandate to serve the needs of women in the home and professional women in the workplace, the Women's Gas Council combined these two types of interwar women's clubs, functioning simultaneously as a pressure group, an educational society and a recreational club. Although the WGC staff and executive hesitated to label their activities 'feminist', for a variety of reasons to be discussed, their agenda fitted that of the interwar 'new' social welfare feminists. This final chapter considers the work of the home service women between the wars, their continued efforts to develop and uphold a professional identity and their contributions to social welfare campaigns aimed at child welfare, housing and nutrition. Studies of interwar feminism have identified corporate home economists with the promotion of a 'cult of domestic consumerism', designed to keep women in the home.[5] This critical interpretation of their work is entirely valid; however, in the case of the gas demonstrators, their pragmatic approach to social activism and their positive interactions with thousands of women in the home warrant a reassessment of that view.

'Goodwill advertising': the Women's Advisory Staff and interwar housewives

When considering the efficacy of advertising, gas men unanimously agreed on the dual importance of attractive prominent showrooms and accomplished lady demons as a means of reaching female consumers.[6] Facing vigorous competition from the electrical industry, they expressed their continued enthusiasm for public cookery demonstrations and instructional visits. Described as 'goodwill advertising', the demonstrators' visits and classes had been the mainstays of gas home service departments since the Victorian and Edwardian years.[7] Showroom expansions and urban housing development meant that the gas companies increased their home service departments accordingly. For example, in 1922, the GLCC maintained a staff of 12 home service advisors. This number had grown to around 50 by the late 1930s.[8]

During the early 1920s, as the number of permanent women employees grew, the company formally established set wage scales for female staff officers, including typists, telephone operators and salespersons. The home service women, classed as 'lady representatives', were categorised as either 'indoor' or 'outdoor' workers, with respective starting salaries of £65 and £80 per annum. Experienced home service women were paid accordingly. Women in the highest wage bracket, Class C 'outdoor lady representatives', made between £200 and £250 per annum.[9] Department heads, a special category, took home even more. Helen Edden, the company's first lady demon and principal of the home service department since its establishment in 1906, received £375 per annum by the time of her retirement in 1922.[10] Helen Edden's salary was exceptional, particularly if one compares her wages with that of her successor, Ethel Willans. In 1906, Edden started her job with the GLCC at £200 per annum, a highly respectable salary for a single woman in a new profession. After the GLCC implemented official pay scales for their female employees in 1922, 17 years later, Edden's starting salary was the maximum wage that an experienced Class C indoor home service woman could expect. Edden's replacement, Ethel Willans, assumed the job of head of home service, not at £375, but at the much lower salary of £175 per annum. This represented a significant saving for her employer.

Why was Willans' pay so much lower than Miss Edden's? It should not have been a matter of job experience as Willans also had a long employment history with the company. She joined the GLCC's home service staff around 1910, the eighth 'lady demon' on Helen Edden's team, at the starting salary of £1 per week with the potential of earning up to 40 shillings per week.[11] Although Willans' pay more than tripled as she moved from demonstrator to showroom attendant and, finally, to principal director of home service, her salary never approached that of Helen Edden. A possible explanation is that, as the numbers of female employees increased, the gas companies attempted to regulate women's wages based on the requirements of the job and relevant work experience. The set wage scales may also have been an attempt to roll back some of the large salaries that had been demanded by female pioneers in the industry, such as Helen Edden. Nonetheless, the post-war generation of women demonstrators were still relatively well paid compared to other positions for women in the industry. Under the company's official pay scales, home service women earned more than the typists and telephone operators who received between £40 and £60 per annum, while starting salaries for skilled shorthand typists compared favourably with the wages of 'indoor' lady representatives, including the showroom saleswomen and receptionists. Home service women undertaking personal home visits and public demonstrations were still the best-paid women in the gas industry, with experienced Class A demonstrators receiving between £90 and £120 per annum.[12] One former demonstrator with the South Suburban Gas Company apparently earned about £2 per week as a home service advisor, about the same wage as women employed by the GLCC. As the junior member of the home service department (she joined the South Suburban Gas Company in 1939 at the age of 19), Pearl Thorpe (then Pearl Brown) did her rounds to customers' homes on a bicycle. The cookery demonstrations were the responsibility of a more senior home service woman, who had been with the company for many years. Apparently, the cookery demonstrator drove herself to her appointments in a company car, in recognition of her higher job status and years of experience.[13]

Interwar writers reporting on female employment noted the inconsistencies in the demonstrating profession. According to Vera

Brittain's 1927 study, demonstration work in showrooms and at trade exhibitions was often part-time, with long and irregular hours. Part-time demonstrators earned around 27 shillings per week, with a commission for goods sold. In contrast to these short-term jobs, 'good firms' offered regular positions to domestic science graduates to demonstrate products like cooking apparatus, household utensils and patent foods. According to Brittain's estimates, these jobs paid around £3 per week. Although the gas industry had employed full-time lady demons since 1906 and expanded its home service departments in the interwar years, Brittain made no specific mention of this pioneer group of female sales professionals. Instead, she described the new prospects for women in the electrical industry, particularly the growing need for trained 'electrical advisors' to visit householders had explain 'the use of new kinds of apparatus, such as electric stoves, refrigerators, bed-warmers and kettles'.[14] By the 1920s, the electrical industry had adopted the gas industry's practice of hiring trained domestic experts to visit consumers in the home.

The job of cookery demonstrator, the first position open to women in the gas industry, remained an important aspect of their home service duties. Between the wars, gas companies increased the frequency of these public demonstrations, probably to counter the effects of electrical advertising by ensuring that housewives remained loyal to gas. Approximately half of the GLCC's new and renovated urban showrooms included special lecture theatres for regular demonstrations. By 1937, the GLCC had 23 fully-equipped lecture theatres.[15] The largest showroom offices – Kensington and Barking, for example – featured lecture halls with platforms and seating for 200 persons. To improve the acoustics at the Kensington theatre, the GLCC went to the extra expense of lining the walls with three-ply quilted fabric to ensure that housewives missed not a word of their demonstrators' talks.[16] Even the smaller gas showrooms – including those in Ilford, Islington, Hackney, Kilburn, Kingsbury and Edgware – featured demonstration theatres with raised platforms, model kitchen equipment and seating for about 40 viewers.[17] The Kingsbury lecture room had the added novelty of gas-controlled air conditioning, while the spacious Acton demonstration hall doubled as a staff dining and recreation room.[18] Like the GLCC, the SMGC's updated gas showrooms also featured lecture rooms for cookery demonstrations with seating for between 40 and 60 persons.[19] The Eltham showroom's theatre, designed by W. Watson, adopted tiered seating which fanned out from a tiled, fully-equipped model kitchen, allowing everyone an 'unrestricted view of the demonstrator at work'. For the purposes of illustrating the number of heat units consumed during each demonstration, many of the new lecture halls, like the GLCC's Kilburn showroom, came equipped with enormous gas gauges, the size of a public wall clock (see Fig. 6.1). Note also the large mirror above the demonstration table that allowed the audience to see exactly what the cook was doing on-stage.

In terms of format, the interwar demonstrations were much the same as the pre-war versions. They were conducted before a live audience and a woman demonstrator cooked something in a gas stove, illustrating the most effective use of the appliance. Throughout the war years, and immediately after, attendance at cookery demonstrations had been a patriotic duty, a way for the housewife to gather information on rationing and fuel conservation. By the mid-1920s, cookery demonstrations resumed their former commercial purpose (see Fig. 6.2). Demonstrations incorporated a variety of marketing techniques, combining the soft-sell methods of customer service and public relations with the more aggressive tactics of flamboyant advertising and the showroom sales pitch. The cookery demonstration was a recognised means of attracting thousands of potential female customers to gas showrooms on a regular basis. The SMGC held daily cookery demonstrations at its six largest showrooms, including Bermondsey, Woolwich and Vauxhall. At four demonstrations per showroom on a weekly basis, with an average attendance of 40 women per demonstration, the SMGC reached approximately 960 customers each week through this promotional method. Another eight demonstrations per week were held in two separate furniture shops, Holdron's and Chiesman's, reaching another 320 women, bringing the total number of customers attending per week to around 1,280.[20] Ethel Willans, at the GLCC's home service department, estimated that, in 1937, over 100,000 women attended

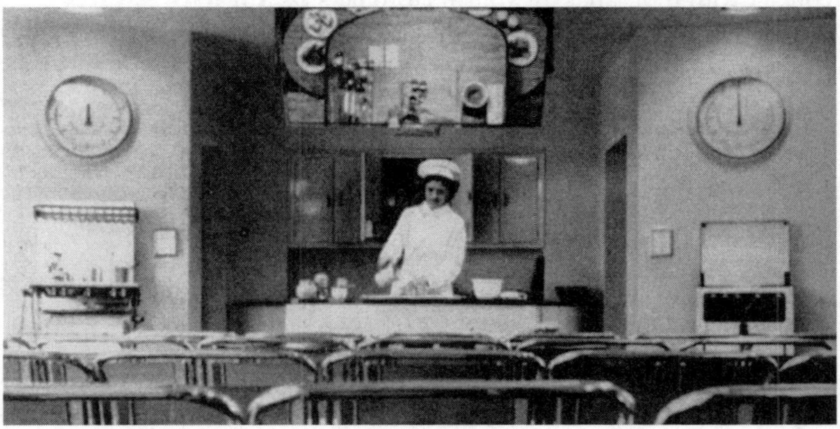

6.1 A cookery demonstrator in the Gas, Light and Coke Company's Kilburn showroom

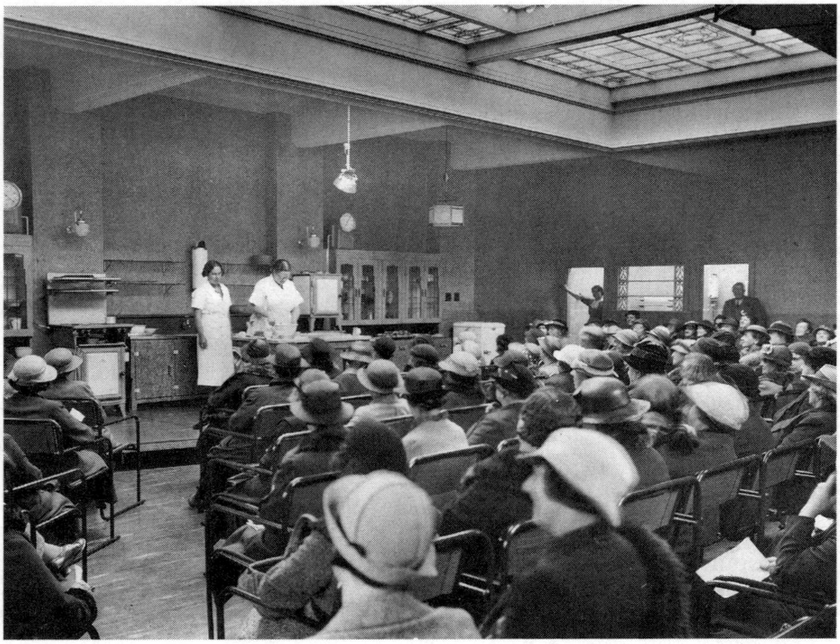

6.2 An afternoon cookery demonstration in a Harrogate Gas Company lecture theatre, c. 1936.

nearly 1,700 demonstrations given by her company's home service staff.[21] Clearly, despite their small numbers, the home service women established contact with thousands of women customers each year.

Demonstrations began at mid-afternoon, usually around three o'clock, which was just when housewives might have a bit of leisure time between preparing and serving the noon-time meal and the children's return from school for their evening tea.[22] Demonstration programmes appeared in the local press and in the showrooms, and usually combined the presentation of several domestic tasks. For example, during the time that the demonstrator cooked a four-course meal on the gas range, she might also have washed a load of clothes in a gas-fuelled washing machine, to fully illustrate how domestic equipment saved both time and energy. Impromptu requests for quick dishes, like omelettes, filled up any extra time, and, throughout the talk, the demonstrator responded to a steady steam of questions from the audience. Reporting on her impressions of a home service presentation in Croydon, Eileen Murphy noted the numerous requests for various special dishes and the demonstrator's willingness to comply, indicating that 'the programme of lectures and demonstrations [was] actually planned by the consumers'.[23]

Evidently, through the demonstrations, home service women created a genuine dialogue between the gas companies and their female consumers. C.M. Croft, the chief engineer at the Wandsworth, Wimbledon and Epsom District Gas Company, described cookery demonstrations as 'one of the cheapest and most intimate forms of effective publicity' because they established a 'sympathetic bond between the consumer and the company'.[24]

Despite the obvious commercial motivations, the cookery demonstrations were also a form of entertainment. Not only were these demonstrations cheap publicity, but they were also cheap outings for housewives. In the late 1930s, Mrs E. Warne of the Walthamstow and Leyton District branch regularly attended cookery demonstrations as a young housewife. She recalled that the hour-long programmes usually ended with tea and biscuits, while the demonstrators raffled off the finished results to audience members, so that several lucky women took home something for the evening meal. Pearl Thorpe confirmed that raffling off the demonstration results to the audience was a common practice and she fondly noted that the serving dishes were always returned clean to the showroom by the following week.[25] Home service women recommended demonstrations of simple, everyday dishes suitable for 'the average household'.[26] Even if one wasn't lucky enough to win the demonstrator's finished result, the assumption was that the dishes would be duplicated at home. Mrs Warne also noted that dropping in at the gas demonstration was a nice diversion from the daily chores of housekeeping and shopping.[27] Historians of working-class women have noted that most interwar housewives had little time or money for personal leisure.[28] Possibly, the gas cookery demonstration offered women of limited means a brief respite from the usual routine of housework, childcare, laundry and meals, with the added bonus of a free cup of tea and the chance to win the family's evening meal.

On some occasions, the gas companies presented their demonstrations in cooperation with large department stores. According to Bill Lancaster, English department store owners experimented with creative methods of attracting business in the 1920s and 1930s within an increasingly competitive retail market. Enthusiasts of such commercial showmanship included Gerrard Bentall, whose south-west London store hosted baby competitions, musical performances and, at Christmas, an in-store circus complete with elephants and a lion. Store owners mounted demonstrations of domestic arts, like sewing and flower arranging, and offered customers free 'mothercraft' and 'fathercraft' classes. Between the wars, 'imaginative manufacturers' used teams of travelling demonstrators to advertise new products in department stores. For example, Colley Shorter, the manufacturer of the colourful hand-painted pottery known as

'Bizarre Ware', sent his head designer Clarice Cliff and a group of female assistants on a tour of department stores to publicly demonstrate their technique.[29] Lancaster observes that electrical appliance manufacturers also used in-store demonstrations to sell new products. This promotional method was first popularised, to a great extent, by the gas industry and by the interwar years had been adopted by a growing number of retailers and manufacturers.

The use of cookery demonstrations in department stores was another promotional tool adopted by the gas industry. Prior to arrangements with the furniture shops, Holdron's and Chiesman's, the SMGC staged regular cookery shows at the Bon Marché in Brixton in a company-sponsored model gas kitchen. In 1935, the GLCC offered a series of demonstrations and hourly talks at Selfridge's Oxford Street store, in the heart of London's busiest shopping district.[30] This high-profile series offered a departure from the everyday menus presented in district showrooms for lower-middle-class wives on a strict budget. At Selfridge's, the company tailored the lectures to appeal to single business girls and more affluent married women by presenting topics that addressed entertaining, dieting and servant management. Topics such as 'Sensible Snacks for Business Girls', 'Sane Slimming' and 'Cooks and Cooking' suggest a younger audience with more disposable income. For working-class girls who might have been window-shopping in Oxford Street, a favourite pastime for young women, attendance at the GLCC's Selfridge's demonstrations offered a glimpse into the world of layer cakes, iced drinks and gourmet menus from famous hotels.[31] Interwar gas demonstrators – working in the showrooms, at trade exhibitions, department stores or, occasionally, in special off-site demonstrations at busy 'tube' stations – offered their audiences a subtle combination of entertainment, public education, domestic advice and advertising.[32]

As the number of lady demons grew, between the wars, the occupational scope of home service departments expanded to include female sales representatives and showroom managers engaged in the sale of gas services and appliances. As noted in the previous chapter, during World War I, the gas companies employed women as showroom attendants to replace enlisted men. For example, Ethel Willans managed the GLCC's showrooms throughout the war until she replaced Helen Edden as head of home service in 1921. Although there are no exact figures to indicate how many women remained in the sales departments after the war, the creation of official pay scales for 'indoor lady representatives' suggests that female salespersons were not the anomaly that they had been prior to the war.[33] In a promotional photograph, a female showroom attendant advises a man and a woman about the merits of a gas heating stove, the assumption being that trained women were accessible and well-informed sales experts, who could address both male and female customers (see Fig. 6.3).

In support of this claim, Francis Goodenough remarked, in 1934:

> More women are employed on the selling side than ever before. In addition to specialist cookery demonstrators, there are women receptionists, showroom managers and saleswomen, many of whom have taken the course in Domestic Gas Salesmanship. The importance of having at least one educated woman of some social standing to meet customers in the showrooms and outside them is widely recognized.[34]

As chief inspector and, later, district controller of sales for the GLCC, Goodenough had consistently supported the role of the women's advisory staffs since their establishment in approximately 1905. On the assumption that female sales personnel were as capable of selling the 'big ticket' consumer durables, like cookers, gas fires and refrigerators, as their male co-workers, he supported the expansion of female occupations in the gas industry. Few of his male colleagues were as generous.

The expansion and modernisation of showroom facilities focused the industry's attention on salesmanship. Although publicity specialists, including the BCGA's Mrs Maud Adeline Cloudesley Brereton and William Mason, had recommended improved training for salesmen since before World War I, it was not until the interwar years and the genuine

6.3 Female showroom attendant demonstrating a gas heating stove for two customers (provenance unknown)

threat of electrical competition that gas men seriously considered the need for professional development in this area. Mrs Cloudesley Brereton had long lamented the erroneous assumption that any young man of neat appearance would suffice as a gas salesman. She deplored the standard image of the showroom attendant as being too incompetent to perform any genuinely useful task within the industry, but she insisted that the 'commercial side should be as much respected and as well paid as the purely professional and technical side'.[35] At the time of her comments in 1913, gas salesmen earned around £100 per annum or about the same salary of an accounts clerk; however, she assured her Edwardian listeners that, in terms of quality and quantity of work, employers never got more than they paid for. Moreover, she added that indifferent and misinformed sales personnel damaged the industry's reputation, resulting in lost sales.[36] Before her assembled audience, Mrs Cloudesley Brereton outlined a plan for the creation of a consulting salesmanship department in connection with the BCGA. She also urged gas managers to encourage consultation between different groups involved in gas sales, by, for example, organising meetings that brought together outdoor canvassers, lady advisors, gas inspectors and showroom attendants, thereby creating a multi-level sales team. Though her suggestions were politely received, respondents to Mrs Cloudesley Brereton's paper feared that the added cost of training and adequately remunerating professional salesmen would be met with hostility from their shareholders.[37] Given the commercial interruptions of World War I, it was not until the early 1920s that gas managers and public relations officers returned to the topic of salesmanship.

In a general sense, Goodenough and Mrs Cloudesley Brereton's preoccupation with salesmanship and staff training had finally come into vogue by the early 1920s. At this time, the difficulties facing British industry and commerce following World War I prompted businessmen to consider the importance of technical and commercial education.[38] Inspired by the American efficiency experts, including Frank Gillbreth and Frederick Taylor, books on 'rationalisation' and 'scientific management' presented strategies to streamline production and reduce selling costs from the factory to the office. Business writer Paul E. Derrick warned British manufacturers against overlooking the crucial element of salesmanship. According to Derrick, the development of an efficient and modern sales department was the greatest challenge that British businessmen faced, and the most vital to their survival. He confidently proclaimed that, in the modern business world, 'everything depends on the Sales Director'.[39] In 1931, a Board of Education inquiry, under the chairmanship of Francis Goodenough, expressed concern at the continued prejudice against commerce, noting that young men of ability and good education were not offered the proper incentives to consider a sales career. The retail trade was treated in a separate section and the report advocated the provision of some type of formal commercial training for retail salesmen, whether in commercial schools or through part-time instruction offered by employers.[40]

Although educators and industrialists focused on salesmanship as a means of rejuvenating British industry, white-collar workers, such as retail salesmen and office clerks, still remained members of a 'marginalised' lower-middle class whose salaries did not support their social and occupational aspirations.[41] Nevertheless, the development of sales associations, training programmes and publications directed at salesmen suggested an improvement in their occupational prestige, a development that was particularly evident in the English gas industry. This same prestige was not automatically extended to female sales personnel, despite their long years of customer service work in the gas showrooms and in customers' homes. By the interwar years, the definition of the perfect salesman linked technical knowledge with smooth-talking persistence, while home service women were apparently left with the soft-sell approach of cookery shows, cake competitions and housekeeping advice.

Although the publicity specialists at the BCGA had recommended improved training for salesmen as early as 1911, it was not until the early 1920s that the majority of gas men accepted the need for professional development in this area. In an effort to better train showroom personnel, beginning in 1922, the BCGA organised regional meetings for salesmen. At these 'Salesmen's Circles', guest speakers delivered papers on a variety of relevant topics, ranging from the latest technical innovations in gas merchandise and showroom design to the benefits of local advertising and the new psychology of selling. Reports on regional meetings and national conferences appeared in a special salesmen's monthly supplement included in the *Gas Journal* after 1922.[42] Business etiquette and tricks of the trade were offered to young salesmen at the regular meetings. William Mason, the general manager of the BCGA, with many years of experience in advertising and sales, warned salesmen against getting a reputation for 'lightness or frivolity'. To protect one's credibility, it was unsafe to get a reputation for telling stories, unless they were of a self-deprecating nature since this induced a smile from the customer and might create a more sympathetic atmosphere. He advised the salesmen to be positive, not argumentative, and always 'stick to the truth' or understate things, but never exaggerate the claims of a particular product. Tact, 'the oil that makes things go smoothly', in Mason's opinion, was the most fundamental quality of the good salesman.[43] To standardise the training of its salesmen, the BCGA organised a certified educational programme for gas salesmen, offered in conjunction with the regular circle meetings.[44]

In addition to Mason's common-sense advice, gas salesmen were also exposed to a new brand of sales literature that combined a smattering of psychology with old-fashioned self-help. In a series of articles entitled 'Salesmanship and the Subconscious', business author H. Ernest Hunt urged salesmen to look inside themselves, to cultivate a positive self-image by picturing themselves as more capable, reliable and successful than they really were. This enabled them to develop a more positive mental picture of themselves that they then strove to achieve. Service to the consumer, not money, should motivate salesmen, but profits were sure to follow. Hunt advised salesmen to focus on their energies 'towards greater service, the reward end of the transaction may be trusted in good time to look after itself'.[45] Having covered the psychology of selling, in a later article, Hunt outlined the importance of good physical health for the salesman by offering advice on diet, sleep and exercise. He asked salesmen to avoid alcohol since it dulled the finer sensibilities and reduced self-control. Hunt even addressed the delicate subject of 'elimination through the digestive tract', recommending that the healthy salesman have a routine of 'at least twice a day' to ensure a 'feeling of marked well-being' and 'keener mental activity'. Hunt's regimes for mental and physical self-improvement, though prudish and even comical, suggest the level of attention directed at the making of the 'perfect salesman', who was not simply a wage-earner but a bridge between the corporation and the consumer.

In addition to the salesmen's circles and the printed supplements, in 1931, the gas industry standardised the training of its salesmen by offering a certified educational programme for gas salesmen.[46] Offered by correspondence through the BCGA at a cost of £6.6s, the graduates received the Gas Salesman's Certificate following the successful completion of the required technical and practical courses. Goodenough, who helped establish the programme, advised gas administrators to encourage their salesmen to enrol in the course and also to attend their local salesmen's circles by paying their employees' travel expenses and, when necessary, providing time off work.[47] By the late 1920s, through this growing network of conferences, training schemes and trade publications, showroom salesmen were granted a new level of professional status within an industry that had traditionally emphasised the masculine role in production, but not marketing.[48] It is difficult to determine whether this new respect went beyond words. For example, the articles on the gas salesmen's course offered no assurance that certificate holders could expect wage increases for their trouble. In addition, despite some suggestions that showroom salesmen receive payment by commission, the majority of gas undertakings retained their sales attendants at salaries of around £160 per annum.[49]

How did this image of the new white-collar professional salesmen affect women employed in gas sales and home service? Although women worked in the showrooms and interacted with consumers through their home service demonstrations, published accounts of the salesmen's circles suggest that women did not regularly attend these information sessions. Instead, women in gas sales were called upon to advise and assist salesmen on the best means of appealing to female consumers, including housewives and career women. For example, an article in the *Gas Salesman* supplement, written by a 'Saleswoman', described for male readers the domestic life of the working bachelor woman and the importance of gas appliances in the female 'family of one'.[50] The author cautioned salesmen never to underestimate the social conscience of the modern educated woman, who was even more aware of contemporary issues, such as fuel conservation and national efficiency, than male consumers because of her 'natural bent towards personal domestic economy'. Writing from the dual perspective of the experienced salesperson and the working woman, the female author provided an insider's report, describing exactly what single women needed and wanted in their compact flats.

In order that gas salesmen might better appreciate the 'woman's point of view', home service women were invited to speak periodically at sales meetings. Miss M.A. Tench, a home service advisor with the Darlington Gas Company, speculated that because of the post-war housing shortage, women were worse off than their grandmothers were, in terms of their 'domestic workshops'. She reflected that improvements to women's working environment were slow because housewives were not organised workers and design changes to the home depended on the goodwill of male architects and builders, who had no experience of domestic work and, therefore, were unsympathetic to the 'woman's point of view'. On another occasion, Miss Tench lectured a group of salesmen on developments in gas cooker design, including the women's positive and negative reactions to various stove features as a result of home use.[51] On a similar theme, lady demons Miss B. Randle of the GLCC and Miss M.A. Chalkley of the Tottenham District Gas Company addressed the Southern District (London Area) Gas Salesmen's Circle. In their respective talks, both women listed the priorities of the 'average woman' shopping for gas appliances. Both Randle and Chalkley noted the merits of regular cookery demonstrations and service calls by demonstrators, especially after a new cooker had been installed. In addition, they mentioned the importance of overlooked details which affected cooking results in various stove models – the sizes and weights of baking pans, the placement of grills and browning shelves – the details of which were familiar to home service women, but not necessarily to their male co-workers. Called upon to address male showroom attendants on the particular needs of female consumers, those responsible for training salesmen clearly recognised the special relationship that home service advisors had with their customers.

Home service women used these speaking engagements to emphasise their longstanding contributions to the industry, their continued importance and the need for greater cooperation between the various departments involved in service and sales, from showroom salesmen to gas fitters and inspectors. A number of gas managers supported their women demonstrators in this respect. In an effort to improve customer service and extend the role of its home service department, W. Clark Jackson, manager of the Neath Gas Company, substituted his male gas inspectors with 'lady inspectors' because, in his view, women established a better relationship with female customers and were generally allowed into consumers' homes with greater ease than male inspectors. Jackson contended that women had successfully undertaken the inspection and repair of gas appliances during the war and, when properly trained, were highly effective in this area of work.[52] In a later address to the Midland District Gas Salesmen's Circle, Mr. A. Forshaw raised the issue of extended training for women advisors to include more technical information. He also suggested that fitters and meter inspectors, on the other hand, might benefit from some basic knowledge of gas cookery. Despite the difficulties of initiating such a plan, Forshaw anticipated that the crossover of duties might eliminate service delays and facilitate interdepartmental cooperation with the ultimate objective of increasing gas sales, despite electrical competition.[53]

Not surprisingly, interwar proposals to extend the lady demons' occupational duties drew mixed reactions. Responding to Jackson's advocacy of 'lady inspectors', a male gas inspector argued that women customers placed more faith in the recommendations of a male inspector than in those of a female inspector, except in the area of cookery.[54] And Forshaw's suggested plan for breaking down the occupational barriers between male fitters and female demonstrators elicited a mixed response. While one editorial saw this scheme as a positive means of overcoming 'the strict demarcation' of vocations, as imposed by the craft unions, another writer warned that, given the 'great number of men unemployed', gas workers would naturally oppose any plans to train women for traditionally male occupations.[55] Despite the home service women's self-confidence in the workplace and professional demeanour, the above debates over work allocation and gender make it easy to understand why women advisors during the interwar period still made their service calls outfitted with handbags, not tool chests. In many respects, they were no better prepared to deal with a mechanical stove malfunction than the pre-war lady demons armed only with their hat-pins.

Gender was an issue in the showroom as well. Although some gas managers, notably Francis Goodenough, fully endorsed women's entrance into a variety of gas sales jobs, other sales experts maintained that women lacked the masculine traits of personal aggressiveness and mechanical ability necessary for the tough job of selling 'hard goods'. Ironically, while gas men considered women's technical knowledge sufficient to demonstrate appliances, these same skills were somehow inadequate when it came to the job of actually selling domestic equipment in the showrooms. A debate on this topic developed at the 1927 annual conference of the BCGA, between the outspoken Mrs Cloudesley Brereton and the conference's guest speaker, Miss Gladys Burlton. Miss Burlton had been invited to address the BCGA membership on the topic of 'Salesmanship in the Showrooms'. As the former director of education and staff controller to Selfridge's department store, and a past chairman of the Association for Education in Industry and Commerce, Burlton was a recognised expert on the art and science of selling. When asked her opinion of saleswomen, she stated that, although experienced lady demons made excellent saleswomen and sales staff trainers and 'a feminine touch' was desirable in the showroom, overall, women's chief weakness in the showrooms was their inability to grasp technical matters, leading female customers to place more confidence in male salesmen.[56] In defence of women showroom attendants, Mrs Cloudesley Brereton responded that women trained in domestic economy made highly effective salespersons, since they possessed a technical familiarity with the appliances, but also because they were more sympathetic to women's domestic issues than the young salesmen were. Women, she argued, were entirely capable of assuming greater responsibility in the industry.

Mrs Eileen Murphy took this same point at the 1934 BCGA annual conference in Sheffield. A speaker at that year's conference, Mrs Murphy had been recently appointed head of the BCGA's home service section. She was instructed to 'coordinate the activities of women engaged in the gas industry and to foster additional contacts between gas undertakings and housewives in the districts'.[57] As the new head of home service, Mrs Murphy did not replace Mrs Cloudesley Brereton as chief editor of the BCGA publications after her retirement in 1932.[58] However, Mrs Murphy did assume her role as the association's female spokesperson, stridently boosting women's contributions to the industry whenever possible. Her remarks to the BCGA in 1934 were reminiscent of Mrs Cloudesley Brereton's inaugural speech some 22 years earlier.

Mrs Murphy recognised the prejudice against women in business, particularly in the 'old days', but she hoped that prejudice did not exist in the gas industry. Slightly chastising her listeners, she stated:

> Remember, there is no gender in brains. It should make no difference in business whether a person's legs are concealed in petticoats or in trousers. In fact, if any difference is made, surely the petticoats should have the advantage in an industry where women form the bulk of the customers![59]

According to Mrs Murphy, 70 per cent of the gas output in Britain was sold for home use, making women responsible for using three-quarters of the total gas made. As the largest industry catering to the daily needs of women, Mrs Murphy declared that 'the gas industry is a women's industry' and therefore they should offer 'brilliant employment prospects for the best women'. She asked gas managers to value and respect the potential of their home service employees, 'to give them more scope and responsibilities' beyond routine cookery demonstrations. Home service, in her view, encompassed a wider range of employment opportunities than ever before, including jobs for demonstrators, saleswomen, showroom managers, receptionists, lecturers and interior architects.[60] But, despite her crusading tone, Mrs Murphy was cautious before her male audience, never mentioning the contentious issue of female gas inspectors, and even stressing that women 'should not try to do men's work'.[61] Although Murphy prudently left undefined exactly what constituted 'men's' and 'women's work', in a later speech, she stated that girls who entered the gas industry 'were taking up a job … that can only be done by women' and therefore the women 'were not competing with men on their ground'. She still considered home service women as 'pioneers' in a new field, the products of a 'new set of social conditions'. Of great importance to the work that home service women did was the personal satisfaction gained from knowing that they helped other women 'to get more interest and pleasure out of life'.[62] As in the pre-war years of the lady demons, the interwar home service women's role still combined a sense of professional duty with social responsibility along with the insistence that the women were not poaching male jobs.

The gas industry's debates over female salesmanship and the expansion of female employment reflect a larger series of contemporary questions. In light of the interwar debates that addressed gender and women's right to work, and in conjunction with the advent of the professional salesman, home service women may have inadvertently undermined the aspiring professionalism of their male co-workers by assuming a more active role in the gas showrooms. Although home service demonstrators were respected as female sales experts, particularly when called upon to explain the feminine perspective, they were not included as equals in the salesmen's circles. Moreover, suggestions for extending the duties of home service women by training them to perform appliance inspections and minor repairs rattled the carefully reconstructed perceptions of male and female occupations within the industry; such perceptions were previously shaken by women's work experiences during the war years. Although a few companies, like the Neath Gas Company, braved the criticism and overlapped duties, the vast majority still maintained that home service women and gas inspectors must undertake entirely separate functions, thereby continuing to observe a hierarchy of skill that the war had only temporarily dismantled.

Despite these ongoing discussions on the role and duties of the women demonstrators, the gas women fared much better than other groups of interwar female white-collar workers in terms of their professional advancement. In the civil service and the teaching profession, women faced an increasing number of professional barriers, most notably the marriage bar.[63] It appears that home service women within the gas industry were not similarly restricted and the industry continued to hire married women for some of its highest-profile positions. The gas women's sales duties in the showrooms and at trade exhibitions continued to enhance their public and professional roles in both the workplace and the community, particularly as gas managers believed that the women fulfilled the dual role of simultaneously representing the needs of the female consumer and the interests of the gas companies. The establishment of the WGC in 1935 officially consolidated the gas industry's female employees. By the late 1930s, under the auspices of this organisation, home service women hosted their own professional conferences, published a journal for women consumers and female gas employees and cooperated with a network of interwar women's service groups in support of several prominent social welfare causes. While the WGC helped create professional networks for women in the gas industry, the development of positive public relations between the gas industry, through its demonstrators, and its female consumers in the home remained its first and foremost objective.

The WGC and interwar housewives

Although the BCGA and the GLCC usually led the way in terms of public relations schemes, the idea for a women's gas club did not originate in London, but in Yorkshire. In January 1934, prior to Mrs Eileen Murphy's promotion at the BCGA and her crusading remarks on behalf of gas women, Mr W.B. McLusky, manager of the Halifax and District Gas Company, offered his showroom as a meeting place for local women interested in public service and community affairs. According to McLusky, the Halifax Gas Development Association started among a group of prominent local women, including the mayoress; however, membership was open to anyone wishing to attend.[64] Monthly meetings were held in the gas showrooms and consisted of an invited speaker's talk followed by a question period or debate. Discussion topics addressed housing, slum clearance, smoke abatement and nutrition or 'dietetics'. On several occasions, the company's home service women addressed the meetings, either lecturing or performing demonstrations. Special outings included 'motor coach' trips to new gas showrooms or tours of large restaurants, shops and factories that used gas services. After one year, the new organisation reported 96 members, including Mrs Eileen Murphy. Mrs Murphy enthusiastically claimed that the group's membership was representative of the community and included young mothers, businesswomen, politicians, civil servants, social and church workers and elderly ladies, without 'distinction of class, age or calling'. According to McLusky, a woman did not have to be a gas consumer to join, although he supposed that all the members used gas, given the 'unforced enthusiasm for gas' among the membership.[65] Mrs Murphy and McLusky anticipated the steady growth of local branches, to be federated in a national union of women's gas associations. The federation would represent the interests of all domestic consumers of gas to outside official bodies, while also maintaining a strong local connection between staff and consumers.

The plan for a gas club for women was probably inspired by a similar organisation affiliated to the electrical industry. In 1924, Mrs M.L. Matthews founded the Electrical Association for Women (EAW), a group designed to collect and distribute information on the use of electricity as it affected women in the home and the community. The organising secretary, Miss Caroline Haslett, was a prominent figure in the electrical industry. A founding member of the professional organisation, the Women's Engineering Society (WES), established in 1919, Haslett joined with other female engineers, trained during the war, who were lobbying for job security and advancement in this field. The WES actively promoted technical education for girls, leading to careers in science, commerce and industry. The EAW, an offshoot of the WES, directed its attentions to customer education with the intention of bridging the gap between electrical consumers and producers, anticipating the labour-saving potential of electrical services for women in the home. To demonstrate electrical labour-saving devices, the EAW constructed a model all-electric house, Bristol House, which they used as a training centre for demonstrators.[66] According to Suzette Worden, neither the Electrical Development Association, the industry's public relations group, or the Institute of Electrical Engineers expressed any initial interest in the EAW; however, the WES lent its immediate support. The EAW flourished. By 1940, it had 85 branches, with 9,000 members, drawn mainly from the middle classes. It published a journal entitled *The Electrical Age for Women*, had members on local electricity and housing committees and instituted training programmes for saleswomen and demonstrators in an attempt to increase women's employment in the electrical industry.[67]

In 1934, with the EAW already a decade old, there can be little doubt that the gas industry drew inspiration from this group. The editors of the *Gas Journal* wrote disparagingly about 'Miss Haslett's girls', describing them as bored middle-class do-gooders whose ultimate crusade in the name of progress was 'to install an electric cooker in every slum kitchen'. Nevertheless, the staff writers conceded that the gas industry might profit from a similar federation, run by professional women.[68] Ironically, although the gas industry felt that it had to catch up in this area, the founders of the EAW had considered themselves to be in a similar position in 1924 because of the promotional work being done by Mrs Maud Adeline Cloudesley Brereton and Francis Goodenough on behalf of the gas industry. In a handwritten memo to Caroline Haslett, a Mrs Messanger noted that Goodenough and Mrs Cloudesley Brereton at the BCGA had 15–20 years of business experience, while 'she had wandered into the Electrical World just two years ago'. Evidently, this EAW organiser was conscious not only that the EAW were inexperienced in matters of public relations and advertising, but also that the BCGA had already established a means of communicating with its female consumers, even if it did not have a separate women's organisation. The author added, however, that the presence of women was a positive development for both industries, as it might reduce the antagonism generated by the men and create a 'new atmosphere' of positive coexistence.[69] The assumption that women's presence in the business world neutralised the worst excesses of the masculine competitive spirit was also upheld by other groups of interwar women who worked closely with men. The presence of women in the office encouraged men to behave like gentlemen, minding their language and their manners. Some businesswomen anticipated that by having women in their staff, men were encouraged to conduct their commercial dealings with a greater sense of fairness.[70] Mrs Eileen Murphy stated that the addition of more women employees had 'humanized' the gas industry.[71] This spirit of cooperation was the sentiment of the WGC founders who were anxious to promote the use of gas in the home, but without directly attacking electricity in the process. In addition, the WGC executive believed that they had a very definite contribution to make to the gas industry's commercial advancement.

In 1934, David Milne-Watson at the GLCC contacted the Women's Employment Federation (WEF) to recruit someone capable of setting up a national women's organisation to represent his industry. WEF executive committee member, educationalist Mary Wolseley-Lewis, recommended Miss Kathleen Halpin for the job.[72] The daughter of an upper-level civil servant, Halpin had lived in Paris for a year after leaving school in 1922; upon her return to London she took secretarial training, eventually working at the *Encyclopaedia Britannica* and the *Architect's Journal*. Between 1932 and 1935, Halpin worked as the personal secretary to the national government's foreign secretary, Sir John Simon. As a member of several post-suffrage feminist groups, including the Fawcett Society, the London Society of Women's Service and the WEF, Halpin met prominent constitutional suffragists, including Pippa and Ray Strachey, whose continued dedication to the women's movement inspired younger women like herself who had not personally campaigned for the vote.[73] Recalling her first interview with David Milne-Watson, where he outlined his ideas for an organisation that would 'do for the gas industry what Caroline Haslett had done for electricity', Miss Halpin reflected that Milne-Watson had pitted her against 'the cleverest woman in London with ten years start'. Nonetheless, in May 1935, Halpin accepted the job of organising secretary for the WGC because it offered her 'a challenge', something she appreciated in her work.[74]

While Milne-Watson and Wolseley-Lewis played leading roles in the establishment of the national WGC, the tough job of opening local branches fell to Halpin. In an unpublished essay on the early years of the WGC, Halpin referred to herself as 'the commercial traveller', selling the idea of a women's organisation to gas undertakings:

> I was forever travelling all over the country, setting up branches and talking at meetings. First I had to persuade local managers and there was a fair amount of male prejudice to overcome. Some of them couldn't see why a women's organisation was needed. If there was a ladies circle already, I had to persuade them to join us. If there wasn't, I'd ask for names of people who might make a suitable president and chairman and talk to them. They'd appoint a committee and get started. We were very lucky having the home service advisors. They generally took a very active part in organising new branches and handling the secretarial side.[75]

After the inaugural meeting with Halpin, the branches were on their own, albeit under the patronage of the local gas undertakings. The clubs met monthly in the gas showrooms, at mid-afternoon, with an average attendance of 150 women, who were mainly housewives. Members paid an annual subscription fee of one shilling, to offset the operating costs of the central office. Within each branch, the secretary was the only salaried member; however, this position was usually held by a home service woman.[76] Topics presented at meetings ranged from domestic themes, such as 'interior decoration' and 'hints on careful buying', to information on family health and welfare, including 'home nursing and first aid' and 'care of the teeth'. Branch members might also recommend topics of interest or suggest club outings. According to one early WGC member, Mrs E. Warne, the local chapters planned their own events and meetings, although they sent the minutes of the meetings to headquarters for official approval. Branch delegates attended the WGC's annual general meetings.[77]

Even within the WGC's first year, Halpin's recruiting skills showed impressive results. At its first annual meeting, in June 1936, Halpin welcomed 400 delegates who represented 7,000 members from 23 branches throughout Great Britain. By 1939, the WGC reported an estimated 13,000 members in 42 branches.[78] As a national body representing the interests of women who used gas in their homes, the WGC aspired to 'bridge the gap between the scientist and the housewife and link the manufacturer with the consumer'. In Halpin's words, the WGC was a 'means of collecting and distributing information about gas as it affected the home, child welfare, hygiene, housing, slum clearance, and smoke abatement – all subjects of vital importance to women'. In addition, through its associations with a network of women's service organisations, including the National Council of Women in Britain, it served as the 'housewife's voice on Government and Public Committees'.[79] Beyond this educational and political mandate, the WGC's meetings and field trips provided a social outlet for middle- and working-class housewives, as did the cookery demonstrations already offered in the company showrooms.[80] That the WGC served 'an educational, professional and social function' was clearly stated at the first annual meeting. But, despite their association's close cooperation with the gas industry, WGC organisers did not consider it a 'commercial concern'.[81]

While the WGC brought together women in the home, it doubled as a professional organisation for those women who were employed in the gas industry. Activities intended for women in the business included conferences for gas demonstrators and regional meetings, modelled on the masculine salesmen's circles. For example, within the first year, Halpin and Wolseley-Lewis organised a 'summer school' for home service women. Held in Cardiff in September 1936, and jointly sponsored by the WGC, the local domestic science college and the Cardiff Gas Light and Coke Company, the first summer school drew 98 delegates. The women attended a series of seminars, delivered by professional businesswomen, on technical subjects, such as refrigeration, and public service issues, including the problems and challenges of the demonstrator's job.[82] The success of this event resulted in annual summer schools and regular conferences for demonstrators; however, political events in late 1939 suspended this activity.[83]

To bring demonstrators together on a more regular basis, the WGC encouraged the formation of 'demonstrators' circles'. At these sessions, home service women met monthly throughout the autumn and winter months to hear guest speakers and to exchange ideas about their work. The London and Home Counties Demonstrators' Circle convened its first meeting in January 1937, under the chairmanship of Ethel Willans. As an organisation dedicated to career development, the London circle immediately affiliated with the British Federation of Business and Professional Women. Within a year, the circle reported having 135 members, which included home service women from gas undertakings in London and the surrounding area. Unfortunately, despite this successful start, the London area circle remained the only demonstrators' club, probably because insufficient numbers elsewhere in the country did not justify the formation of additional groups.[84]

As a means of serving its educational mandate, the WGC executive cooperated with technical schools to establish certified training programmes for gas demonstrators. In 1937, the WGC and the Battersea Polytechnic School offered a course in gas cookery demonstration for domestic science students interested in pursuing careers in the gas industry. Miss Godfrey, a home service advisor with the GLCC, instructed her students twice-weekly for two terms. The syllabus extended beyond cookery lessons, including methods for successful public speaking, demonstration practice and the mechanical workings of gas apparatus. After a final examination, successful candidates received a certificate awarded by the WGC Training Sub-Committee, an educational advisory group consisting of home service women and representatives from the domestic science colleges and the Board of Education.[85] For students interested in the gas industry but not wishing to have a career in home service, the WGC offered the elementary gas homecraft course, which was held in company showrooms.

In addition to its training programmes and the conferences, the WGC fulfilled its educational and social mandate through the publication of a quarterly magazine. Under the editorship of Miss A.M.S. Wilson, *Fanfare* first appeared in December 1936, at a cost of sixpence per issue.[86] In its first edition, Wilson enthusiastically outlined the journal's intent:

> As women introduce modern methods into running their homes, they find themselves with more leisure. How to fill this time for their own happiness and the benefit of others? Only the individual can answer for herself, but it is up to *Fanfare* to help her with suggestions and to keep her in touch with the work and pleasures of the modern world.[87]

By 'work', Wilson referred to contemporary efforts to eradicate slums, improve urban air quality and eliminate useless toil through modern technology. Brimming with optimism, despite 'wars and rumours of wars', Wilson saw the present age as a step towards 'a brave new life', with women contributing as never before. 'We have a right to herald it with a fanfare of trumpets and sometimes we will blow our own!'[88] This editorial characterised the tone of the magazine: moderately feminist with an emphasis on maternal issues, of no definite political party bias, but stressing social progressivism and women's responsibilities and rights as citizens, at every level of government (local, national or international). Halpin described *Fanfare* as 'a magazine for women who were beginning to think for themselves'.[89]

Fanfare aimed to satisfy a broad readership, be they home service advisors or lower-middle-class housewives. The content was mixed, including the usual offerings found in interwar women's weeklies: recipes, fiction and housecraft.[90] Despite these similarities to other popular women's magazines, Wilson also ran articles on more serious topics, including women and public service, employment opportunities for women, the ongoing problem of poor infant and maternal health and the need for increased provisions for children's school meals.[91] Not surprisingly, *Fanfare*'s lead articles often contained an endorsement for the gas industry. For example, a story on the history and current work of 'The Invalid Kitchens of London', a type of meals-on-wheels service that dated back to 1912, mentioned that the 85,585 meals served annually to needy adults and children were prepared in gas cookers.[92] Given the industry's sponsorship, such a boost could be expected; however, it was neither obvious nor offensive. Wilson avoided direct negative comparisons between gas and electricity, never alluding to the fierce and ongoing competition between the two industries. Such professional restraint may have been a combination of editorial good taste coupled with the realisation that it was hard to argue with the growing popularity of electrical lighting;[93] her restraint may also have been influenced by the current sentiment that women brought a humanising quality to business and were therefore above crass comparisons.

Keeping the membership abreast of the WGC's activities was another important function of the publication. In a regular news section, the secretaries and chairmen of local branches reported on the events of their respective clubs. Additional columns addressed the WGC's activities at the London headquarters in Grosvenor Place, emphasising the WGC connections to other prominent women's and public service organisations, including the National Federation of Women's Institutes, the National Union of Townswomen's Guilds, the National Smoke Abatement Society and the National Women Citizens Association.

Realising that not all of its members subscribed to *Fanfare*, which was an expensive publication compared to popular women's magazines, such as *Woman* and *Woman's Own*, Wilson printed the WGC news as a special supplement that was free to every member.[94] This gesture indicated that the WGC dedicated itself to keeping its members informed of its local, national and international activities, connecting its readers to ongoing political and social issues of relevance to women.

From its inception, the WGC aligned itself with a network of women's organisations that were attempting to improve women's working lives in the public and private sphere. This action on the part of the WGC was entirely understandable, since its list of vice presidents included a number of women who were active in several organisations whose leadership and committee membership tended to overlap. Consider the example of one WGC vice president, Lady Emmott. The widow of Lord Alfred Emmott, Liberal MP for Oldham, she chaired the Women's Sub-committee on Housing (1918) and later ran as an Independent Liberal in 1922 for Oldham. She lost, partly on a Liberal split, as voters rallied behind either the Lloyd George or the Asquith supporters, but also because of her forthright support for women's issues. In her campaign, Lady Emmott openly endorsed the political programme of the National Union of Societies for Equal Citizenship, an amalgamation of women's suffrage groups, headed by Eleanor Rathbone.[95] Hereafter, Lady Emmott dedicated her time to women's concerns, serving as a vice president of the NCW of Great Britain, an organisation whose agenda addressed political, educational, legal and employment-related issues of importance to women in the workplace and the home.[96] An active member of the WEF, she supported the feminist efforts to end the marriage bar and broaden employment prospects for educated women. In April 1933, Lady Emmott attended a conference on the cooperation among societies, concerned with the employment of educated women, along with better-remembered feminist activists, including Ray and Pippa Strachey and Marjory Spring-Rice.[97] Lady Marjorie Nunburnholme, another WGC vice president, was a committed WEF member whose 'generous donations' to the WEF kept the organisation afloat in the 1930s.[98] Lady Nunburnholme became president of the NCW in 1934. She and Lady Emmott served together on numerous sub-committees, such as an investigative committee on public health and child welfare, whose membership included Eva Hubback. In addition, Lady Emmott chaired the WEF's committee for parliamentary and legislative activity.[99]

WGC executive members and staff were similarly active within a circle of interwar women's groups. We have already noted Wolseley-Lewis's joint participation in the WEF and the NCW. WGC executive member, Lady Ruth Balfour, was also a member of the WEF. She and

Wolseley-Lewis co-wrote a series of newspaper articles on careers for women in non-traditional sectors, such as agriculture, advertising, 'auctioneering' and photography.[100] In 1936, Lady Balfour acted as the WGC delegate to the Ladies' Public Health Committee of the International Council of Women, which met in Yugoslavia that year.[101] Halpin and Wolseley-Lewis represented the WGC at meetings of the WEF. Executive member Ethel Willans attended her share of affiliate meetings, representing the WGC at gatherings of the Association of Teachers of Domestic Science, the National Safety-First Association and the British Federation of Business and Professional Women, an organisation headed by Caroline Haslett of the WES. Evidently, under the auspices of the WGC, women from the gas industry became part of an overlapping network of organisations dedicated to furthering women's employment and educational opportunities.[102]

Serving the interests of women in the home, the WGC participated in a number of projects intended to draw attention to the continuing problems of domestic architecture. In 1937, Halpin was elected to the executive of Lady Sanderson's Women's Advisory Housing Council, a revived version of the 1917 reconstruction committee that had been chaired by Lady Emmott. A Labour Party representative on that earlier committee, Lady Sanderson's, and her group resumed the work of the former organisation, that being to undertake and collate research on working-class housing and report its findings to the Ministry of Health. In addition, Lady Sanderson's committee urged the appointment of more women to local housing committees, because of their knowledge of women's needs and problems.[103] In a show of support, the WGC passed a resolution 'that branches should consider the question of getting more women councillors put up for municipal elections and other committees'.[104] Thus, the WGC assumed the agenda of housing activism that had been in place since the Edwardian years when women in the gas industry, notably Mrs Cloudesley Brereton, had first lectured on the topic of efficient domestic design and the labour-saving possibilities gas technology offered for heating and cooking.[105]

Prior to the establishment of the WGC, home service women participated in research groups aimed at improving working conditions for women in the home. As a member of the NCW, Ethel Willans undertook research for the Council Of Scientific Management In The Home (COSMITH). An ad hoc committee of the NCW Household Service Committee, COSMITH was formed in 1931, under the chairmanship of Mrs Ethel M. Wood, the advertising executive and consultant on domestic design who had worked on the gas industry's 1924 British Empire exhibit. COSMITH had been established by the NCW to ascertain whether methods of scientific management currently revolutionising industrial processes could simplify domestic work in the home.[106] Willans's home service staff presented gas customers with a questionnaire dealing with the 'common faults found in kitchens'. Her staff then collected and analysed the data, and thereby arrived at recommendations for ideal heights for kitchen work surfaces and shelves.[107] In conjunction with the surveys, COSMITH members, most of whom were domestic economy experts, used time and motion theories to evaluate equipment, interior layouts, natural and artificial lighting and sequences of work.[108] The COSMITH group included Caroline Haslett, from the EAW, A.C. Franklin of the Townswomen's Guilds, women lecturers from King's College, London and the London School of Economics, as well as professional female designers, such as Dorothy Braddell.[109] For the female investigators involved in this research, domestic scientific management meant anything related to the operation of the home. This included the establishment of minimum standards for nutrition and family household budgets, and also support for family allowance payments to women as a way of bringing all families up to a national minimum standard of living. COSMITH members were particularly concerned that their research should benefit lower–middle income families earning around £3 per week, and, to this purpose, they promoted their housing designs to borough councils and speculative builders, setting up 'Kitchen Clinics' to advise builders and consumers. In addition, they presented their findings at International Scientific Management Conferences in Amsterdam (1932) and London (1935).[110]

COSMITH members sought ways of integrating labour-saving methods into the homes of middle- and working-class women not simply through the purchase of kitchen equipment, but through efficient design principles. With the benefit of hindsight, historians point out that the concept of 'labour-saving technology' was highly problematic as it was often associated with the exacting demands of domestic science or persuasive and sexist advertising. Its proponents, including the Americans Christine Frederick and Lillian Gilbreth, used the promise of technology and management theories in a bid to professionalise housework, reassuring women that housewifery constituted a satisfying and respectable career. Unlike the Edwardian material feminists, Charlotte Perkins Gilman or even Mrs Cloudesley Brereton, they did not consider technology as a tool for female social and economic emancipation. In addition, the propaganda of domestic scientists and product advertisers altered women's perceptions of minimum standards of domestic hygiene, so housewives spent more time than ever before cooking, cleaning and washing clothes. The time that women saved in one area they were expected to reallocate to another task, and some household chores, such as cleaning carpets and washing windows, were undertaken with greater frequency.[111]

That said, it is easy to be critical of the home service women, such as Ethel Willans and Eileen Murphy. They were corporate domestic scientists, selling cookers and gas services, teaching women how to be professional housewives, despite their own careers in the business world. It must be remembered, however, that the work of the home service women and the WGC had several dimensions, including the recognition of specific design problems as they affected women in the home and the development of strategies for change. They recognised that, although the majority of young women between the wars worked before marriage, most of them would spend a great deal of their adult lives in the home, as mothers and housewives. Working as mediators that moved between their customers and the corporations that produced gas appliances and services, between government housing authorities and builders, the home service women and the WGC acted as a pressure group, representing the interests of ordinary housewives.

As an executive member of the WGC, Ethel Willans continued her work with COSMITH. Both Willans and Halpin represented the WGC on Lord Leverhulme's British Management Council, a think-tank of businessmen and industrialists searching for ways to improve the efficiency of British industry using scientific management. In September 1938, Willans and Halpin were among 100 British delegates who attended the Seventh International Management Congress, held in Washington DC. Although the conference focused on business concerns related to administration, distribution, personnel and agriculture, for the first time, the organisers included a separate home management section. Lillian Gilbreth, well known for her applications of time and motion studies to household routines, was a principal organiser of these meetings.[112]

Representing COSMITH, Ethel Willans read a paper that directly challenged Mrs Gilbreth's work. Miss Willans noted the difficulties of applying time and motion studies to domestic labour as daily housework entailed fewer repetitions per task than industrial jobs and women usually undertook several tasks at one time.[113] The theories of work efficiency that might apply in the workshop and the factory were not readily transferable to the home. Willans explained that, by using its contacts with a number of women's organisations, COSMITH had polled housewives about their flats and houses. While they wanted to learn of chronic design problems, even in relatively new homes, COSMITH strove to impress upon women that, as workers whose input was vital to the national economy, they had the right to expect better, more efficient workplaces. Willans expressed the need to help women find a means of articulating their dissatisfaction with poorly designed houses that contributed to excessive walking, lifting and cleaning. She stated: 'If only she [the housewife] can be incited to make herself heard and refuse to buy or live in a badly planned house, our work will be half done.'[114] Willans emphasised the critical importance of cooperation with manufacturers, and, in the interests of developing labour-efficient designs for kitchen equipment, COSMITH passed its findings on to the British Standards Institution, a group that was slowly turning its attention to domestic equipment. In the ensuing discussion, delegates noted how high the proportion of income spent on renting or purchasing a home was, which, in turn, diminished the family food budget.

As an example of how this problem might be addressed, Halpin presented a paper on Kensal House, a modern working-class housing complex that actually saved its tenants money because of its efficient gas services.[115] Although this housing solution was heavily subsidised by private industry, namely the GLCC, borough councils in the metropolitan area – including Chelsea, Paddington, Stoke Newington, Islington and Westminster – were also studying Kensal House as a possible model for their public housing estates.[116]

As conference delegates and as visitors to a foreign country, Willans and Halpin were struck by the differences between America and Britain, differences they recorded on their return to London in late 1938. Both women agreed that American women fared better than British housewives, the former possessing far more 'mechanical contrivances' that theoretically minimised housework. This being her first visit to America, Halpin marvelled at American plumbing by describing the 'joy of constant and efficient hot water' and the attractions of central heating, which were luxuries that the majority of English homes still lacked.[117] Both women expressed their surprise at the number of American papers presented at the conference that addressed family dynamics and the psychological role of the home, and not the actual physical improvements to domestic architecture, as if the problems of urban housing had already been solved. Reflecting on her impressions of the Washington conference, Halpin recalled that she initially believed that the Americans were 'further ahead on the issues of home management'. She later decided that British women's groups were undertaking similar projects, but did not express it in the same manner. Halpin decided that, in America, it was 'all talk, but not necessarily more action'.[118]

This summation of America seems sadly ironic considering the final note of Halpin's trip to America. The Munich crisis of October 1938 raised fears of imminent war with Germany. Halpin and Willans cut short their stay, returning home on the *Empress of Britain* to find London's streets filled with cheering crowds as Neville Chamberlain announced his successful appeasement of Nazi aggression in Europe.[119] By the end of the following year, however, the WGC had once more turned its agenda to wartime activities on the 'domestic front'. It also experienced the loss of its organising secretary when in January 1939 the British government commandeered Halpin to begin arrangements for the possible evacuation of thousands of children from London to Wales.[120] Scientific management and affordable fully-equipped kitchens would have to wait for yet another post-war reconstruction era.

Gas women and interwar feminism

In terms of the interwar women's movement, where should we place the home service women and the WGC? According to both contemporary and historical accounts, the women's movement divided along class and party lines after women aged 30 and above received the vote under the Representation of the People Act (1918). These divisions were exacerbated by disputes over priorities and tactics for implementing feminist change, debates which pitted 'equal rights' feminists against the 'new' social welfare feminists. Party politics complicated matters further, particularly for women in the Labour Party. For those women, feminist commitment was constrained by a party that stressed the importance of unified class interests over potentially divisive gender issues.[121]

Writing at the end of the 1920s, Vera Brittain identified three main divisions within feminist societies. The 'equal rights' feminists aimed specifically at obtaining equal status, pay and opportunities for women. Lady Rhondda's Six Point Group and its publication, *Time and Tide*, focused on issues of female equality, particularly the amendment of the 1918 legislation to enfranchise all women on the same terms as men, at age 21, and not 30. Elizabeth Abbott's Open Door Council campaigned against any form of government measure that imposed prohibitionary regulations on women's employment, regardless of the trade, profession or working conditions. Brittain identified the second group of feminists as members of the National Union of Societies for Equal Citizenship (NUSEC). Descendants of Millicent Fawcett's constitutional suffragists, the NUSEC directed its energies to reforms necessary to achieve equality of 'liberties, status and opportunity' between men and women. Brittain noted that the organisation's executive had divided over priorities.[122] The dispute erupted between equality feminists, who favoured adherence to a strict equal rights agenda, such as the opposition of protective work legislation, and the 'new feminists' led by Eleanor Rathbone, who gave priority to social welfare reforms aimed specifically at women. This included, for example, family allowance payments to women and free access to birth control information. Although the equality feminists did not oppose these reform measures, they believed that, by focusing on maternal issues, the new feminists reinforced traditional notions of sexual difference and female essentialism.[123] Vera Brittain's third group, the most moderate feminists in the women's movement, preferred to work 'for women's social welfare' rather than engage in 'heated battles for equality of status'.[124] This group included the NCW of Great Britain and its affiliate, the International Council of Women.[125] In many respects, the NCW aligned itself with the 'new feminists' by supporting campaigns for family allowances, nursery schools and increased public funding for maternal and child health.[126]

By the late 1930s, the divisions within the women's movement were even more pronounced than they had been a decade earlier. According to Eleanor Rathbone, the 'old' equal rights feminists, pejoratively nicknamed 'Me-too-feminists', remained preoccupied with a narrow agenda that aimed for equal legal and political rights with men. In contrast, she maintained that the greatest legislative challenges had been won, and women should no longer measure their wants by men's achievements. If the women's movement was to survive, feminists had to recruit support from ordinary women, wives and mothers in the home, finding support through moderate organisations, such as the Townswomen's Guilds and Women's Institutes. 'New feminism' recognised that women were equal, but different, from men, and, therefore, did not share identical needs and aspirations. She attributed this difference to women's maternal qualities and their compassion for 'human suffering'. Whether gender differences were the product of biology or social conditioning she hesitated to say. Rathbone firmly believed, however, that it was women's responsibility to address the social and economic injustices which mothers and children faced, notably by supporting plans for ameliorative legislation.[127]

In addition to these tactical and political divisions, historians have described a general anti-feminist sentiment that undermined popular support for the post-suffrage women's movement. Throughout the suffrage campaign, increasing numbers of women endorsed the feminist demand for the vote. Later, as a result of World War I, women challenged the proscribed social and physical limitations of their capabilities by performing non-traditional work; for example, they worked as retort stokers in the gas industry. Both the suffrage campaign and women's wartime employment undermined traditional demarcations between male and female spheres. After the war and women's partial enfranchisement, however, the popular vision of social reconstruction presented by politicians, intellectuals and the media stressed a return to pre-war social relations and traditional sex roles, but with a slight difference. Marriage manuals stressed the virtues of conjugal harmony as a means of ensuring normalisation after the war, putting a final end to the 'sex antagonism' of the Edwardian suffrage campaign. Moreover, according to Susan Kingsley Kent and Sheila Jeffreys, the new rhetoric of sexual liberation and companionate marriage served to reinforce the interwar adulation of modern domesticity and motherhood. In contrast, feminism became associated with old-fashioned spinsterhood or the malaise of misguided women trying to imitate men.[128]

Although the discourse of sexual reform, as promoted by psychologist Havelock Ellis, may have reflected a post-suffrage anti-feminist backlash, the images of domestic bliss and contented maternal womanhood found in the popular press more effectively eroded post-suffrage feminism. Immediately after the war, the press urged women to leave their factory jobs and return to their normal duties at home. Ray Strachey noted that those women who remained in the workforce were branded 'blacklegs', although, only months before, the press had praised female workers for their wartime contributions.[129] The images in advertising also pressurised women to resume cheerfully their wifely duties: the advertisers of cocoa, cosmetics and cleaning products stressed the importance of feminine beauty and the virtues of domesticity. Deirdre Beddoe notes the changing images of women in advertising between 1918 and 1920, as slogans about factory girls were replaced by those of happy young wives. In addition, she observes the enormous growth of women's magazines during the 1920s and 1930s, including *Good Housekeeping*, *Woman's Own* and *Woman's Illustrated*. All of these publications extolled the importance of motherhood, elevated domesticity to a science and equated femininity with consumption. With the focus on homemaking and romantic fiction, and few references to feminism or professional women, popular women's interwar magazines, by implication, reinforced a negative stereotype of emancipated women.[130]

Brian Harrison cites several factors that combined to erode the feminist movement between the wars. According to his analysis, the post-suffrage feminist demise was a natural slump, resulting from the elimination of its primary 'cause', female enfranchisement, a shifting focus to international feminist causes, like peace and anti-fascism, and the advance of domestic political conservatism under Baldwin's governments. Feminism was no longer fashionable among younger women. The grey-haired activists increasingly dominated the established offshoots of the old suffragist societies while interwar women MPs 'often publicly disclaimed feminist views'.[131] In this environment, it is not surprising that Eileen Murphy, the director of home service for the BCGA, was cautious, even contradictory, in her attitude to the women's movement. She at one time stated: 'I can never be accused of being a feminist, but after seeing the good work women are doing in the gas industry, I have automatically become the woman's champion.'[132]

Within the gas industry press, which addressed an overwhelmingly male readership, Eileen Murphy distanced herself from feminist politics, despite her outspoken advocacy for women's employment opportunities in the gas industry through the WGC and the WEF.[133] Her predecessor, Mrs Cloudesley Brereton, had adopted a similarly evasive position vis-à-vis the suffrage movement. Recall that in 1913, before the gathered members of the BCGA, Mrs Cloudesley Brereton had been equally hesitant to support openly women's suffrage despite her friendship with the active suffragist Alice Zimmern.[134] Instead, Mrs Cloudesley Brereton had urged her male colleagues to consider how gas technology could emancipate women from the home, releasing them for more important work in the public sphere. It is entirely conceivable that both Murphy and Mrs Cloudesley Brereton appreciated that women in the business world could easily antagonise their male colleagues by strident pronouncements of support for feminist activities.

Describing her feminist politics, the countess of Limerick also chose the middle ground. A vice president of the WGC with a number of other public offices, including a seat on the LCC and a membership to the Kensington Housing Association, she did not consider herself a 'full-blooded feminist'. She celebrated the 'enormous change' in women's position over the past 50 years, and, although she insisted that women had an important role to play in public life, she did not personally support 'any widespread agitation to have women admitted to every occupation where men are employed'.[135] Instead, the countess favoured 'organizing our institutions in such a way as to include the variety and freshness of ideas and outlook characteristic of the two sexes rather than encouraging everyone to have the same conventional outlook and training'. She praised women's valuable contributions to municipal affairs, maternity and child welfare work, education, social work and health services. Moreover, she stated that women were eminently suitable for the management of housing estates because of their practical knowledge of domestic requirements. For the countess, it was no longer simply a matter of 'Women's Rights', but 'Women's Duty' as citizens to make the best use of their qualifications and political voice in the interests of national life. She reflected that there was 'no more need of a feminist movement' now that women had the vote, had proved their capabilities during the war, and were entering new areas of public work, not because they were women but because they had the necessary education and expertise.

The countess's audience might have come away with contradictory impressions of women's potential for public office and professional employment. By her account, the old equal rights feminist movement had run its course. Women were able to succeed in the public sphere not because they demanded it, as a right to redress past wrongs against their sex, but because they deserved it, based on the merits of their education and training. Nevertheless, as women, they were still inherently better suited to some types of work, for example in the social services and teaching. Women's essential maternal qualities made them more sympathetic to domestic issues related to housing and child welfare. However contradictory this line of reasoning might appear, it was quite in keeping with the contemporary ongoing discussions among interwar women's groups, as they debated their priorities, politics and strategies, now that women had equal access to the political system.

The countess of Limerick's separation of 'Women's Rights' from 'Women's Duty' becomes clear in light of these debates. Demands for 'rights' was the agenda of the 'old' feminists, still associated with radical egalitarian principles that resisted special treatment for women. In contrast, 'duty' alluded to 'new' feminism, the social welfare concerns taken up by Rathbone's NUSEC, and its offshoot group, the National Council for Equal Citizenship. 'Women's duty', to reform society in accordance with the particular needs of women and children, was the moderate reformist agenda of the NCW.

Evidence suggests that the WGC adhered to this latter brand of interwar feminism. For example, in 1938, the WGC invited Rathbone to address their fourth annual meeting on the topic of family allowances. She was unable to attend, but her colleague, Marjorie Green, the secretary of the Family Endowment Society, delivered a lengthy address on the topic, and a transcript of this talk appeared in the WGC's annual report for members who did not attend the conference.[136] In addition, the WGC was affiliated to the NCW, an organisation associated with a moderate 'new' feminist perspective. When questioned about feminism in the 1930s, Halpin recalled: 'The Women's Gas Council was not overtly feminist but they agreed that society needed changing. They supported family allowances, school meals and nursery schools.'[137] Finally, the WGC's close association with the WEF, an offshoot of the NUSEC, confirms its ties to interwar 'new' feminist activity.[138] Hesitant though they were to identify themselves as 'full-blooded' feminists, the WGC leadership supported various strategies designed to assist women in the home and the workplace, including plans for affordable working-class housing and higher safety standards for clerical workers.

Conclusion

Between the wars, equal rights feminists criticised the maternalist social reform agenda, fearing that it marginalised more important issues related to equality in the public sphere, such as the elimination of the marriage bar from the civil service and the teaching profession, or the demand for equal pay. In addition, equal rights feminists maintained that social reforms directed specifically at women, to assist wives and mothers, merely conformed to the traditional social-sexual dichotomy of the working male bread-winner and female nurturer in the home. From this perspective, the WGC and the home service advisors were part of the problem, particularly given their close association with the domestic science colleges and the manufacturers of household technology.

In a sense, this is true. As experts in household matters, the gas companies' home service women promoted a feminine ideal that was traditional, maternal and domestic. Ethel Willans stated that her department expanded to meet the needs of its female clientele 'because we found that women wanted to know how to run their homes in the best way'.[139] In their efforts to bring to women the information that they 'wanted to know', home service women engaged in wider social debates concerning modern technology, the physiological and social impact of rational design and the importance of 'planning', arguably the greatest reorganising concept of the 1930s.[140] Although home service women promoted an ideal of domestic efficiency, through gas technology and rationalised design, they did not assume that all women were housewives and mothers, or, indeed, that married women did not work outside of the home. There were enough examples of professional women, some married with children, within their own ranks, for home service women to dismiss so narrow a view of women's work. Being practical women, the home service advisors recognised that women assumed the responsibility for the care and management of the home, a situation that was unlikely to change despite egalitarian legislation or improved work opportunities for women. Mrs Cloudesley Brereton understood the difficulty of running a busy household while at the same time meeting the demands of full-time employment. In her letters to Marie Stopes, her references to her work and their shared interest in public health education appear alongside news about family, domestic obligations, and, occasionally, complaints of illness from exhaustion and overwork.[141] Women in public life, including Eva Hubback and Dora Russell, expressed the frustrations of balancing both domestic life and employment, while socialist feminist Margaret Cole, looked to rationalisation and technology, not radical sex role reversals, as the agents of women's release from domestic drudgery.[142]

Among the numerous women's service clubs that catered to housewives between the wars, the WGC most resembled the Townswomen's Guilds, the Electrical Association for Women and the Women's Institutes. It differed from these groups, however, because the WGC also functioned as a professional association for women in the industry, forging contacts through local meetings and national conferences. Its magazine, *Fanfare*, reflected this dual role by running articles of joint relevance to women in the industry as well as the rank and file membership. In this respect, the WGC sustained and developed the home service women's sense of professional achievement and solidarity amid a social and economic climate that was not sympathetic to women's work outside of the home.

Granted, men were not competing for these jobs, and, in all but a few companies, home service advisors were not eligible for gas inspector positions or upper level sales and service jobs. The lady demons had developed their own code of feminine professional conduct and successfully proven their managerial skills during World War I. With the advent of the professional white-collar salesman in the 1920s, however, business authors questioned women's ability to compete in the tough world of modern retail. According to the literature, the supposedly masculine qualities of competitiveness, sportsman-like fairness, personal resolve and quiet confidence were essentials for the professional salesman.[143] While some marketing experts questioned whether women had sufficient drive to succeed in sales, they still agreed that, in the showrooms, the women demonstrators were invaluable for reaching female customers. Evidently, by the interwar years, gendered definitions of occupational ability affected the gas industry, just as they affected other business sectors. Despite this frustrating demarcation, home service women remained active professionals, offering advice to male salesmen on the best means of appealing to female customers, while they also mediated between the gas industry and women in the home. Spokeswomen like Eileen Murphy asserted that women in the gas industry were capable of accepting greater responsibility, and the lady inspectors at Neath Gas Company successfully confirmed that position.

Notes

1 Eileen Murphy, 'Women's activities in the gas industry: home service as it exists today', *GJ*, 29 September 1934, 304.

2 'Mrs. M.A. Cloudesley Brereton relinquished the editor's chair', *GJ* , 29 June 1932, 864–865; 'Changes at the BCGA', *GJ*, 29 June 1932, 864. General manager, William Mason, also retired in 1932; J.C. Walker became general secretary of the BCGA to oversee the office administration; however, Mason continued to work as a consultant for the organisation as did executive chairman Sir Francis Goodenough, following his retirement from the GLCC in 1931.

3 Pugh, *Women and the Women's Movement in Britain*, chap. 4; Beddoe, *Back to Home and Duty*, 135–145; Harold L. Smith, 'British Feminism in the 1920s', in Harold L. Smith (ed.), *British Feminism in the Twentieth Century* (Amherst, Mass., 1990), 47–65.

4 Harrison, *Prudent Revolutionaries*, 317.

5 Horsfield, *Biting the Dust*, 147–148.
6 For example, see Philip G.G. Moon, 'Sales and service', *GJ*, 20 June 1923, 733–735; T.P. Ridley, 'The present position of the gas industry', *GJ*, 30 April 1924, 308–310; 'Local publicity', *GJ*, 3 November 1926, 295–297; 'BCGA district conference at Buxton: local publicity', *GJ*, 25 May 1927, 502–503; Francis Goodenough, 'Essential factors in the promotion of gas sales', *GJ*, 28 November 1934, 687–688.
7 'Local publicity', *GJ*, 3 November 1926, 297.
8 'Lady demons: the story of a pioneer corps', *Co-partners' Magazine* (January–February 1942), 17.
9 LMA/B/GLCC 49/*Minutes of the Board of Directors*, 13 January 1922, 27.
10 LMA/B/GLCC 48/*Minutes of the Board of Directors*, 17 December 1920, 361.
11 LMA/B/GLCC 44/*Minutes of the Board of Directors*, 23 November 1906, 150; GLCC/*Minutes of the Board of Directors*, 30 June 1911, 394. The board agreed to increase the maximum weekly wages paid to demonstrators from 30 shillings to 40 shillings, to be 'attained after five years in service, by annual increases according to merit'.
12 LMA/B/GLCC 49/*Minutes of the Board of Directors*, 22 June 1922, 84.
13 Interview with Pearl Thorpe (née Brown), Bexley Heath, Kent, 23 March 2001.
14 Vera Brittain, *Women's Work in Modern England* (London, 1928), 41–42, 48–49.
15 'Lady demons: the story of a pioneer corps', *Co-partners' Magazine* (January–February 1942), 17.
16 'Opening of the new Kensington showrooms', *Co-partners' Magazine* (November 1926), 311; 'Gas service in East London', *GJ*, 30 July 1930, 254.
17 'Opening of the company's new showrooms in Islington', *Co-partners' Magazine* (December 1925), 310; 'New showrooms at Kilburn', *Co-partners' Magazine* (June 1927), 224; 'Our new premises at Hackney', *Co-partners' Magazine* (August 1931), 312; 'New showroom at Ilford', *Co-partners' Magazine* (March 1936), 100; 'New showrooms at Edgware', *Co-partners' Magazine* (April 1936), 151–152; 'New showrooms at Kingsbury', *Co-partners' Magazine* (July 1936), 321.
18 'Our fiftieth showroom', *Co-partners' Magazine* (July 1932), 263.
19 'Local offices and showrooms', *Co-partnership Journal* (May 1929), 125–127; 'Gas in the market-place', *Co-partnership Journal* (June 1937), 181; 'New showrooms at Eltham', *Co-partnership Journal* (July 1938), 153.
20 'Cookery demonstrations', *Co-partnership Journal* (March 1929), 73.
21 'Lady demons: the story of a pioneer corps', *Co-partners' Magazine* (January–February 1942), 17.
22 Spring-Rice, *Working-Class Wives*, 96–99.
23 Eileen Murphy, 'Free advice for housewives', *Gas World*, 15 September 1934, 657.
24 C.M. Croft, 'BCGA Eastbourne conference: service to the consumer', *GJ*, 13 November 1929, 281.
25 Interview with Mrs E. Warne, Walthamstow, London, 28 October 1995. Interview with Pearl Thorpe (née Brown), Bexley Heath, Kent, 26 May 2003.
26 Miss B. Randle and Miss M.A. Chalkley, 'Gas service from the feminine point of view', *Gas World*, 12 April 1930, 392.

27 'Lady demons: the story of a pioneer corps', *Co-partners' Magazine* (January–February 1942), 17.
28 Beddoe, *Back to Home and Duty*, 115–131.
29 Lancaster, *The Department Store*, 96–102.
30 'A demonstration of gas cookery', *Co-partnership Journal* (October 1922), 294; 'A super gas demonstration: week's display at Selfridges', *GJ*, 6 November 1935, 362.
31 Sally Alexander, 'Becoming a Woman in London in the 1920s and 1930s', in David Feldman and Gareth Stedman Jones (eds), *Metropolis London: Histories and Representations Since 1800* (London, 1989), 263–264.
32 'A gas appliance display at an underground railway station', *Gas World*, 9 January 1932, 38.
33 LMA/B/GLCC 49/*Minutes of the Board of Directors*, 13 January 1922, 27.
34 Goodenough, 'Essential factors in the promotion of gas sales', 688.
35 M.A. Cloudesley Brereton, 'The study of the problems of domestic economy as a factor in salesmanship', *Second Annual General Meeting and Conference Proceedings of the BCGA*, 109.
36 Ibid., 117; M.A. Cloudesley Brereton, 'Addresses to gas salesmen: exact knowledge as a factor in salesmanship', *GJ: Gas Salesman Supplement*, 29 May 1929, 137–138.
37 M.A. Cloudesley Brereton, 'The study of the problems of domestic economy as a factor in salesmanship', *Second Annual General Meeting and Conference Proceeding of the BCGA*, 120; for the response to Cloudesley Brereton's paper, see 130–139.
38 G.A.N. Lowndes, *The Silent Revolution: An Account of the Expansion of Public Education in England and Wales 1895–1965* (Oxford, 1970), 150–152; Sarah King, 'Technical and Vocational Education for Girls: A Study of the Central Schools of London, 1918–1939', in Eric Evans and Penny Summerfield (eds), *Technical Education and the State Since 1850: Historical and Contemporary Perspectives* (Manchester, 1990), 84–85.
39 Paul E. Derrick, *How to Reduce Selling Costs* (London, 1919), 34–35.
40 Board of Education, *Education for Salesmanship: Final Report of the Committee on Education for Salesmanship* (London, 1931), 10–12, 130–131.
41 Crossick, 'The Emergence of the Lower Middle Class in Britain: A Discussion', in Crossick (ed.), *The Lower Middle Class in Britain*, 12; Bailey, 'White collars, gray lives?', 273–290.
42 'The circles', *GJ, Gas Salesman Supplement*, 25 April 1923, 58; 'Annual meeting of gas salesmen', *GJ*, 12 October 1927, 119.
43 W.M. Mason, 'Some BCGA experiences in salesmanship, BCGA Gas Salesmen's Circles – Southern District – London Area', *GJ, Gas Salesman Supplement*, 26 November 1924, 210–211.
44 NGA/ British Commercial Gas Association, *Building Up Gas Sales: The Domestic Gas Salesmanship Course* (London, n.d.).
45 H. Ernest Hunt, 'Addresses to gas salesmen: salesmanship and the subconscious', *GJ*: *Gas Salesman Supplement*, 1 April 1929, 117–118; H. Ernest Hunt, 'The higher salesmanship: the right start', *GJ: Gas Salesman Supplement*, 16 October 1929, 237–238.
46 Francis Goodenough, 'The education of gas salesmen', *GJ*, 5 November 1930, 410–413; Francis Goodenough, 'Education and training of gas salesmen', *GJ*, 15 April 1931, 147–148.

47 Francis Goodenough, 'Salesmen's circles – an appeal to administrators', *GJ*, 12 October 1927, 120.

48 Benson, *Counter Cultures*, fn. 265. Retail historians have overlooked the occupational development of the sales profession for men. Within the American context, Benson suggests that, by the interwar years, a growing number of salesmen worked for commission selling the 'big ticket' items, notably cars and appliances. Selling these 'hard goods' was extremely competitive and depended on the salesmen's technical expertise, or at least his ability to convince the customer of his competency in this regard.

49 LMA/B/GLCC, *Minutes of the Board of Directors*, 22 January 1923, 168; 'BCGA southern district conference at Bournemouth', *GJ*, 20 June 1923, 734.

50 A Saleswoman, 'The home of the bachelor woman', *GJ, Gas Salesman Supplement*, 11 January 1928, 5–6.

51 Miss M.A. Tench, 'The domestic workshop from a woman's point of view', *Gas World*, 26 July 1924, 90–91. Miss M.A. Tench, 'Improvements in gas cookers', *GJ: Gas Salesman Supplement*, 30 October 1929, 261–262.

52 'Politic service – lady inspectors', *GJ*, 25 May 1927, 483–484.

53 'Gas from a woman's point of view: Mr. A. Forshaw gives some second-hand impressions', *Gas World*, 7 February 1931, 133–134.

54 'Lady inspectors', *GJ*, 1 June 1927, 578; 'The BCGA conference: home management in the machine age – appendix II – maintenance service by women', *GJ*, 9 October 1935, 135.

55 'The proper blend', and 'two illustrations', *Gas World*, 21 February 1931, 173–174.

56 'Salesmanship in the showrooms by Miss Gladys Burlton', *GJ*, 12 October 1927, 122, 124; also see 'By a woman for women: saleswomen necessary in the showroom', *Gas Salesman* (January 1931), 25. The anonymous author offered a plea to gas managers to employ more women in their showrooms, but not to take away jobs from salesmen, to improve relations with female customers who were their families' 'purchasing agents'.

57 'Interesting BCGA appointment', *Gas World*, 25 August 1934, 178.

58 'Mrs. M.A. Brereton relinquishes the editor's chair', *GJ*, 29 June 1932, 864–865; 'Changes at the BCGA', *GJ*, 29 June 1932, 864. BCGA general manager William Mason and editor Maud Adeline Cloudesley Brereton retired from the BCGA in 1932. After this, all the BCGA's advertising work was handled by the advertising agency, the London Press Exchange, although the BCGA retained its editorial staff for the association publications.

59 Murphy, 'Women's activities in the gas industry', 304.

60 Ibid., 304–305; Eileen Murphy, 'Home service', *GJ*, 3 October 1934, 67–68; 'Our home service section "At Home"', *News: A Newspaper for the Employees of the Gas, Light and Coke Company* (February 1938), 6–7.

61 'Home service: Mrs. Eileen Murphy champions women in the gas industry', *Gas World*, 6 October 1934, 304.

62 'The BCGA conference: home management in the machine age – Mrs. Eileen Murphy', *GJ*, 9 October 1934, 134.

63 On the marriage bar in teaching and the civil service, see Beddoe, *Back to Home and Duty*, 5, 42, 82–85; Lewis, *Women in England*, 197–200; Smith, 'British Feminism in the 1920s', in Smith (ed.), *British Feminism in*

the Twentieth Century, 53; Pugh, *Women and the Women's Movement in Britain*, 90–100.

64 'G.D.A', *GJ*, 17 October 1934, 200–201; 'Gas development association for women: Halifax Centre visit Bradford', *GJ*, 12 December 1934, 809.

65 'Under the lens: this woman business', *GJ*, 12 December 1934, 811.

66 'Bristol House', *The Woman Engineer* (December 1935), 70–71.

67 Worden, 'Powerful Women' in Attfield and Kirkham (eds), *A View from the Interior*, 134–148; Sparke, *As Long As It's Pink*, 153–154; Electrical Association for Women, *The Electrical Association for Women: A Introduction to the Society* (London, n.d.).

68 'Under the lens: this woman business', *GJ*, 12 December 1934, 811.

69 Institution of Electrical Engineers/Dame Caroline Haslett Papers/Personal Files/NA Est 33/17.1.11. Gas and Electricity memo, 19 November 1926.

70 Kwolek-Folland, *Engendering Business*, 235–238.

71 'The BCGA conference: home management in the machine age – Mrs. Eileen Murphy', *GJ*, 9 October 1935, 134.

72 Women's Library/6/Women's Employment Federation/493/Women's Employment Federation: National Federation of Organisations concerned with the Employment and Training of Educated Women, 1. Milne-Watson was a vice president of this organisation.

73 Video-taped interview with Kathleen Halpin, Chagford Street, London, 23 September 1996; Second interview with Kathleen Halpin, Chagford Street, London, 27 September 1996. Collection of the author.

74 First interview with Kathleen Halpin, 23 September 1996.

75 Business papers of Kathleen Halpin, 'Women's Gas Federation: our story 1935–1985', unpublished paper presented at the Fiftieth Annual General Meeting of the Women's Gas Federation, Westminster, 1985, 2.

76 Miss M. Goggin, 'My day's work in the gas industry', *Fanfare*, 2, 7 (1938–39), 6, Miss Goggin, head of home service for the Commercial Gas Company, London had one staff demonstrator assigned to the company's two WGC branches in Poplar and Bow. Branch News, *Fanfare*, 2, 5 (1938), 51, Mrs E.M. Menhinick, a demonstrator with the GLCC since the lady demon era, served as secretary to the West Ham branch of the WCG until she retired in 1938 due to failing eyesight.

77 Interview with Mrs E. Warne, Walthamstow, 2 November 1995; see also NGA/Women's Gas Council papers/Women's Gas Council, Walthamstow and Leyton District Branch, *Formation of the Branch and Report of the First Meeting*, 3 November 1938; WGC, Walthamstow and Leyton Branch, *Agenda for First Meeting*, 3 November 1938; WGC, Lea Bridge District Branch, *Aims and Objectives*, November 1938; WGC, Lea Bridge District Gas Company, *Invitation Letter to New Members*, September 1938.

78 NGA/Women's Gas Council, *First Annual Report, June 1935–March 1936* (London, 1936), 5; WGC, *Fourth Annual Report, June 1938–March 1939* (London, 1939), 3.

79 'Women's Gas Council hold first annual meeting', *GJ*, 3 June 1936, 733.

80 NGA/WGC, *First Annual Report*, 10. A party of 200 women visited the British Industries Fair, Birmingham and 140 WGC members attended the 1936 Ideal Home Exhibition, London; 'The Women's Gas Council', *Co-partnership Journal* (April 1937), 107; 'Woolwich Women's gas circle and Downham Women's gas circle, visit to Ascot Water Heaters, Ltd.', *Co-partnership Journal* (November 1937), 347.

81 NGA/WGC, *First Annual Report*, 6.
82 NGA/WGC, *Second Annual Report, 1936–37* (London, 1937), 5–6; 'The Women's Gas Council', *GJ*, 28 April 1937, 209; also see Branch News, 'Summer school for demonstrators', *Fanfare*, **1**, 3 (1937).
83 NGA/WGC, *Third Annual Report 1937–1938* (London, 1938), 9–10; NGA/WGC, *Fourth Annual Report*, 11–12; NGA/WGC, *Fifth Annual Report 1939–1940* (London, 1940), 4–5. The summer school for demonstrators, planned for August 1939 in Manchester, was cancelled because of the threat of war.
84 NGA/WGC, *Second Annual Report 1936–1937* (London, 1937), 7–8; NGA/WGC, *Third Annual Report*, 10; NGA/WGC, *Fourth Annual Report*, 10–11; 'Women's Gas Council: demonstrators' circle meets', *GJ*, 24 March 1937, 778; 'London and Home Counties demonstrators' circle', *Fanfare*, **2**,4 (1938), 42; 'Demonstrators' circle', *Fanfare*, **2**,3 (1938), 39; 'Demonstrators' circle', *Fanfare*, **2**, 6 (1938–39), 40.
85 NGA/WGC, 'Training of demonstrators', *First Annual Report*, 13; NGA/WGC, 'Training of demonstrators', *Second Annual Report*, 7; NGGA/WGC, 'Training of demonstrators', *Third Annual Report*, 8–9; NGA/WGC, 'Training of demonstrators', *Fourth Annual Report*, 10.
86 'Fanfare: first issue of the new WGC magazine', *GJ*, 16 December 1936, 788; NGA/WGC, *Second Annual Report*, 9.
87 Miss A.M.S. Wilson, 'Editorial', *Fanfare*, **1**, 1 (1936), 3.
88 Ibid., 3.
89 Video-taped interview with Kathleen Halpin, 23 September 1996.
90 For a discussion of interwar women's magazines see Pugh, *Women and the Women's Movement in Britain*, 209–219.
91 For example see Lilian Baker, 'Schools for mothers', *Fanfare*, **2**, 2 (1938), 11–13; 'Model kitchens for modern houses', *Fanfare*, **2**, 2 (1938), 20–23; 'More women in public life?' *Fanfare*, **2**, 3 (1938), 9–11; Dorothy Norman, 'Feeding the nursery folk', *Fanfare*, **2**, 4 (1938), 13–15.
92 'The invalid kitchens of London', *Fanfare*, **1**, 1 (1936), 11–17.
93 Video-taped interview with Kathleen Halpin, 23 September 1996. Miss Halpin noted with irony that she lived in an all-electric flat, despite her continued preference for gas cookers and heaters.
94 NGA/WGC, *Second Annual Report*, 11; Irene Dancyger, *A World of Women: An Illustrated History of Women's Magazines* (Dublin, 1978), 142–144; Beddoe, *Back to Home and Duty*, 14–15.
95 Pugh, *Women and the Women's Movement*, 186.
96 National Council of Women, *Women in Council: National Council of Women News*, **1** (January 1931), 3.
97 Women's Library/WEF/ *Executive Meeting Minutes*, 19 December 1933, 2.
98 Ibid., 2.
99 National Council of Women, *NCW Handbook 1931–32* (London, 1932), 2, 28; Harrison, *Prudent Revolutionaries*, 278–287. Eva Hubback, a former suffragist and close associate of Eleanor Rathbone and Ray Strachey, was active in the NUSEC promoting a feminist agenda that included family allowance payments to mothers, women's access to free birth control information and the prevention of child poverty.
100 WL/WEF/*Executive Meeting Minutes*, 26 February 1935, 5; 10 October 1935, 1–2.
101 NGA/WGC, *Second Annual Report*, 15–18.

102 NGA/WGC, *Fourth Annual Report*, 4–5. The British Federation of Business and Professional Women concerned itself with the treatment of business and professional women in countries controlled by fascism. Had the war not redirected their activities to labour concerns at home, the federation planned to send delegates to the International Labour Conference at Geneva.
103 NGA/WGC, *Third Annual Report*, 4–5, 12–13.
104 Ibid., 12.
105 For example see 'Labour-saving bungalows', *ATAOUFG* (April 1924), 34–43; 'Home equipment on common sense lines', *ATAOUFG* (July 1925), 58–64; 'Labour-saving kitchens', *ATAOUFG* (February 1928), 11–16; 'Gas in London's homes', *ATAOUFG* (April 1930), 36–42.
106 WL/WF/Box 563/Letter from Mrs E.M. Wood to Mrs Stocks, LCC (March 1939), regarding the history and purpose of the Council of Scientific Management of the Home.
107 WL/WF/Box 563/Letter from Ethel Willans to Mrs Raphael at the Institute of Industrial Psychology, 9 March 1934; COSMITH, *List of Subcommittees*, 1 June 1937. Mrs Raphael was also a COSMITH member.
108 WL/WF/Box 563/*National Council of Women Household Service Sectional Committee*, 1–5.
109 WL/WF/B4/Box 563/COSMITH, *Membership list*, 1 June 1937.
110 WL/WF/Box 563/ 'Scientific household management', copy of paper presented at Fifth International Scientific Management Congress, Amsterdam, July 1932; National Council of Women of Great Britain, 'Household Service Committee: the research sub-committee', *NCW Handbook 1936–37* (London 1937), 82–83.
111 Hoy, *Chasing Dirt*, 153–157; Horsfield, *Biting the Dust*, 140–150; Cowan, *More Work for Mother*, 173–187; Davidson, *Woman's Work*, 204.
112 Business papers, Kathleen Halpin, *Program of The Seventh International Management Congress*, Washington, DC, 19–23 September 1938, 46, 49.
113 WL/WF/563/COSMITH, *Report by Miss E. Willans*, 12 December 1938, 8–9.
114 WL/WF/Box 563/COSMITH, 'Planning of kitchens and developments in time and motion study', paper delivered by Ethel Willans at Seventh International Management Congress, Washington, DC, 19–23 September 1938, 3.
115 Business Papers, Kathleen Halpin, *Report of Seventh International Congress of Management*, Washington, DC, 19–23 September 1938, Home Management Session, 13.
116 'Kensal House scheme', *GJ*, 15 February 1939, 440.
117 Kathleen Halpin, 'To America and back', *Fanfare*, 2, 5 (1938–39), 7–10; WL/WF/Box 563/Miss E. Willans, COSMITH, 1.
118 Video-taped interview with Kathleen Halpin, 23 September 1996.
119 Ibid. Halpin recalled that because of the threat of war in Europe, they had no trouble changing their reservations due to the great number of cancellations. Halpin was amused because she, her mother and Ethel Willans were given the bridal suite for their return voyage to England.
120 Katherine Bentley Baumann, *Green Sleeves: The Story of the Women's Voluntary Service/Women's Royal Voluntary Service* (London, 1977), 16, 166; 'Women's Gas Council: Miss Halpin to assist work on evacuation', *GJ*, 4 January 1939, 39. Kathleen Halpin joined the Women's Voluntary Service in late 1938; she did eventually oversee the mass evacuations of

children from London to Wales, taking a wartime leave from her duties at the WGC.

121 Jane Lewis, 'Beyond suffrage: English feminism in the 1920s', *Maryland Historian*, **6** (Spring 1975), 1–17; Smith, 'British Feminism in the 1920s', in Smith (ed.), *British Feminism in the Twentieth Century*, 47–65; Harold L. Smith, 'Sex vs. class: British feminists and the labour movement, 1919–1921', *Historian*, 8 (November 1985), 19–37; Pat Thane, 'The Women of the British Labour Party and Feminism 1906–45', in Harold L. Smith (ed.), *British Feminism in the Twentieth Century* (Amherst, Mass., 1990), 124–143; Pamela M. Graves, *Labour Women: Women in British Working-Class Politics, 1918–1939* (Cambridge, 1994), chap. 4.

122 Brittain, *Women's Work in Modern England*, 173–174.

123 Smith, 'British Feminism in the 1920s', in Smith (ed.), *British Feminism in the Twentieth Century*, 60–61; Pugh, *Women and the Women's Movement in Britain*, 236–240.

124 Brittain, *Women's Work in Modern England*, 174.

125 Ibid., 173–174.

126 National Council of Women, *NCW Handbook 1931–32* (London, 1932), 7–20.

127 Eleanor Rathbone, 'Changes in Public Life', in Ray Strachey (ed.), *Our Freedom and Its Results* (London, 1936), 57–74.

128 Sheila Jeffreys, 'Sex Reform and Anti-Feminism in the 1920s', in The London Feminist History Group (ed.), *The Sexual Dynamics of History* (London, 1983), 179; Susan Kingsley Kent, 'Gender Reconstruction After the First World War', in Harold L. Smith (ed.), *British Feminism in the Twentieth Century* (Amherst, Mass., 1990), 71–75.

129 Ray Strachey, *The Cause: A Short History of the Women's Movement in Great Britain* (Port Washington, 1928), 79.

130 Beddoe, *Back to Home and Duty*, 9–33.

131 Harrison, *Prudent Revolutionaries*, 309–310; for international feminist activity, see Beryl Haslam, *From Suffrage to Internationalism: The Political Evolution of Three British Feminists, 1908–1939* (New York, 1999), chap. 5.

132 'Home service: Mrs. Eileen Murphy champions women in the gas industry', *Gas World*, 6 October 1934, 340.

133 Murphy, 'Women's activities in the gas industry', 304–305; NGA/WGC, *Fourth Annual Report*, 6. Murphy gave several talks to the Women's Employment Federation, including 'Careers for women in the gas industry' and 'Openings for women in the gas industry'.

134 Anne Clendinning, 'Gas and water feminism: Maud Adeline Brereton and Edwardian domestic technology', *Canadian Journal of History*, **33**, 1 (April 1998), 12–13.

135 NGA/WGC, 'Address by the Countess of Limerick', *Third Annual Report*, 17–18.

136 'Fourth annual general meeting', *Fanfare*, **2**, 6 (1938–39), 40; NGA/WGC, *Fourth Annual Report*, 18–22.

137 Video-taped interview with Kathleen Halpin, 23 September 1996.

138 Pugh, *Women and the Women's Movement*, 242.

139 '100,000 women want to know', *Modern Living* (Summer 1939), 21.

140 Ritschel, *The Politics of Planning*, 1–17.

141 BL, Stopes Papers, Add. 568699 f. 80, 1926.
142 Harrison, *Prudent Revolutionaries*, 317–18; Dyhouse, *Feminism and the Family in England*, 144.
143 Arthur Coe, *The Scientific Promotion of Gas Sales*, 2nd edition (London, 1926), 347–348.

Conclusion

> The significance of domestic technology lies in its location at the interface of public and private worlds. The fact that men in the public sphere of industry, invention and commerce design and produce technology for use by women in the private domestic sphere, reflects and embodies a complex web of patriarchal and capitalist relations.[1]

As this book has shown, the gas industry's lady demons and home service women mediated between the private and public worlds of domestic consumption and industrial production. As home economists, saleswomen and publicists employed by the gas industry and promoting the benefits of domestic technology to women in the home, they were positioned at the very centre of this 'complex web of patriarchal and capitalist relations'. An occupational group dedicated to increasing domestic sales through personalised customer service, the lady demons assured housewives that gas technology would eliminate useless toil, improve housing standards and help them to raise healthier children. From a contemporary perspective in an age of media saturation and corporate mistrust, it is easy to be sceptical of the lady demons' sincerity and their sense of social purpose. However, the gas women must be considered within the historical context of Edwardian progressivism, the child welfare movement, the reconstruction plans for building 'homes fit for heroes', and the practical agenda of interwar 'new feminism'. Their work intersected with this broad agenda of social welfare reforms. The dedication of these women – notably Mrs Maud Adeline Cloudesley Brereton, Helen Edden, Ethel Willans, Eileen Murphy and Kathleen Halpin – indicate that their interest in the health and well-being of the gas industry's female customers extended beyond corporate rhetoric. These women did believe that affordable gas technology substantially improved women's working lives in the home and the workplace, reducing the time spent on household labour and making possible new standards of domestic comfort and hygiene.

Feminist critics have pointed out the fallacy of their position, noting that domestic mechanisation actually increased the time that women spent on housework.[2] The professional home economists, particularly those who worked in advertising and sales, have been singled out for criticism, accused of colluding with the corporations by promoting unrealistic domestic standards and the costly equipment to achieve them.[3] These home economists' false consciousness served to prolong women's domestic enslavement; they were part of the problem, not the solution.

Sociologist Judy Wajcman questions this school of thought, particularly its brand of 'technological determinism, where technology is said to have resulted in social changes'. Instead, she considers the historical, material and social circumstances under which women promoted and accepted domestic technology, thereby positing an alternative feminist analysis of the mechanisation of the British home. Wajcman writes:

> The common feminist stress on the negative effects of domestic technology has contributed to the view that women have been duped. There is a tendency among some feminist scholars to assume an unqualified anti-technological stance and to imply that modern housewives are worse off than their grandmothers. This tendency is evident in those authors who stress the increasing isolation of the domestic worker and see domestic labour as having lost much of its creativity and individuality. Once we recognize that the mechanization of the home did bring substantial improvements to women's domestic working conditions, even while it also introduced new pressures, women seem less irrational.[4]

Continuing Wajcman's argument, it follows that English women's growing acceptance of, and reliance upon, gas technology from the late Victorian era to the interwar years was entirely rational, as was the home service women's promotion of this domestic infrastructure.

Wajcman also questions the caricature of the consuming housewife duped by the avaricious adman and the corporate home economist. She admits that people were often taken in by false promises, and that housewives have been 'susceptible' to 'well-targeted' advertising campaigns. She also believes, however, that the 'effectiveness of the professional experts in imposing new notions of domestic life has been overestimated'. The acquisition of domestic capital goods, including cookers, heating systems and washers, enabled English women to rationalise domestic labour by allocating less time to repetitive heavy jobs, such as hauling coal and water. In addition, domestic technology compensated for the declining numbers of working-class women entering domestic service, as more middle-class women undertook their own domestic chores.[5] If domestic technology failed to deliver women from household toil because it elevated housekeeping standards, the problem lies not with the technology but with the social relations within which it operates. Sociologists Christine Bose and Philip Bereano have arrived at similar conclusions, positing that while the 'technological infrastructure of the household' eliminated the physical difficulty of some routine tasks, technology did not free women from the domestic sphere because 'its goal was never to decrease male power in the home'.[6]

After 30 years of feminist scholarship, it should come as no surprise that the promise of domestic mechanisation was undermined by social constructions of work and gender. Still, even if domestic technology did not challenge 'male power in the home', it had the potential to raise women's status within the domestic sphere, while simultaneously asserting the interconnections between women's work that was unpaid and private versus paid and public employment. As shown in Chapter 3, Mrs Cloudesley Brereton's Edwardian 'gas and water feminism' anticipated that domestic mechanisation would minimise the heaviest household chores, thereby allowing women time for more valuable pursuits, such as paid work and community service or leisure time spent with their friends, husbands and children. Mrs Cloudesley Brereton urged her male colleagues at the BCGA to recognise that women's time was of real value, and should not be wasted even on routine unpaid domestic work. Her assertion that 'home life' was the foundation of society echoed the maternal feminist arguments of her contemporaries then lobbying for political representation.

During the interwar years, Ethel Willans and Kathleen Halpin reasserted the need to modernise the domestic sphere, working through a network of women's organisations that included the WGC, the NCW, COSMITH and the WEF. Promoting domestic gas technology to rationalise 'women's workshops', they also hoped to elevate women's status in the home. By focusing attention on housing standards, nutrition, working-class budgets and women's health, the basis of the 'new feminist' programme between the wars, the home service women asserted women's equal status as citizens, while recognising their different social responsibilities and needs as mothers. Nevertheless, as women working in business, sales and advertising, the gas industry's home service women demonstrated, by their own example, that domestic technology enabled women to have choices. In Chapter 6, I noted how Ethel Willans described her work with COSMITH as a form of consumer consciousness-raising. She wanted to make women aware that they had the right to voice their housing needs and preferences to builders, architects and housing authorities. As a means of female empowerment, Willans's advocacy efforts asserted the value of women's time and labour, and their right to control their working environment. In so doing, both Willans and Mrs Cloudesley Brereton continued the efforts of an earlier group of Victorian women who had identified the connection between domestic architectural reform and the 'woman question'.

It can be argued that by stressing women's domestic contributions, the home service women, and indeed the 'new feminists', reinforced gender differences, including the sexual division of labour. Like their female colleagues working in advertising, despite the good intentions of either group, the gas demonstrators offered 'an essentially conservative approach to domestic reform'. However, as Alison Ravetz notes, a broad spectrum of women activists and organisations welcomed labour-saving domestic technology and architectural designs. It was not until the late 1950s that women grew dissatisfied with their 'mechanized isolation' in the suburban new towns, and, even then, the extent of this social malaise remains open to question.[7]

How then, should historians of women, work and business interpret Eileen Murphy's 1934 assertion to the BCGA that the gas industry was a 'women's industry'? The employment of women as company lecturers and demonstrators in the late 1880s was an integral aspect of the gas industry's evolving promotional strategies, as outlined in Chapter 2. While the recruitment of demonstrators created a new form of employment for the graduates of the cookery colleges, the lady demons consciously worked to define their new profession, in terms of its public image, duties, pay and job status. The letters from lady demons that appeared in the *Journal of Gas Lighting* in 1905 indicate that the women were quite aware that they were establishing the parameters of a new occupation that crossed sales and social work with teaching. Ideologies of class and gender informed these discussions, as gas managers and lady demons all debated the women's occupational status and their required levels of technical expertise.

The professional status of male salesmen and showroom attendants received less consideration, at least until after World War I, despite the concerns of BCGA spokesmen, like William Mason and Francis Goodenough. In addition to the contract cookery ladies taken on for exhibitions, the gas showrooms were staffed by young men, often clerks, with no sales training and no technical background. With the expansion of the electrical industry after the war, Goodenough initiated training programmes and salesmen's workshops to remedy the shortcomings of their gas salesmen in this increasingly competitive atmosphere. The training initiatives coincided with a general trend towards the professionalisation of the male retail worker in a new recognition of the alleged importance of sales to business success and the association of that success with identifiably masculine qualities, such as confidence and assertiveness. Despite the home service women's certified training and their longstanding sense of professional solidarity, by the 1920s they were no longer the unrivalled sales experts, even though they were often called upon to offer advice and assistance on sales techniques. While this attests to the women's importance in the industry, it also indicates that the gas industry management was demarcating what jobs women could and should hold. This movement is clearly reflected in the introduction of specific pay scales for different female jobs.

Debates about women's mechanical skills, physical strength and stamina were common throughout World War I as female replacement workers undertook a wide range of industrial jobs at the gasworks, even the dangerous job of stoking the retort houses. Given proper training and equipment, women successfully performed previously designated masculine forms of employment, thereby challenging existing assumptions regarding the categories of skilled male versus unskilled female work. Although the demobilisation of men in service relieved the women 'gassies' of their wartime posts in 1919, the women's example had definite repercussions for the gas industry. During the 1920s, with the outdoor female replacement workers gone, male white-collar workers raised their concerns about job designations. Some managers, inspectors and showroom personnel questioned the ability of female employees to undertake anything beyond cookery demonstrations, based on the women's assumed poor understanding of technical matters. Although women like Mrs Cloudesley Brereton and Murphy understood the fallacy of such a claim, few women performed traditionally 'male' jobs, such as inspecting appliances or collecting money from the gas meters after the war. Although the gas industry had praised the technical abilities of its nearly 5,000 female replacement workers during the war years, those same skills were deemed inadequate once the war was over and skilled women workers represented a perceived threat to men's jobs within the industry.

Nevertheless, Eileen Murphy's claim warrants some consideration. The lady demons successfully created a new profession for women within the gas industry. Despite their small numbers – there were about 150 home service women active 'in and around London' by the late 1930s – they reached thousands of female consumers through public lectures, club meetings, department store displays, public exhibitions and home visits. As a result of this activity, the gas industry's home service women established a connection between the corporations and their reported 11 million female consumers.

By some accounts, women's relationship to the 'world of goods' has been entirely negative. Female shoppers, blinded by opulent displays and seduced by advertising, have abandoned themselves to the pleasures of the marketplace, recklessly running up debts, shoplifting or coveting what they cannot afford.[8] But, in a recent history of the department store, Bill Lancaster presents a more positive assessment of women's experiences in late Victorian consumer culture. While the expansion of the mass market created employment for women in shops, journalism and advertising, the proprietors' growing awareness of customer relations fostered the merchandising ethos that the 'customer is always right', and the customer was usually a woman. In addition, women's participation in consumer activity subverted a number of traditionally 'feminine' qualities, including 'dependence, passivity, religious piety and domestic inwardness'. According to Lancaster, women were not 'passive consumers', but used their purchasing power to make rational and calculated choices, even applying those choices to political ends, in terms of store boycotts and support for the cooperative movement. Lancaster's description of the rational female consumer aided by the professional woman of business aptly characterises the gas industry's home service women and their female customers.

This is not to suggest that gas managers were indifferent to the commodity spectacle, be it a trade exhibition or a tastefully appointed showroom. As this study has shown, Edison's 1882 Crystal Palace exposition and the resulting public enthusiasm for electricity influenced gas men in their decision to actively promote their own product, perceiving themselves as 'tradesmen' and not simply 'monopolists'. Despite the early predictions of the popular press, the commercial showdown between gas and electricity did not occur until the interwar period. To modernise the industry's public image and cultivate the loyalty of female customers, the London gas companies expanded their promotional practices, by, for example, making documentary films and building an expensive model housing complex, Kensal House, despite the questionable return on those costly projects. Showroom reconstruction was another important aspect of this modernisation scheme. Using contemporary design, be it tasteful Georgian classicism or streamlined art deco, the GLCC and the SMGC created inviting public spaces for female recreation and leisure, where women might relax, meet for a chat, have a cup of tea, hear a talk, watch a free film or see a demonstration. Like the managers of the large department stores described by Lancaster, gas company managers and home service women designed programmes to attract, entertain and educate female consumers.[9]

When compared to its rival, electricity, business historians and scholars of industrial design have often characterised the English gas industry as conservative and old-fashioned. By most accounts, the gas industry's technological advancements and its commercial diversification away from lighting into heating and cooking were motivated by the fear of electrical competition, and not by any independent entrepreneurial spirit. Yet, this characterisation rests to a large extent on a late Victorian popular discourse which prematurely predicted the demise of gas and the rapid ascendancy of electricity. A lingering popular mistrust of the gas industry based on consumers' apprehensions about the soot, smell, toxicity and inflated cost of coal gas also stimulated public interest in the new alternative fuel source. As this study has shown, gas men were motivated to promote their own industry in response to this popular discourse and the flamboyant showmanship of Edison and Swann, not simply because they sincerely believed in the imminent demise of gas lighting. The use of women demonstrators proved to be a highly effective strategy for remaking the popular image of the gas industry. As the women demonstrated with their cookery shows and home visits, coal gas could be used with relative safety and ease in the home, if appliances were properly installed, operated and maintained. Gas cookers, fires and water boilers no longer represented the intrusion of the industrial revolution into the Victorian home, but represented modernity and women's new familiarity with domestic technology. As a reflection of the rehabilitated image, by the Edwardian years, the gas industry easily outdistanced electricity in terms of advertising and public spectacle, thanks to the collective efforts of the BCGA under the guidance of Francis Goodenough, William Mason, David Milne-Watson and Mrs Maud Adeline Cloudesley Brereton.

Curiously, studies of the electrical industry have offered more positive assessments of the continued commercial success of the English gas industry.[10] In their discussion of interwar electrification and its failure to radically transform the English home, Sue Bowden and Avner Offer noted that, despite a preference for electric lighting, British housewives still adhered to coal gas for cooking, heating and washing because gas was familiar and less expensive.[11] Cost and fuel efficiency were important determining factors, as indicated by the consumer protests discussed in Chapter 5. Bowden and Offer suggested that, in comparison to electricity, the gas industry directed its advertising to women of all classes and not simply to an affluent urban minority.[12] According to Hannah, gas cookers were 'better designed and easier to use', although, for reasons that remain unclear, design historians have shown a limited interest in domestic gas technology.[13] Perhaps they too have been swayed by the sort of 'technological determinism' that equates new technology with social advancement.

Finally, there remains the issue of commercial motivations. Although the gas managers continuously reiterated the ameliorative potential of gas technology, their fundamental interest was not social welfare, but increasing consumer demand for coal gas which resulted from the sale of appliances and consumers' dependence on utility services. Despite some connections to progressive circles, gas engineers were practical businessmen, not philanthropists or politicians. Although the advocates of domestic economy preached the benefits of healthy cooking, gas men commissioned women demonstrators to sell stoves based on the latter's ability to appeal to other women. Despite the obvious labour-saving merits of gas technology for Edwardian housewives, as extolled by enthusiasts like Mrs Cloudesley Brereton, gas managers hired lady demons to decrease the maintenance and repair costs on company-owned rental cookers. While Kensal House offered a solution to the housing problems associated with slum clearance, the GLCC only really needed to build just one such complex to convey this message. During the interwar years, the home service women and the industry-sponsored WGC were important adjuncts to the gas industry's sales departments and valuable components of its comprehensive advertising campaigns. Although the success of the women's home service departments was always difficult for gas managers to measure in terms of profits, the women's apparent corporate contributions lay in the areas of public relations, customer service and the developing field of market research.

This sales agenda was often at variance with the home service women's own perceptions of their professional duties and their responsibilities to women in the home. While it remains true that the women recognised the commercial basis of their work, the lady demons and the home service women added a sense of social purpose to their corporate mandate. In debating their professional duties, the early lady demons compared themselves to social workers and district visitors, identifying strongly with these public welfare professionals. The essential difference, however, was that the lady demons were bound to respect the rights and privacy of their customers, the women they were hired to assist and to serve. Thus, the lady demons lacked the official status of the health visitors whose interventionist policies often aroused working-class resentment. Mrs Cloudesley Brereton elected to work for the private sector, specifically the gas industry, because she believed in that industry's potential to fundamentally improve the living conditions of working families while relieving women from the worst aspects of domestic labour. Between the wars, home service women used the WGC to establish professional networks with other working women and to publicly represent the needs of women in the home in campaigns for housing reform and social welfare programmes. While it can be argued that theirs was an essentially conservative message which represented and reaffirmed women's domestic roles, their support for school meals, day nurseries and family allowance payments reflected the overall demands of interwar new feminists.

Unfortunately, it remains unclear exactly how gas managers interpreted this limited feminist involvement. When David Milne-Watson recruited Kathleen Halpin to do for them what Caroline Haslett had done for the electrical industry, did he also realise that the former suffragist Caroline Haslett actively promoted the entrance of women into the engineering trades, despite male opposition from within her industry?[14] Given the general reluctance of gas managers to employ women in traditionally male jobs, such as gas inspection, it appears unlikely that Milne-Watson advocated that particular aspect of Caroline Haslett's work. As might be expected, the commercial agenda of the gas managers, the aspirations of its female employees and the needs of their female customers in the home were not always complementary. For example, in a letter to the *Evening Standard*, a London housewife suggested that gas cookers should be elevated on legs, to relieve women of excessive stooping when using and cleaning their stoves. Responding to this letter, the *Gas Journal*'s editor jeered that women had to bend down to scrub their floors anyway, and that domestic labour gave women that 'lithe figure' so fashionable in the 1920s.[15] Nevertheless, the following year, the BCGA praised a new line of 'modern gas equipment for kitchens', including an enamelled, elevated gas range with the hot plates and the oven 'on the same level' to eliminate unnecessary stooping.[16] Although the gas industry may not have been the 'women's industry' that Eileen Murphy hoped for, male industrial designers and gas managers could no longer afford to ignore the concerns of the majority of their customers.

Notes

1 Judy Wajcman, 'Domestic Technology: Labour-saving or Enslaving?' in Gill Kirkup and Laurie Smith Keller (eds), *Inventing Women: Science, Technology and Gender* (Milton Keynes, 1992), 254.
2 Ruth Schwartz Cowan, 'The "Industrial Revolution" in the home: household technology and social change in the twentieth century', *Technology and Culture*, **17** (1976), 1–23; Davidson, *Woman's Work*, 190–192; Hardyment, *From Mangle to Microwave*, 187–189.
3 Susan Strasser, *Never Done: A History of American Housework* (New York, 1982), 245; Laird, 'Progress in Separate Spheres', 43–45.
4 Wajcman, 'Domestic Technology: Labour-saving or Enslaving?', 250; Wajcman critically cites K. Reiger, 'At home with technology', *Arena*, 75 (1986), 109–110.
5 Wajcman, 'Domestic Technology: Labour-saving or Enslaving?', 240–241.
6 Christine E. Bose and Philip L. Bereano, 'Household Technologies: Burden or Blessing?' in Jan Zimmerman (ed.), *The Technological Woman: Interfacing With Tomorrow* (New York, 1983), 92.
7 Ravetz, 'A View from the Interior' in Attfield and Kirkham (eds), *A View from the Interior*, 200–201; Judy Attfield, 'Inside Pram Town: A Case Study of Harlow House Interiors, 1951–61', in Judy Attfield and Pat Kirkham (eds), *A View from the Interior: Feminism, Women and Design* (London, 1989), 215–216; Frank Schaffer, *The New Town Story* (London, 1972), 192–194. In contrast to Attfield's conclusions, Schaffer contends that female dissatisfaction in the new towns was overstated.
8 Bowlby, *Just Looking*, 31–32; Beetham, *A Magazine of Her Own?*, 154.
9 Lancaster, *The Department Store*, chap. 6.
10 Hannah, *Electricity Before Nationalization*, 204.
11 Bowden and Offer, 'The Technological Revolution that Never Was' in de Grazia with Furlough (eds), *The Sex of Things*, 267.
12 Ibid., 267.
13 Hannah, *Electricity Before Nationalisation*, 206; Forty, *Objects of Desire*, chap. 8; Sparke, *Introduction to Design and Culture*, 125.
14 Worden, 'Powerful Women', 134–137.
15 'She stoops to conquer', *GJ*, 2 November 1927, 309.
16 'Labour-saving kitchens', *ATAOUFG* (February 1928), 14.

Bibliography

Business records and manuscript sources

Committee of Scientific Management in the Home, Women's Library (WL), WF/Box 563.
CPGC, London, London Metropolitan Archives (LMA), B/CPGC.
Electrical Association for Women, Institute of Electrical Engineers (IEE), London EAW/2.1–8.
GLCC, London, LMA, B/GLCC.
GLCC, London, Capital Housing Corporation, Minute Book, National Gas Archive (NGA), Warrington.
Heal and Son's Corporate Records, Archive of Art and Design (AAD), Victoria and Albert Museum, London AAD/1978/2/2.
King's College for Women, King's College, University of London, Minutes KWA/M; KWA/GPF11, Miss Baddley, unpublished speech.
LCC Subcommittee on Cookery, LMA, LCC 1.
LCC, Report re: Gas versus Electricity in the Home, LMA, LCC.
National Society's Training College of Domestic Subjects, Hampstead, LMA, ACC 900.
SMGC, London, LMA, B/SMet.
WEF, Executive Meeting Minutes, WL.
WGC, Minute Book, Leyton and Walthamstow Branch, NGA.
WGC, Annual Reports and Conference Proceedings, 1935–1939, NGA.
Women and Work Collection, microfilm, Imperial War Museum.

Unpublished papers

Braddell, Dorothy, business papers, Archive of Art and Design, AAD/2/8/1980–AAD2/110/1980.
Halpin, Kathleen, business papers, estate of Kathleen Halpin.
Haslett, Dame Caroline, Institute of Electrical Engineers, London, NA Est 33, 17.1. 11–14.
Horrigan, Pauline, business papers, collection of Pauline Horrigan.
Matthews, Wendy, business papers, collection of Wendy Matthews.
Stopes, Marie Carmichael, Manuscripts Collection, BL, Add. 58447–58770.
Warne, Elizabeth, NGA, Warrington.

Ephemera

BCGA and Associated Realist Films, *Housing Problems* (1935); *Enough to Eat* (1936); *Children at School* (1937), British Film Institute, London.

John Johnson Collection, Bodleian Library, Oxford, Gas and Electrical Appliances, Box 1–2.

Government publications

Board of Education, Education for Salesmanship (1931), *Final Report of the Committee on Education for Salesmanship*, London: HMSO 1931.

Board of Trade Committee on Deaths from Gas Poisoning (1930), *Report to the Board of Trade on the Departmental Committee on Deaths from Gas Poisoning*, London: H.M. Stationery Office.

Board of Trade (1918); *Report of the Board of Trade on the Increased Employment of Women in the UK during the War; Statistics to April 1917*, (d. 9164; xiv. 783.

Board of Trade, *Report on the State of Employment in the United Kingdom at the end of March 1918*, Imperial War Museum, Women at Work Collection, Reel 40, Emp. 25.

Board of Trade, Report of Departmental Committee, *Methods of Charging for Gas on a Thermal Basis* (1923), Cmd. 1825 xi 569, 7; 3–16.

Board of Trade, *Reports of the Standing Committee (and Subcommittees) on the Investigation of Prices, Gas Apparatus – Sale and Hire*, May 1921, Cmd. 1381, xvi: 717.

Central Committee, *Women's Employment, Interim Report 1914–1916* (1915), Cd. 7848, xxxvii: 669.

Health of Munitions Workers Committee, *Final Report*, Cd. 9065, vol. 5, part 3, 34, 161.

Home Office (1919), *Substitution of Women for Men During the War, Reports by H.M. Inspectors of Factories showing the Positions in Certain Industries at the end of 1918*, London: H.M. Stationery Office.

Home Office, *Substitution of Women for Men, Tabular Reports by H.M. Inspectors Showing the Present Position in Industries Other Than Munitions Industries* January 1917.

Home Office (1919), *Substitution of Women in Non-Munition Factories During the War*, London: H.M. Stationery Office.

Interdepartmental Committee on Physical Deterioration, *Report of the Interdepartmental Committee of Physical Deterioration*, Cd. 2210, minutes of evidence (1904) 1, 13–47, 299–324.

Ministry of Munitions, 1920–24 (1976), *History of the Ministry of Munitions*, 7, i, iv, ix; Appendix vii, Microfilm, 1976.

Ministry of Reconstruction, Women's Employment Committee, *Ministry of Reconstruction, report of the Women's Employment Committee*, 1919. Cmd. 9239, xiv: 767.

Ministry of Reconstruction, Women's Housing Sub-Committee, *First Interim Report of the Women's Housing Subcommittee of the Advisory Council of the Ministry of Reconstruction* (1918), Cmd. 9166, x: 629.

Ministry of Reconstruction, Women's Housing Sub-Committee, *Final Report* (1918), Cmd. 9232, x: 637.

War Cabinet Committee (1919), *Women in Industry: report of the War Cabinet Committee on women in industry*, Cmd. 167 xxxi: 593.

War Cabinet Committee, *Report of the War Cabinet Committee on Women in Industry* (Physical Sub-Committee), Cmd. 135, 21 November, 1918.

War Cabinet Committee, *Report of the War Cabinet Committee on Women in Industry Appendices*, Summaries of Evidence, (1919). Cmd 167, xxxi: 593.

War Office, *Women's War Work*, issued by the War Office, London, September 1916.

Newspapers, trade and other journals

Architectural Review
A Thousand And One Uses For Gas
British Medical Journal
Bulletin
Cassell's Family Magazine
Chamber's Journal
Common Cause
Cookery Teacher
Co-Partnership Journal
Co-Partners' Magazine
Daily Express
Daily Mail
Daily Mail Ideal Homes Catalogue
Daily Telegraph
Expositor
Fanfare
Gas Journal (known as the *Journal of Gas Lighting* until January 1917)
Gas World
Graphic
Home and Hearth

Illustrated London News
Journal of Gas Lighting (renamed the *Gas Journal* in January 1917)
Journal of the Royal Institute of British Architects
Lancet
Macmillans' Magazine
Modern Living
New Statesman
Nineteenth Century
Pall Mall Gazette
Picture Post
Punch
The Times
T.P.'s Weekly

Contemporary publications

Abercrombie, Patrick (ed.) (1939), *The Book of the Modern House*, London: Hodder & Stoughton.

Apprenticeship and Skilled Employment Association (ed.) (1909), *Trades for London Girls and How to Enter Them*, London: Longmans.

Atkinson, Amy and Grace Holyroyd (1909), *Practical Cookery with an Introduction on Cookery by Gas*, Leeds: Nutt & Co.

Barnett, Henrietta O. (1884), 'Women as Philanthropists', in Theodore Stanton (ed.), *The Woman Question in Europe*, London: G.P. Putnam, 108–138.

Bateson, Margaret (1895), *Professional Women Upon Their Professions*, London: Hozace Cox.

Bayley, Edric (1888), *Education in London Board Schools*, London: London Board of Education.

Bennett, Arnold (1949), *Buried Alive*, reprint of 1912 edition, London: Methuen.

Beveridge, Sir William H. (1928), *Economic and Social History of the World War: British Food Control*, London: Humphrey Milford, Oxford University Press.

Bird, M. Mostyn (1911), *Women at Work: A Study of Different Ways of Earning a Living Open to Women*, London: Chapman & Hall.

Black, Clementina (1907), *Sweated Industry and the Minimum Wage*, London: Duckworth.

Bondfield, Margaret G. (1905), 'The effects on health of women's employment in shops', *Journal of the Royal Sanitary Institute*, transactions, **25**, 738–745.

Bosanquet, Helen (1899), *Rich and Poor*, London: Macmillan.

Bosanquet, Helen (1914), *Social Work in London, 1869–1914*, London: Macmillan.

Bowley, A.L. and Margaret H. Hogg (1985), *Has Poverty Diminished?*, reprint of 1925 edition, New York and London: Garland.

Bowley, Marian (1985), *Housing and the State, 1919–1944*, reprint of 1945 edition, New York and London: Garland.

Brackett, C.F. and F.L. Pope (1891), *Electricity in Daily Life*, London: Kegan Paul.

Brereton, M.A. Cloudesley (1909), *The Mother's Companion*, London: Mills & Boon.

Brereton, M.A. Cloudesley (1914), *First Annual Report and Conference Proceedings of the British Commercial Gas Association, Manchester, October 8–9, 1912*, London: British Commercial Gas Association.

Brereton, M.A. Cloudesley (1916), *The One Maid House*, London: British Commercial Gas Association.

Brereton, M.A. Cloudesley (ed.) (1914), *Papers Delivered to the Second Annual General Meeting and Conference of the British Commercial Gas Association, London, October 21–22, 1913*, London: British Commercial Gas Association.

Brereton, M.A. Cloudesley and Helen Edden (1913), *Cooking By G.A.S*, London: British Commercial Gas Association.

British Commercial Gas Association (1939), *Modern Films on Matters of Moment*, London: British Commercial Gas Association.

Brittain, Vera (1928), *Women's Work in Modern England*, London: Noel Douglas.

Brittain, Vera (1978), *Testament of Youth*, reprint of 1933 edition, London: Virago.

Buckmaster, J.C. (1906), 'Cooking as a Branch of Domestic Education', in C. Herman Senn (ed.), *The Universal Cookery and Food Association Annual*, London: Universal Cookery and Food Association, pp. 52–55.

Burlton, Gladys (1935), *Staff-Service and Training*, Manchester: Co-operative Union.

Coe, Arthur (1926), *The Scientific Promotion of Gas Sales*, 2nd edition, London: Ernest Benn.

Denby, Elizabeth (1938), *Europe Re-housed*, London: George Allen & Unwin.

Derrick, Paul E. (1920), *How to Reduce Selling Costs*, 3rd edition, London: George Newnes.

Devonport-Hill, Rosamond (1884), 'Cookery teaching in the London School Board', *Macmillans' Magazine*, **50**, June, 99–104.

Edden, Helen and N. Moser (1908), *The Household Gas Cookery Book*,

London: Gas Publications.

Evans, E.M. (1904), 'The essential qualifications of a lady health visitor', *Journal of the Sanitary Institute*, transactions, **24** (1903), 306–18.

Fabian Women's Group (1989), *Women's Fabian Tracts*, ed. Sally Alexander, London: Routledge.

Francatelli, Charles Elme (1861), *A Plain Cookery Book for the Working Classes*, London: Winstable & Walsall.

Fraser, Percival M. (1912), 'Suggestions for a well-planned house', *Journal of Royal Institute of British Architects*, 23 March, 390.

Freeden, Michael (ed.) (1989), *Minutes of the Rainbow Circle, 1894–1924*, London: Royal Historical Society.

Fry, Maxwell (1975), *Autobiographical Sketches*, London: Elek Books.

Gas Journal (1937), *New Kitchens for Old*, foreword by Dorothy Braddell, London: Walter King Ltd.

Gas, Light and Coke Company (1930), *Record of Progress*, London: Gas Publications.

Gibbon, Sir Gwilym and Reginald W. Bell (1939), *History of the London County Council, 1889–1939*, London: Macmillan.

Gilman, Charlotte Perkins (1966), *Women and Economics: The Economic Factor Between Men and Women as a Factor in Social Evolution*, reprint of 1898 edition, foreword by Carl Degler, New York: Harper Torchbooks.

Glasier, Katherine (1920), 'The Labour woman's battle with dirt', in Marion Phillips (ed.), *Women and the Labour Party*, New York: B.W. Huebsch, 86–93.

Hamilton, Cicely (1909), *Just to Get Married*, London: Chapman & Hall.

Hardy, Forsyth (ed.) (1979), *Grierson on Documentary*, reprint of 1947 edition, London and Boston: Faber & Faber.

Jeune, Lady Mary (1895), 'The ethics of shopping', *Fortnightly Review*, **57**, January–June, 125–132.

Leslie, S.C. (1937), *Kensal House: The Case for Gas is Proved*, London: British Commercial Gas Association.

Lloyd, E.M.H. (1924), *Experiments in State Control: A Study of Economic Control and Social Response at the War Office and the Ministry of Food, 1914–1918*, Oxford: Clarendon.

Lyttleton, M.K. (1901), *Women and Their Work*, London: Methuen.

Marshall, Mrs A.B. (n.d.), *Mrs. A.B. Marshall's Cookery Book*, London: Simpkin, Marshall, Hamilton, Kent & Co.

Moran, Clarence (1905), *The Business of Advertising*, London: Methuen.

Morley, Edith J. (ed.) (1914), *Women Workers in Seven Professions: A Survey of Their Economic Conditions and Prospects*, London: Routledge.

Morrison, Lord (1960), *Herbert Morrison: An autobiography*, London:

Odhams.

National Council of Women (1932), *NCW Handbook 1931–32*, London: National Council of Women.

Peel, Mrs C.S. (1920), *The Daily Mail Cookery Book*, 4th edition, London: Associated Newspapers.

Peel, Mrs C.S. (1929), *How We Lived Then, 1914–1918*, London: John Lane.

Phillips, Nora and Marion Edwardes, Janet Tuckey and E. Dixon (eds) (1898), *A Dictionary of Employments Open to Women*, London: The Women's Institute.

Phillips, Dr Marion (n.d.), *The School Doctor and the Home*, London: Women's Labour League.

Political and Economic Planning Group (1936), *Report on the Electrical Industry in Great Britain*, London: PEP.

Political and Economic Planning Group (1939), *Report on the Gas Industry in Great Britain*, London: PEP.

Reeves, Maud Pember (1980), *'Round About a Pound A Week*, reprint of 1913 edition, New York and London: Garland.

Richards, William (1877), *The Gas Consumer's Handbook*, London: E. & F.N. Spon.

Richmond, Lillie (1903), *Cookery Recipes with Special Hints on Gas Cooking*, London: Richmond Gas Stove and Meter Co.

Romanes, George (1887), 'The capacity of women', *Nineteenth Century*, May, 664–667.

Sanderson Furniss, Averil D. (1920), 'The working woman's house', in Marion Phillips (ed.), *Women and the Labour Party*, New York: B.W. Huebsch, pp. 74–85.

Scott, Walter Dill (1905), *The Theory of Advertising: a Simple Exposition of the Principles of Psychology in Their Relation to Successful Advertising*, Boston: Small, Maynard &Co.

Senn, C. Herman (ed.) (1894), *Cookery Souvenir of the Universal Cookery and Food Association*, London: Universal Cookery and Food Association.

Senn, C. Herman (ed.) (1902), *Cookery Annual: Yearbook of the Universal Food and Cookery Association*, London: Universal Cookery and Food Association.

Sillitoe, Helen (1933), *A History of the Teaching of Domestic Subjects*, London: Methuen.

Simon, Viscount (1952), *Retrospect*, London: Hutchinson.

Sims, George R. (1889), *How the Poor Live and Horrible London*, reprint of 1889 edition, London and New York: Garland.

Smiles, Samuel (1958), *Self Help: With Illustrations of Conduct and Perseverance*, centenary introduction by Asa Briggs, London: J. Murray.

Snowden, Ethel (1913), *The Feminist Movement*, London: Collins.

Spring-Rice, Margery (1939), *Working-Class Wives: Their Health and Conditions*, Harmondsworth: Penguin.

Standing Joint Committee of Industrial Women's Organisations (1928), *Protective Legislation and Women Workers*, London: Labour Party.

Strachey, Ray (1927), *Women's Suffrage and Women's Service*, London: National Society for Women's Service.

Strachey, Ray (ed.) (1936), *Our Freedom and Its Results*, London: Hogarth.

Sugg, Marie Jenny (1890), *The Art of Cooking By Gas*, London: Cassell & Co.

Sugg, William (1882), 'Gas as an illuminating agent compared with electricity', paper read before the Institution of Civil Engineers, London: Walter King.

Sugg, William (1884), *Domestic Uses of Coal Gas*, London: Walter King.

Thompson, Henry (1891), *Food and Feeding*, 6th edition, London: Frederick Warne.

Unwin, Raymond (1911), 'Housing and town planning', *Journal of British Architects*, **9**, December, 110–11.

Williams, Miss Constance (1913), '"The Silent Servant": How the Use of Gas Can Minimize Life's Little Worries', in British Commercial Gas Association (ed.), *Gas in the Home, Office and Factory – Hints on Health, Comfort, Economy: A Series of Popular Lectures at the National Gas Congress and Exhibition*, London: British Commercial Gas Association, pp. 66–72.

Women's Labour League (1912), *The Needs of Little Children: report on a conference on the care of babies and young children*, London: Women's Labour League, Central London Branch.

Zimmern, Alice (1900), 'Ladies dwellings', *Contemporary Review*, 77, January, 96–104.

Zimmern Alice (1909), *Women's Suffrage in Many Lands*, London: The Athenaeum Press.

Books, articles and theses

Abelson, Elaine (1989), *When Ladies Go A-Thieving: Middle-class Shoplifters in the Victorian Department Store*, New York: Oxford University Press.

Adams, Annmarie (1996), *Architecture in the Family Way: Doctors, Houses and Women, 1870–1900*, Montreal and Kingston: McGill-Queen's University Press.

Adams, R.J.Q and Philip P. Poirier (1987), *The Conscription Crisis in Great Britain, 1900–1918*, Basingstoke: Macmillan.

Adburgham, Alison (1975), *Liberty's: A Biography of a Shop*, London:

Allen & Unwin, 1975.

Adburgham, Alison (1979), *Shopping in Style: London from the Restoration to Edwardian Elegance*, London: Thames & Hudson.

Aldcroft, Derek H. (1987), *The British Economy: vol. 1, The Years of Turmoil 1920–1951*, Atlantic Highlands, NJ: Humanities International.

Alexander, Sally (1984), 'Women, class and sexual differences in the 1830s and 1840s: some reflections on the writing of a feminist history', *History Workshop Journal*, **17**, Spring, 125–149.

Alexander, Sally (1989), 'Becoming a Woman in London in the 1920s and 1930s', in David Feldman and Gareth Stedman Jones (eds), *Metropolis London: Histories and Representations Since 1800*, London: Routledge, pp. 245–271.

Altick, Richard (1978), *The Shows of London*, Cambridge, Mass.: Belknap.

Anderson, Olive (1987), *Suicide in Victorian and Edwardian England*, Oxford: Clarendon.

Attar, Dena (1987), *A Bibliography of Household Books Published in Britain, 1800–1914*, London: Prospect.

Attar, Dena (1990), *Wasting Girls' Time: The History and Politics of Home Economics*, London: Virago.

Attfield, Judy (1989), 'Inside Pram Town: A Case Study of Harlow House Interiors, 1951–61', in Judy Attfield and Pat Kirkham (eds), *A View from the Interior: Feminism, Women and Design*, London: Women's Press, pp. 215–238.

Auerbach, Jeffrey A. (1999), *The Great Exhibition of 1851: A Nation on Display* (New Haven/London: Yale University.

Bailey, Peter (1990), 'Parasexuality and glamour: the Victorian barmaid as cultural prototype', *Gender and History* **2** (2), Summer, 148–172.

Bailey, Peter (1998), *Popular Culture and Performance in the Victorian City*, Cambridge: Cambridge University Press.

Bailey, Peter (1999), 'The politics and poetics of modern British leisure', *Rethinking History* **3** (2), 131–175.

Bailey, Peter (1999), 'White collars, gray lives? The lower-middle class revisited', *Journal of British Studies*, **38** (3), July, 273–290.

Barnett, L. Margaret (1985), *British Food Policy During the First World War*, Boston: George Allen & Unwin.

Barnouw, Eric (1993), *Documentary: A History of the Non-fiction Film*, 2nd revised edition, Oxford: Oxford University Press.

Barty-King, Hugh (1984), *New Flame: The Illustrated History of Piped Gas and the Way It Changed the Lives of Those Who Lived and Worked in Britain, 1783–1984*, Tavistock, Devon: Graphmitre.

Baumann, Katherine Bentley (1977), *Green Sleeves: The Story of the Women's Voluntary Service/Women's Royal Voluntary Service*,

London: Seeley Service.

Baxendale, John and Christopher Pawling (1996), *Narrating the Thirties: A Decade in the Making*, Basingstoke: Macmillan.

Beaver, Patrick (1970), *The Crystal Palace 1851–1936*, London: Hugh Evelyn.

Beddoe, Deirdre (1989), *Back to Home and Duty: Women Between the Wars, 1918–1939*, London: Pandora.

Beetham, Margaret (1996), *A Magazine of Her Own? Domesticity and Desire in the Woman's Magazine, 1800–1914*, London and New York: Routledge.

Benson, John (1994), *The Rise of Consumer Society in Britain*, London and New York: Longman.

Benson, Susan Porter (1986), *Counter Cultures: Saleswomen, Managers, and Customers in American Department Stores, 1890–1940*, Urbana: University of Illinois Press.

Berghoff, Hartmut (1991), 'British businessmen as wealth-holders, 1870–1914: a closer look', *BH*, **33**, April, 222–240.

Berghoff, Hartmut (1992), 'A reply to W.D. Rubinstein's response', *BH*, **34**, April, 82–85.

Best, Geoffrey (1971), *Mid-Victorian Britain, 1851–1875*, London: Weidenfeld & Nicolson.

Blakestad, Nancy L. (1992), '"What's the use of learning this?" Careers for KCHSS graduates, 1910–1949', *Envoy: Queen Elizabeth Old Student's Paper*, Autumn, 7–14.

Bose, Christine E. and Philip Bereano (1983), 'Household Technologies: Burden or Blessing?' in Jan Zimmerman (ed.), *The Technological Woman: Interfacing With Tomorrow*, New York: Praeger.

Bowden, Sue and Avner Offer (1996), 'The Technological Revolution That Never Was: Gender, Class and the Diffusion of Household Appliances in Interwar England', in Victoria de Grazia, with Ellen Furlough (eds), *The Sex of Things: Gender and Consumption in Historical Perspective*, Berkeley: University of California Press, pp. 244–274.

Bowlby, Rachel (1985), *Just Looking: Consumer Culture in Dreiser, Gissing and Zola*, New York: Methuen.

Branca, Patricia (1975), *Silent Sisterhood*, London: Croom Helm.

Braybon, Gail (1981), *Women Workers in the First World War: The British Experience*, London: Croom Helm.

Braybon, Gail (1995), 'Women and the War', in Stephen Constantine, Maurice W. Kirby and Mary B. Rose (eds), *The First World War in British History*, London: E. Arnold.

Brewer, John and Roy Porter (eds) (1993), *Consumption and the World of Goods*, London and New York: Routledge.

Briggs, Asa (1963), *Victorian People: A Reassessment of Persons and*

Themes, 1851–67, 2nd edition, New York: Harper & Row.

Burnett, John (1986), *A Social History of Housing, 1815–1985*, 2nd edition, London: Methuen.

Burnett, John (1994), *Plenty and Want: A Social History of Food in England From 1815 to the Present Day*, 3rd edition, London and New York: Routledge.

Burstyn, Joan N. (1980), *Victorian Education and the Ideal of Womanhood*, London: Croom Helm.

Chandler, Jr, Alfred D. (1977), *The Visible Hand: The Managerial Revolution in American Business*, Cambridge, Mass.: Belknap.

Chatterton, D.A. (1972), 'State control of public utilities in the nineteenth century: the London gas industry', *BH*, **14** (2), July, 166–178.

Cherry, Deborah (1993), *Painting Women: Victorian Women Artists*, London: Routledge.

Clark, Anna (1995), *The Struggle for the Breeches: Gender and the Making of the British Working Class*, Berkeley: University of California Press.

Clendinning, Anne (1998), 'Gas and water feminism: Maud Adeline Brereton and Edwardian domestic technology', *Canadian Journal of History*, **33**, April, 1–24.

Clendinning, Anne (2000), '"Deft Fingers" and "Persuasive Eloquence": the "lady demons" of the English gas industry, 1888–1918', *WHR*, **9** (3), 501–537.

Coffin, Judith (1996), *The Politics of Women's Work: The Paris Garment Trades, 1750–1915*, Princeton: Princeton University Press.

Collette, Christine (1989), *For Labour and for Women: The Women's Labour League, 1906–1918*, Manchester: Manchester University Press.

Condrell, Diana and Jean Liddiard (1987), *Working for Victory? Images of Women in the First World War, 1914–18*, London: Routledge and Kegan Paul.

Coombs, Annie (1994), *Reinventing Africa: Museums, Material Culture and Popular Imagination in Late Victorian and Edwardian England*, New Haven and London: Yale University Press.

Corley, T.A.B. (1987), 'Consumer marketing in Britain, 1914–60', *BH*, **29** (4), October, 65–83.

Corr, Helen (1983), 'The Schoolgirls' Curriculum and the Ideology of the Home', in Glasgow Women's Studies Group (eds), *Uncharted Lives: Extracts from Scottish Women's Experiences*, 1850–1982, Glasgow: Press Gang.

Cowan, Ruth Schwartz (1976), 'The "Industrial Revolution" in the home: household technology and social change in the twentieth century', *Technology and Culture*, **17**, 1–23.

Cowan, Ruth Schwartz (1983), *More Work for Mother: The Ironies of Household Technology from the Open Hearth to the Microwave*,

New York: Basic Books.

Cowen, Bruce and Pat Hodgeson (1991), *Eric Fraser: An Illustrator of Our Time*, exhibition catalogue, London: British Gas.

Cross, Gary (1993), *Time and Money: The Making of Consumer Culture*, London and New York: Routledge.

Crossick, Geoffrey (1977), 'The Emergence of the Lower Middle Class in Britain: A Discussion', in Geoffrey Crossick (ed.), *The Lower Middle Class in Britain*, London: Croom Helm.

Cunningham, Hugh (1980), *Leisure in the Industrial Revolution c. 1780–1880*, London: Croom Helm.

Dancyger, Irene (1978), *A World of Women: An Illustrated History of Women's Magazines*, Dublin: Gill & Macmillan.

Davidoff, Leonore (1995), *Worlds Between: Historical Perspectives on Gender and Class*, London and New York: Routledge.

Davidson, Caroline (1982), *A Woman's Work Is Never Done: A History of Housework in the British Isles 1650–1950*, London: Chatto & Windus.

Davies, Celia (1988), 'The health visitor as mother's friend: a woman's place in public health, 1900–14', *Social History of Medicine*, **1** (1), April, 39–59.

Davin, Anna (1975), 'Imperialism and motherhood', *History Workshop Journal*, **5**, Spring, 9–65.

Davis, Belinda (1994), 'Food Scarcity and the Empowerment of the Female Consumer in World War I Berlin', in Victoria de Grazia with Ellen Furlough (eds), *The Sex of Things: Gender and Consumption in Historical Perspective*, Berkeley: University of California Press, 287–310.

de Grazia, Victoria (1996), 'Empowering Women as Citizen-Consumers', in Victoria de Grazia with Ellen Furlough (eds), *The Sex of Things: Gender and Consumption in Historical Perspective*, Berkeley: University of California Press, pp. 275–286.

de Vries, Leonard and James Laver (1968), *Victorian Advertisements*, London: John Murray.

Dean, David (1983), *The Thirties: Recalling the English Architectural Scene*, London: Trefoil Books, 1983.

Dodd, Dianne (1990), 'Women in Advertising: The Role of Canadian Women in the Promotion of Domestic Electrical Technology in the Interwar Period', in Mariann Gosztonyi Ainley (ed.), *Despite the Odds: Essays on Canadian Women and Science*, Montreal: Vehicule, pp. 134–151.

Dowling, Robin, 'Femininity, place and commodities: a retail case study', *Antipode*, **25**, 4 (1993), 295–319.

Dyhouse, Carol (1989), *Feminism and the Family in England, 1880–1939*, New York: Basil Blackwell.

Everard, Stirling (1992), *The History of the Gas, Light and Coke*

Company, 1812–1949, reprint of 1949 edition, London: A & C Black.

Finnegan, Margaret (1999), *Selling Suffrage: Consumer Culture and Votes for Women*, New York: Columbia University Press.

Forty, Adrian (1995), *Objects of Desire: Design and Society Since 1750*, 2nd edition, London: Thames & Hudson.

Fraser, W. Hamish (1981), *The Coming of the Mass Market, 1850–1914*. Hamden, Conn.: Archon Books.

Freeden, Michael (1979), 'Eugenics and progressive thought: a study in ideological affinity', *Historical Journal*, **22** (3), 645–671.

Freeman, Sarah (1977), *Isabella and Sam: The Story of Mrs. Beeton*, London: Victor Gollancz.

Freeman, Sarah (1989), *Mutton and Oysters: The Victorians and Their Food*, London: Victor Gollancz.

Gamber, Wendy (1998), 'A gendered enterprise: placing nineteenth-century businesswomen in history', *BH Review*, **72**, Summer, 188–217.

Gaskell, S. Martin (1987), *Model Housing from the Great Exhibition to the Festival of Britain*, Oxford: Alexandrine.

Glucksmann, Miriam (1990), *Women Assemble: Women Workers and the New Industries in Interwar Britain*, London: Routledge.

Glynn, Sean and John Oxborow (1976), *Interwar Britain: A Social and Economic History*, London: Allen & Unwin, 1976.

Goodall, Frances (1999), *Burning to Serve: Selling Gas in Competitive Markets*, Ashbourne, Derbyshire: Landmark.

Gorham, Deborah (1982), *The Victorian Girl and the Feminine Ideal*, London: Croom Helm.

Graves, Pamela M. (1994), *Labour Women: Women in British Working-Class Politics, 1918–1939*, Cambridge: Cambridge University Press.

Hall, Catherine (1979), 'The Formation of Victorian Domestic Ideology', in Sandra Burman (ed.), *Fit Work for Women*, London: Croom Helm.

Hall, Catherine (1980), 'The History of the Housewife', in Ellen Malos (ed.), *The Politics of Housework*, London: Allison & Busby, pp. 44–71.

Hall, Catherine and Leonore Davidoff (1987), *Family Fortunes: Men and Women of the English Middle-Class, 1700–1850*, Chicago: University of Chicago Press.

Hannah, Leslie (1979), *Electricity Before Nationalisation: A Study of the Development of the Electricity Supply Industry in Britain to 1948*, London: Macmillan.

Hardyment, Christina (1988), *From Mangle to Microwave: The Mechanization of Household Work*, Cambridge: Polity.

Harris, Jose (1994), *Private Lives, Public Spirit: Britain 1870–1914*, 2nd edition, Harmondsworth, Middlesex: Penguin.

Harris, Neil (1990), *Cultural Excursions: Marketing Appetites and Cultural Tastes in Modern America*, Chicago: University of

Chicago Press.

Harrison, Brian (1987), *Prudent Revolutionaries: Portraits of British Feminists Between the Wars*, Oxford: Clarendon.

Harrison, J.F.C. (1957), 'The Victorian gospel of success', *Victorian Studies*, **1**, December, 155–164.

Harrison, Royden (1971), 'The War Emergency Workers' National Committee, 1914–18', in Asa Briggs and John Saville (eds), *Essays in Labour History, 1886–1923*, London: Macmillan.

Hartcup, Guy (1988), *The War of Invention: Scientific Developments, 1914–18*, London and Washington: Brassey's Defense Publications.

Haslam, Beryl (1999), *From Suffrage to Internationalism: The Political Evolution of Three British Feminists 1908–1939*, New York: Peter Lang.

Hayden, Dolores (1981), *The Grand Domestic Revolution: A History of Feminist Designs for American Homes, Neighbourhoods and Cities*, Cambridge, Mass. and London: MIT Press.

Hennessey, R.A.S. (1971), *The Electric Revolution*, Newcastle upon Tyne: Oriel.

Hindley, Diana and Geoffrey Hindley (1972), *Advertising in Victorian England, 1837–1901*, London: Wayland.

Holcombe, Lee (1973), *Victorian Ladies at Work*, Hamden, Conn.: Archon.

Hollis, Patricia (1987), *Ladies Elect: Women in English Local Government, 1865–1914*, Oxford: Clarendon.

Holton, Sandra (1986), *Feminism and Democracy: Women's Suffrage and Reform Politics in Britain*, Cambridge: Cambridge University Press.

Horn, Pamela (1975), *The Rise and Fall of the Victorian Servant*, Dublin: Gill and Macmillan.

Horsfield, Margaret (1997), *Biting the Dust: The Joys of Housework*, London: Fourth Estate.

Hosgood, Christopher P. (1994), '"The Knights of the road": commercial travellers and the culture of the commercial room in late-Victorian and Edwardian England', *Victorian Studies*, **37** (4), Summer, 525–547.

Hosgood, Christopher P. (1999), '"Mercantile monasteries": shops, shop assistants and shop-life in late-Victorian and Edwardian Britain', *Journal of British Studies*, **38** (3), July, 322–352.

Houghton, Walter E. (1959), *The Victorian Frame of Mind, 1830–1870*, second printing, New Haven: Yale University Press.

Hoy, Suellen (1995), *Chasing Dirt: the American Pursuit of Cleanliness*, New York: Oxford University Press.

Hunt, Felicity (1987), 'Divided Aims: The Educational Implications of Opposing Ideologies in Girls' Secondary Schooling, 1850–1940', in Felicity Hunt (ed.), *Lessons for Life: The Schooling of Girls and Women, 1850–1950*, Oxford: Basil Blackwell, pp. 3–21.

Hurwitz, Samuel J. (1968), *State Intervention in Great Britain: A Study of*

Economic Control and Social Response, 1914–1919, London: F. Cass.

Jeffreys, Sheila (1983), 'Sex Reform and Anti-Feminism in the 1920s', in The London Feminist History Group (eds), *The Sexual Dynamics of History*, London: Pluto, pp. 177–202.

Johnson, Paul Barton (1968), *Land Fit for Heroes: The Planning of British Reconstruction, 1916–19*, Chicago: University of Chicago Press.

Kapp, Yvonne (1989), *The Air of Freedom: The Birth of New Unionism*, London: Lawrence & Wishart.

Kent, Susan Kingsley (1990), 'Gender Reconstruction After the First World War', in Harold Smith (ed.), *British Feminism in the Twentieth Century*, Amhurst: University of Massachusetts Press, pp. 66–83.

King, Sarah (1990), 'Technical and Vocational Education for Girls: A Study of the Central Schools of London, 1918–1939', in Eric J. Evans and Penny Summerfield (eds), *Technical Education and the State Since 1850: Historical and Contemporary Perspectives*, Manchester: Manchester University Press.

Klein, Dan, Nancy McClelland and Malcolm Haslam (1987), *In the Deco Style*, London: Thames & Hudson.

Kwolek-Folland, Angel (1994), *Engendering Business: Men and Women in the Corporate Office, 1870–1930*, Baltimore and London: John Hopkins University Press.

Laird, Pamela Walker (1996), 'Progress in Separate Spheres: Selling Nineteenth-Century Technologies', *Knowledge and Society*, 10, 19–49.

Lancaster, Bill (1995), *The Department Store: A Social History*, London: Leicester University Press.

Laybourn, Keith (1990), *Britain on the Breadline: A Social and Political History of Britain Between the Wars*, Gloucester: Alan Sutton.

Laybourn, Keith (1995), *The Evolution of British Social Policy and the Welfare State*, Keele: Keele University Press.

Lewis, Jane (1975), 'Beyond suffrage: English feminism in the 1920s', *Maryland Historian*, 6, Spring, 1–17.

Lewis, Jane (1980), 'In Search of Real Equality: Women Between the Wars', in Frank Gloversmith (ed.), *Class, Culture and Social Change: A New View of the 1930s*, Sussex: Harvester, pp. 208–239.

Lewis, Jane (1984), *Women in England, 1870–1950: Sexual Divisions and Social Change*, Sussex: Wheatsheaf and Bloomington: Indiana University Press.

Lewis, Jane (1986), 'The Working-Class Wife and Mother and State Intervention, 1870–1918', in Jane Lewis (ed.), *Labour and Love: Women's Experiences of Home and Family, 1850–1940* Oxford: Basil Blackwell, pp. 99–122.

Light, Alison (1991), *Forever England: Femininity, Literature and*

Conservatism Between the Wars, London and New York: Routledge.

Loeb, Lori Anne (1994), *Consuming Angels: Advertising and Victorian Women*, London and New York: Oxford University Press.

Low, Rachael (1979), *The History of the British Film, 1929–1939: Documentary and Educational Films in the 1930s*, London: Allen & Unwin.

Lowndes, G.A.N. (1970), *The Silent Revolution: An Account of the Expansion of Public Education in England and Wales, 1895–1965*, Oxford: Oxford University Press.

Lury, Celia (1996), *Consumer Culture*, New Brunswick, NJ: Rutgers University Press.

MacKenzie, John M. (1984), *Propaganda and Empire: The Manipulation of British Public Opinion, 1880–1960*, Manchester: Manchester University Press.

Mappen, Ellen (1985), *Helping Women at Work: the Women's Industrial Council, 1889-1914,* London and Dover, N.H.: Hutchinson in association with Explorations in Feminism Collective.

Marchand, Roland (1985), *Advertising the American Dream: Making Way for Modernity, 1920–1940*, Berkeley: University of California Press.

Marsh, Neville (1986), *The History of Queen Elizabeth College: One Hundred Years of University Education in Kensington*, London and Dover: King's College Press.

Matthew, H.C.G. (1988), *Gladstone, 1809–1874*, 2nd edition, Oxford: Oxford University Press.

Matthews, Derek (1986), 'Laissez-faire and the London gas industry in the nineteenth-century: another look', *EHR*, **39** (2), February, 244–263.

Matthews, Derek (1988), 'Profit-sharing in the gas industry, 1889–1949', *BH*, **30** (3), July, 306–328.

McBride, Theresa (1976), *The Domestic Revolution*, London: Croom Helm.

McClelland, Keith, 'Masculinity and the "Representative Artisan" in Britain, 1850–80', in John Tosh and Michael Roper (eds), *Manful Assertions: Masculinities in Britain Since 1800*, London: Routledge, pp. 74–91.

McCord, Norman (1995), *British History, 1815–1906*, 3rd edition, Oxford: Oxford University Press.

McKendrick, Neil, Colin Brewer and J.H. Plumb (1982), *The Birth of a Consumer Society: The Commercialization of Eighteenth-Century England*, Bloomington: Indiana University Press.

McKibbin, Ross (1990), *The Ideologies of Class: Social Relations in Britain, 1880–1950*, Oxford: Clarendon.

McKibbin, Ross (1998), *Classes and Cultures, England 1918–1951*,

Oxford: Oxford University Press.

Meggs, Philip B. (1998), *A History of Graphic Design*, 3rd edition, New York and London: A. Lane.

Melling, Joseph (1979), 'Industrial strife and business welfare philosophy: the case of the South Metropolitan Gas Company from the 1880s to the war', *BH*, **21** (2), July, 163–179.

Merrett, Stephen (1979), *State Housing in Britain*. London: Routledge/Kegan Paul.

Middlemas, Keith (1979), *Politics in Industrial Society: The Experience of the British System Since 1911*, London: Andre Deutsch.

Milan, Sarah (1999), 'Refracting the Gaselier: Understanding Responses to Victorian Gas Lighting, in Inga Bryden and Janet Floyd (eds), *Domestic Space*: *Reading the Nineteenth-Century Interior*, Manchester and New York: Manchester University Press, pp. 84–102.

Mills, Mary (1988), 'The gasworkers of South London', *South London Record*, **3**, 6–16.

Mills, Mary (1991), 'George Livesey and profit sharing: a comment on some recent literature', *BH*, **33** (4), October, 87–90.

Millward, Robert and Robert Ward (1993), 'From private to public ownership of gas undertakings in England and Wales, 1851–1947: chronology, incidence and causes', *BH*, **35** (3), July, 1–21.

Morgan, Kenneth O. (1979), *Consensus and Disunity: The Lloyd George Coalition Government, 1918–22*, Oxford: Clarendon.

Moss, Christopher (1983), 'Industrial copartnership, labour unrest and the last Liberal government', unpublished MA thesis, McMaster University.

Mowat, Charles Loch (1983), *Britain Between the Wars, 1918–1940*, Cambridge: Methuen/Cambridge University Press.

Munich, Adrienne (1996), *Queen Victoria's Secrets*, New York: Columbia University Press.

Nava, Mica (1992), *Changing Cultures: Feminism, Youth and Consumerism*, London and Newbery Park, California: Sage.

Nava, Mica (1996), 'Modernity's Disavowal: Women, the City and the Department Store', in Mica Nava, and Alan O'Shea (eds), *Modern Times: Reflections on a Century of English Modernity*, London: Routledge, pp. 38–76.

Nevett, T.R. (1982), *Advertising in Britain: A History*, London: Heinemann.

Nye, David (1994), 'Electrifying Expositions, 1880–1939', in Robert W. Rydell, Nancy E. Gwinn and James B. Gilbert (eds), *Fair Representations*, Amsterdam: Vu University Press, pp. 140–156.

Opie, Robert (1985), *Rule Britannia: Trading on the British Image*, Harmondsworth: Viking.

Pederson, Susan (1993), *Family, Dependence and the Origins of the Welfare State: Britain and France, 1914–1945*, Cambridge:

Cambridge University Press.

Peebles, Malcolm (1980), *Evolution of the Gas Industry*, London: Macmillan.

Peiss, Kathy (1994), 'Making Up, Making Over: Cosmetics, Consumer Culture and Women's Identity', Victoria de Grazia with Ellen Furlough (eds), *The Sex of Things: Gender and Consumption in Historical Perspective*, Berkeley: University of California Press, pp. 311–336.

Peiss, Kathy (1998), '"Vital industry" and women's ventures: conceptualizing gender in twentieth century business history', *Business History Review*, **72**, Summer, 219–241.

Peiss, Kathy (1998), *Hope in a Jar: The Making of America's Beauty Culture*, New York: Metropolitan Books.

Perkin, Harold (1991), *The Origins of Modern English Society, 1780–1880*, reprint of 1969 edition, London and New York: Routledge.

Perkin, Harold (1991), *The Rise of Professional Society: England Since 1880*, London and New York: Routledge.

Pollard, Sidney (1989), *Britain's Prime and Britain's Decline: The British Economy, 1870–1914*, London: E. Arnold.

Pollard, Sidney (1992) *The Development of the British Economy, 1914–1990*, 4th edition, London: E. Arnold.

Prochaska, F.K. (1980), *Women and Philanthropy in Nineteenth-Century England*, Oxford: Clarendon.

Pugh, Martin (1992), *Women and the Women's Movement in Britain, 1914–1959*, London: Macmillan.

Purvis, June (1991), *A History of Women's Education in England*, Milton Keynes: Open University Press.

Rappaport, Erika Diane (1996), '"A Husband and His Wife's Dresses": Consumer Credit and the Debtor Family in England, 1864–1914', in Victoria de Grazia with Ellen Furlough (eds), *The Sex of Things: Gender and Consumption in Historical Perspective*, Berkeley: University of California Press, pp. 163–187.

Rappaport, Erika Diane (1996), '"The halls of temptation": gender, politics, and the construction of the department store in late Victorian London', *Journal of British Studies*, **35** (1), January, 58–83.

Rappaport, Erika Diane (2000), *Shopping For Pleasure: Women in the Making of London's West End*, Princeton, NJ: Princeton University Press.

Ravetz, Alison (1989), 'A View from the Interior', in Judy Attfield and Pat Kirkham (eds), *A View from the Interior: Feminism, Women and Design*, London: Women's Press, pp. 187–205.

Ravetz, Alison (1995), *The Place of Home: English Domestic Environments, 1914–2000*, London: E & F.N. Spon.

Rendall, Jane (1987), '"A Moral Engine"? Feminism, Liberalism and the English Woman's Journal', in Jane Rendall (ed.), *Equal or Different?*

Women's Politics, 1800–1914, Oxford: Basil Blackwell, pp. 112–138.

Richards, J.M. (1959), *An Introduction to Modern Architecture*, London: Penguin.

Richards, Thomas (1990), *The Commodity Culture of Victorian England: Advertising and Spectacle, 1851–1914*, Stanford: Stanford University Press.

Ritschel, Daniel (1997), *The Politics of Planning: The Debate on Economic Planning in Britain in the 1930s*, Oxford: Oxford University Press.

Roberts, Elizabeth (1995), *Women and Families: An Oral History, 1940–1970*, Oxford: Basil Blackwell.

Rose, June (1992), *Marie Stopes and the Sexual Revolution*, Boston and London: Faber & Faber.

Ross, Ellen (1993), *Love and Toil: Motherhood in Outcaste London, 1870–1918*, New York: Oxford University Press.

Rotha, Paul (1973), *Documentary Diary: An Informal History of the British Documentary Film, 1928–1939*, New York: Hill and Wang.

Rubenstein, W.D. (1992), 'British businessmen as wealth-holders: a response', *BH*, **34** (2), April, 69–81.

Schaffer, Frank (1972), *The New Town Story*, London: Paladin.

Schaffer, Simon (1993), 'The Consuming Flame: Electrical Showmen and Tory Mystics in the World of Goods', in John Brewer and Roy Porter (eds), *Consumption and the World of Goods*, London: Routledge, pp. 489–526.

Schmiechen, James A. (1984), *Sweated Industries and Sweated Labour: the London Clothing Trades, 1860–1914*, Urbana, Ill.: University of Illinois Press.

Scott, Joan W. (1987), 'On language, gender and working-class history', *International Labor and Working-Class History*, **31**, Spring, 1–13.

Scott, Joan W. (1998), 'Conceptualizing gender in American *BH*', *Business History Review*, **72**, Summer, 242–249.

Scott, Margaret E. (1967), *The History of F.L. Calder College of Domestic Science 1875–1965*, Liverpool: Calder College of Education for Domestic Subjects.

Searle, G.R. (1971), *The Quest for National Efficiency: A Study in British Politics and Political Thought, 1899–1914*, Berkeley: University of California Press.

Service, Alistair (1979), *The Architects of London and Their Buildings from 1066 to the Present Day*, London: The Architectural Press.

Simmons, Christina (1989), 'Modern Sexuality and the Myth of Victorian Repression', in Christina Simmons and Kathy Peiss (eds), *Passion and Power: Sexuality in History*, Philadelphia: Temple University Press.

Simms, T.H. (1979), *Homerton College, 1695–1978*, Cambridge:

Trustees of Homerton College.

Skidelsky, Robert (1970), *Politicians and the Slump: The Labour Government of 1929–1931*, Harmondsworth Penguin.

Smith, Harold L. (1978), 'The issue of "equal pay for equal work" in Great Britain, 1914–18', *Societas*, **8** (1), Winter, 39–52.

Smith, Harold L. 'British Feminism in the 1920s', in Harold Smith (ed.), *British Feminism in the Twentieth Century*, Amherst: University of Massachusetts Press, pp. 47–65.

Smith, Michael Pearson (1979), 'Republicanism in Victorian Britain', unpublished PhD thesis, McMaster University.

Soloway, Richard (1990), *Demography and Degeneration: Eugenics and the Declining Birthrate in the Twentieth-Century*, Chapel Hill: University of North Carolina Press.

Sparke, Penny (1986), *An Introduction to Design and Culture in the Twentieth Century*, London: Allen & Unwin.

Sparke, Penny (1995), *As Long As It's Pink: The Sexual Politics of Taste*, London: Pandora.

Sparke, Penny, Felice Hodges, Emma Dent-Coad and Anne Stone (1986), *Design Source Book*, London: Macdonald.

Spender, Dale (ed.) (1987), *The Education Papers: Women's Quest for Equality in Britain, 1850–1912*, New York: Routledge and Kegan Paul.

Spurr, Geoffrey (2001), '"Those who are obliged to pretend that they are gentlefolk": the construction of a clerking identity in Victorian and Edwardian London', unpublished PhD thesis, McMaster University.

St John, Diana E. (1994), 'Educate or domesticate? Early twentieth-century pressures on older girls in elementary school', *WHR*, **3** (2), 191–218.

Stearns, Peter N. (1997), 'Stages of consumerism: recent work on the issues of periodization', *Journal of Modern History*, **69**, March, 102–117.

Stedman Jones, Gareth (1989), *Languages of Class, 1832–1982*, Cambridge: Cambridge University Press.

Stedman Jones, Gareth (1992), *Outcaste London: A Study in the Relationship Between Classes in Victorian Society*, Harmondsworth: Penguin.

Stevenson, John (1990), *British Society, 1914–45*, Harmondsworth: Penguin.

Stevenson, John and Chris Cook (1994), *Britain in the Depression: Society and Politics, 1929–39*, London: Longman.

Stone, Dorothy (1976), *The National: The Story of a Pioneer College, National Training College of Domestic Subjects*, 2nd edition, London: Hale.

Strasser, Susan (1982), *Never Done: A History of American Housework*, New York: Pantheon.

Sussex, Elizabeth (1975), *The Rise and Fall of British Documentary*,

Berkeley: University of California Press.

Swann, Paul (1989), *The British Documentary Film Movement*, Cambridge: Cambridge University Press.

Swenarton, Mark (1981), *Homes Fit for Heroes: The Politics and Architecture of Early State Housing in Britain*, London: Heinemann.

Taylor, Barbara (1983), *Eve and the New Jerusalem: Socialism and Feminism in the Nineteenth-Century*, London: Virago.

Thane, Pat (1990), 'The Women of the British Labour Party and Feminism', in Harold L. Smith (ed.), *British Feminism in the Twentieth Century*, Amherst: University of Massachusetts Press, pp. 123–133.

Thom, Deborah (1988), 'Women and Work in Wartime Britain', in Richard Wall and Jay Winter (eds), *The Upheaval of War: Family, Work and Welfare in Europe, 1914–1918*, Cambridge: Cambridge University Press, pp. 299–326.

Thompson, Paul (1975), *The Edwardians: The Remaking of British Society*, London: Weidenfeld & Nicolson.

Tickner, Lisa (1988), *The Spectacle of Women: Imagery of the Suffrage Campaign, 1907–1914*, Chicago: University of Chicago Press.

Tosh, John (1991), 'Domesticity and Manliness in the Victorian Middle Class: the Family of Edward White Benson', in John Tosh and Michael Roper (eds), *Manful Assertions: Masculinities in Britain Since 1800*, London: Routledge, pp. 44–73.

Turnbull, Annmarie (1987), 'Learning Her Womanly Work: The Elementary School Curriculum, 1870–1914', in Felicity Hunt (ed.), *Lessons for Life: The Schooling of Girls and Women, 1850–1950*, Oxford and New York: Basil Blackwell, pp. 83–100.

Turnbull, Annmarie (1994), 'An isolated missionary: the domestic subjects teacher in England, 1870–1914', *WHR*, **3** (1), 81–100.

Turner, E.S. (1965), *The Shocking History of Advertising*, Harmondsworth: Penguin.

Valverde, Mariana (1989), 'The love of finery: fashion and the fallen woman in nineteenth-century social discourse', *Victorian Studies*, **32** (2), Winter, 169–188.

Vernon, Keith (1995), 'Science and Technology,' in Stephen Constantine, Maurice W. Kirby and Mary B. Rose (eds), *The First World War in British History*, London: E. Arnold.

Vicinus, Martha (1985), *Independent Women: Work and Community for Single Women, 1850–1920*, London: Virago.

Vorspan, Rachael (1976), 'Vagrancy and the New Poor Law in late-Victorian and Edwardian England', *English Historical Review*, **92** (362), January, 59–81.

Wacjman, Judy (1992), 'Domestic Technology: Labour-saving or Enslaving?' in Gill Kirkup and Laurie Smith Keller (eds),

Inventing Women: Science, Technology and Gender, Milton Keynes: Polity and Open University Press, pp. 238–254.

Walkowitz, Judith R. (1992), *City of Dreadful Delight: Narratives of Sexual Danger in Late Victorian London*, Chicago: University of Chicago Press.

Walsh, Margaret (2000), *Making Connections: The Long Distance Bus Industry in the United States*, Aldershot and Burlington, VT: Ashgate.

Ward, J.T. and Hamish Fraser (eds) (1980), *Workers and Employers: Documents on Trade Unions and Industrial Relations in Britain Since the Eighteenth Century*, London: Macmillan.

Wigham, Eric (1976), *Strikes and the Government, 1894–1974*, London: Macmillan.

Wilkinson, James and H. Stuart Hughes (1995), *Contemporary Europe*, 8th edition, Englewood Cliffs, NJ: Prentice Hall.

Williams, Rosalind H. (1982), *Dream Worlds: Mass Consumption in Late Nineteenth-Century France*, Berkeley: University of California Press.

Williams, Trevor I. (1981), *A History of the British Gas Industry*, Oxford: Oxford University Press.

Wilson, J.F. (1991), 'Ownership, management and strategy in early north-west gas companies', *BH*, **33** (2), April, 203–221.

Wilson, John T. (1991), *Lighting the Town: A Study of Management in the North West Gas Industry, 1805–1880*, London: Paul Chapman.

Windsor, Alan (1992) *Handbook of Modern British Painting, 1900–1980*, Aldershot and Brookfield, VT.: Scolar.

Winston, Brian (1995), *Claiming the Real: The Griersonian Documentary and its Legitimations*, London: The British Film Institute.

Woollacott, Angela (1994), *On Her Their Lives Depend: Munitions Workers in the Great War*, Berkeley: University of California Press.

Worden, Suzette (1989), 'Powerful Women: Electricity in the Home, 1919–40', in Judy Attfield, and Pat Kirkham (eds), *A View from the Interior: Feminism, Women and Design*, London: Women's Press, pp. 128–146.

Wright, Cynthia (1992), '"Feminine Trifles of Vast Importance": Writing Gender Into the History of Consumption', in Franca Iacovetta, and Marianna Valverde (eds), *Gender Conflicts: New Essays in Women's History*, Toronto: University of Toronto Press, pp. 229–260.

Wrigley, Chris (1976), *David Lloyd George and the British Labour Movement*, Hassocks, England and New York: Harvester Press.

Yarwood, Doreen (1981), *The British Kitchen: Housewifery Since Roman Times*, London: B.T. Batsford.

Zimmeck, Meta (1986), 'Jobs for Girls: the Expansion of Clerical Work for Women, 1850–1914', in Angela John (ed.), *Unequal Opportunities: Women's Employment in England 1800–1918*, New York: Basil Blackwood, pp. 153–177.

Index